Studies of Brain Function Vol. 17

Robert Miller

Cortico-Hippocampal Interplay

and the Representation of Contexts in the Brain

With 51 Figures

Springer-Verlag
Berlin Heidelberg GmbH

Dr. ROBERT MILLER
Department of Anatomy
University of Otago Medical School
P.O. Box 913
Dunedin, New Zealand

This book was produced with the assistance of the Computing Services Centre, University of Otago, New Zealand, using the LaTeX document preparation system with Apple Laser Writer output.

ISBN 978-3-662-21734-4 ISBN 978-3-662-21732-0 (eBook)
DOI 10.1007/978-3-662-21732-0

Library of Congress Cataloging-in-Publication Data. Miller, Robert, 1943 Aug. 29– Cortico-hippocompal interplay and the representation of contexts in the brain / Robert Miller. p. cm. – (Studies of brain function; vol. 17) Includes bibliographical references. Includes Index. 1. Hippocampus (Brain) 2. Cerebral cortex. 3. Theta rhythm. I. Title. II. Series: Studies of brain function; v. 17. [DNLM: 1. Cerebral Cortex – physiology. 2. physiology. W1 ST937KF v. 17 / WL 307 M649c] QP383.25.M55 1990 599'.0188 – dc20 DNLM/DLC for Library of Congress 90-10397

Originally published by Springer-Verlag Berlin Heidelberg New York in 1991.
Softcover reprint of the hardcover 1st edition 1991

31/3145 (3011)-543210 – Printed on acid-free paper

Acknowledgements

I would like to thank Jeff Wickens for his part in the generation of the ideas in this book. He was the first to whom I explained the concepts advanced in Part 3, and in subsequent discussions, he made significant contributions to the way in which I presented the ideas, and to some of the more distinctive terminology I have adopted. I would also like to thank Valentino Braitenberg for unfailing support in this project. I thank Gunther Palm for some very helpful comments on a seminar I gave at Tubingen based on this work, thus extending some of my own ideas. These extensions have been incorporated here. I would also like to thank a much larger number of experimentalists, few of whom I have met, for providing experimental results from which I could construct my own ideas. Theoretical brain research would be a futile endeavour if we did not have such a rich literature of good experimental findings as now exists. While it is perhaps unfair to single out individuals, recent papers of L. S. Leung and G. Buzsaki and older work of W. R. Adey and the late E. Grastyan and their co-workers were most influential in shaping my ideas. I would like to thank Robbie McPhee for all artwork. I thank the Otago Medical Library, especially the interlibrary loans department, for their efficiency in supplying all the scientific literature I needed. I thank Graeme McKinstry for help in producing camera ready copy of the manuscript. Finally, I would like to thank Gareth Jones for providing an academic environment in which production of a work such as this was possible.

R. Miller
Dunedin, New Zealand
May, 1990.

Contents

Part 3 The Theory of Resonant, Self-Organizing Phase-Locked Loops 157

List of Figures

* Figures published in other works re-lettered to ensure clarity.

List of Tables

Part I

Statement of the Problem

1 Historical Introduction

1.1 Contexts

The principal issue with which this monograph deals is the role of the hippocampus in establishing and using representations of contexts for information processing. However, before this issue can be addressed directly, it is necessary to ask "what is meant by the word 'context' ?". The first answer which comes to mind is likely to be something along the following lines: "A context is a framework (or background) of information with respect to which more specific 'items' of information can be identified and manipulated". This answer may be correct, but it begs a fundamental question. Why should it be necessary to subdivide information into specific "items" of information, and the more global backgrounds, or frameworks? This question is especially pertinent if we are thinking of information representation in the brain, since neuroscientists (or at least the vast majority of them) believe that the basic way in which patterns of information are encoded in the brain is as combinations of connections, selected in a variety of ways. Since both "items" of information and "contexts" are just such patterns, apparently differing only in size, it is far from clear why there should be a categorical division between the two.[1]

This question is relatively new in the neurosciences. However, in a somewhat different guise it has been alive for a long time, since the publication of Immanuel Kant's *Critique of Pure Reason*. In the first chapter of this monograph, the older history of this subject will be traced. First of all it will be traced in the development of post-Renaissance philosophy. Then, a closely parallel debate will be described, which took place in the first half of the present century, amongst psychological learning theorists. In the second chapter, a biological parallel will be drawn between these philosophical and psychological issues and the nature of the connectivity in the cerebral hemisphere. These three strands of the argument will set the scene for the remaining chapters in this book. These later chapters deal with the same issue again, but from the viewpoint of contemporary neuroscience, and advance a theory of the functional relationship between the hippocampus and the neocortex in the mammalian brain.

1.2 "Context" as a Philosophical Concept in Analysis of Logical Relations

1.2.1 The Cartesian View of Logic

Rene Descartes (1596–1650), the first modern philosopher, was a mathematician. He thought he had a method for ascertaining absolute truth. Two of his maxims (Descartes 1637/1964) were: (1) Use only the secure deductive steps of mathematicians;

[1] In terms more relevant to the practicalities of behavioural experiments a related question is also posed: Can one, in *practice*, determine when a pattern of information is an "item" and when it is a "context"? However, this question will not be given a definite answer in this monograph.

(2) reject all premises on which there could be the slightest doubt. He thus believed that, given correct explicit premises, and valid inference, the correctness of the conclusions followed inevitably. This method may be termed that of "Pure Reason".

Some while later, David Hume (1711–76) used Descartes' method to reduce this conclusion to ridiculous nonsense. He argued that all matters of fact were based on experience. He further argued that any such empirical "facts", if they had the least claim to generality (such as the fact that the sun rises every day), were, in deductive terms, questionable. That the sun had risen every morning in the experience of a particular enquirer had (from the point of view of Cartesian deductive inference) no bearing on whether it would rise tomorrow. By extension, all other empirical statements were also dubitable. Thus, those apparently self-evident empirical premises upon which a Cartesian philosopher would most hope to base his deductions were always insecure. In place of reason, Hume suggested that the human mind has a rather unreasonable tendency to believe that constant conjunctions of events in the past are likely to continue into the future. Hume called this principle "custom" or "habit". "All inferences from experience therefore are effects of custom, not of reasoning" (Hume 1748/1977). This principle gave to the human mind a sense of special significance to chains of events which were consistently linked, so that it would be believed that the future would be conformable to the past, despite the unprovability of such a belief.

It is possible to read Hume's arguments in two contrasting ways. Either he can be seen as offering a usage of reason which totally destroys the possibility of our apprehending anything meaningful in our universe, or alternatively he can be read as providing an argument for abandonment of reason as a whole in deference to accepted, traditional, but unprovable beliefs about the empirical behaviour of the universe. If we adopt the former view of Hume's argument, he can be considered as providing a pre-emptive, and rather destructive *Critique of Pure Reason*, in advance of Immanuel Kant's work with this title. If the latter is taken as Hume's primary objective, he was really writing not about logic, but about the human mind. He can then be credited with being an early pioneer of psychology. It was with this background that Immanuel Kant (1724–1804) produced **the** *Critique of Pure Reason*, (1781/1934), undoubtedly a far more subtle and constructive philosophical work than had previously been written on this subject, and one which perhaps makes Kant the greatest of post-Renaissance philosophers.

1.2.2 Kant's Critique of Pure Reason

Kant classified "statements" in two ways (see Table 1). Firstly they could be either *analytic* or *synthetic*. An *analytic* statement was one where the predicate added nothing that was not latent in the subject(s). In other words, an analytic statement was a tautology (either obviously so, or sometimes tautologous in a more subtle way). Kant's example of an analytic statement was:

"All bodies are extended".

A more obvious example is:

"No black cats are green".

A *synthetic* statement is one where the predicate constructs something new, that is not latent in the subjects. Kant's example was:

"All bodies are heavy".

Another example would be:

"All cats like chasing mice".

The second way in which Kant classified statements was a more familiar dichotomy. Statements could be either *empirical* (matters of experience) or *a priori* (given prior to any experience).

Table 1. Kant's classification of statements

	Synthetic	Analytic
Empirical	Yes	?
A priori	*Yes*	Yes

"Yes" indicates that this category of statements exists.

What was the correspondence between the two ways of classifying statements? In so far as this issue had been considered at all, earlier philosophers would probably have assumed that synthetic statements were also empirical, while analytic statements were *a priori*. The subtlety of Kant's approach was to suggest that there could be such a class of statements as the "synthetic *a priori*". This class included all of the mathematical statements which so influenced Descartes' philosophy. He illustrated this class of statement with respect to geometry and number. The following were examples Kant used of *synthetic a priori* statements:

"A straight line between two points is the shortest".

"$7 + 5 = 12$".

Neither statement requires experience to demonstrate its truth, hence they are both *a priori* statements. In neither case, Kant argued, could it be claimed that the subjects alone contained the predicate in latent form. The key to this riddle, according to Kant, was that, in order to make these apparently simple assertions, one required *a priori* concepts, of Euclidean space and natural number theory, respectively. If (and only if) these general concepts were concurrently in mind, could the predicate be synthesized from the subject in the above statements. (For instance, the first of Kant's examples is *untrue* in non-Euclidean space, the latter may be untrue in modular number theory).

This raises two issues. Firstly, Kant suggests that number and space are innate conceptions, rather than ones acquired through experience. Second, regardless of the correct answer on the first issue, the overt and specific premises of a deductive inference require, in addition, some underlying premise of more general background nature. The present work is concerned with both issues. Nevertheless, they should be treated separately. We may find that we agree with one and reject the other.

In the two examples of Kant's synthetic *a priori* statements quoted above, the underlying background premises appeared to define the set of relationships possible within the relevant class of statements (the "ground rules" by which a particular class of ideas was to be interrelated). Henceforth the word "context" will be used to denote this Kantian requirement in deductive inference. Employing this terminology, Kant's message may be summed up in the following way: *"There are many important examples of deductive inference which require an appropriate context if the conclusion is to be derived from the overt premises"*. One might perhaps regard the context for such examples of deductive inference as just another premise. However, it differs from normal premises in that it can be completely hidden in the statement of a syllogism. Moreover, it may in fact be a far more complex set of ideas than one would guess from looking at the synthetic *a priori* statement itself. Both Euclidean space and natural number theory are concepts of some complexity. There is here a major contrast with the view advocated by Descartes.

1.3 Developments of Kant's Ideas up to 1950

In the nineteenth century, the discipline of psychology was in its gestation. The British empiricist school of philosophers, initiated by Hume, was influential in ways running counter to Kant's ideas: first, there was the emphasis on experience as the sole source of knowledge. Second, there was the primordium of describing that experience as having its influence by constant conjunction of different sense impressions. Combining the two, one has the seed which later led to the associationist tradition in learning theory. This tradition became ripe for development at the start of the present century, when it was realized that the nervous system was, anatomically speaking, a complex network forming an ideal substrate for associative operations.

Although around the turn of the century the tide was in favour of the associationist/empiricist approaches to psychology, one of the founders of modern psychology, William James, was aware of Kantian philosophy. In *The Principles of Psychology* (1890/1950, p. 661, footnote) James discusses the issue of whether *a priori* truths are analytic or synthetic. However, he does not address the issue of whether premises for deductive steps require a context before the inference can be made. The present author is aware of no other psychologist from the first half of the present century in the Anglo-American tradition who was familiar with Kant's *Critique*.

1.4 "Context" as a Psychological Concept in the Analysis of Learning Processes

As the neuron theory gained acceptance, matching psychological theories of learning were developed. These theories viewed learning processes as associative in nature, and, from the start, were thought of as occurring in a biological substrate of neural networks. This was seen as an ideal medium for representation of those "constant conjunctions" which Hume and his successors envisaged to underlie learning processes. There was, of course, the additional assumption about the biological basis of learning, that the neural connections were *modifiable*. It was well beyond the techniques of the day to provide experimental support for the latter supposition, though nowadays the investigation of modifiable connections is a major theme in neuroscience.

There followed what has been described as "the great age of learning theory", dominated by familiar names such as Thorndike, Pavlov, Watson, Hull and Skinner. Two particular logical forms emerged, as descriptors of the associations presumed to underlie learning. Firstly, there was the Pavlovian model, in which associations between two stimuli (CS and UCS) came to be encoded if they were repeatedly paired. This form may be designated as "learning on the basis of signal contiguity". In slightly more complex guise this form can be seen as capable of underlying pattern learning. The second form was the Thorndikian one, in which responses, emitted in relation to specific stimuli, were strengthened or weakened according to whether they produced favourable or unfavourable effects for the animal.

The two descriptive forms, though rather different in their detailed implications, had two assumptions in common. First, it was assumed necessary to consider only specific stimuli and responses, rather than the totality of sensory-motor information which an animal may concurrently be processing. Secondly, given that the appropriate conjunctions of stimuli, responses (and reinforcements, in the Thorndikian paradigm) had been presented to allow learning to take place, input would then dictate output *unconditionally*.

In the first half of the twentieth century, the associationist tradition may have dominated learning theory. However, it was not without challenge. E. C. Tolman (1932) could not regard learning as merely the conjunction of representations of stimuli and responses (a *molecular* approach). He insisted that there was a more global (or *molar*) representation of the environment. After all, the explicit stimuli and responses in a learning experiment were arguably only a tiny fraction of the information which could potentially be assimilated by the animal, especially if consideration was given to situations which were ecologically more realistic than the simplified ones of classical or instrumental conditioning. In contrast to the associationists, he was not framing his psychological theories specifically to fit the idea of central nervous tissue as a network. Probably his overriding concern was to be realistic about animal behaviour. He may have had no specific biological basis in mind, though he did not specifically deny the "connectionism" of the associationists.

Tolman preferred to describe learning as the formation of a "cognitive map" of the environment (1932), so that not only did the animal respond appropriately to specific stimuli in that environment, but to some extent "knew what to expect" even if the specific stimuli at a particular time had not been experienced before. The clearest example of this was shown in maze problems. Rats could guide themselves through a somewhat familiar maze from a known starting point to a known goal, but using pathways they had never traversed before (Tolman *et al.* 1946). This result could only be possible if the rat had an overall map, not just a series of responses to specific stimuli. It also seemed to imply that rats knew the set of relationships possible in Euclidian space.

Tolman was apparently unaware of the relation between these implications of his learning experiments and the Kantian idea that there is required a representation of the "ground rules" or "context" for information processing. He never cited Kant's *Critique*. Nor did other important reviewers of mid-century learning theory (Kimble 1964; Hilgard and Bowers 1966). Those psychologists of the middle years of this century who were aware of the philosophical roots from which their discipline sprang cite Kant with reference to the other aspect of Kant's *Critique*, namely the view that space, and by implication other concepts, were innate, *a priori*, rather than acquired (see for instance Turner 1965). Not surprisingly, Kant's *a priori* knowledge did not meet with the approval of such students of learning theory. The only major psychologist of the middle years of the present century who had properly assimilated Kant's ideas was, not surprisingly, a European, Jean Piaget.

He developed, on the basis of many empirical studies and shrewd theoretical ideas, a scheme for the intellectual development of the human infant and child. This scheme consisted of a number of stages, each of which was characterized by particular forms of information processing. Over the time of a child's development, these specific forms became interrelated in the same way as subroutines of a computer programme become interrelated in a more complex programme. Individual subroutines became operational first, but when they later came to be combined, much more complex emergent properties appeared (Piaget 1955). For instance, in linguistic development, a child may acquire first the ability to use naming words (nouns), then, separately, to use verbs, after which the more powerful capacity to form sentences can be developed. Each subroutine (or "schema") was rather like a Kantian *a priori* concept underlying certain types of information processing.

Returning to the animal learning literature, there is a second strand of evidence which challenges the view that a learned response was one where stimuli had acquired the capacity to dictate responses unconditionally. This is evidence that learning may progress via two parallel developments. On the one hand, stimulus-response links may be forged, as the associationists believed. On the other hand, repeated exposure to the learning situation could lead to the improved ability to use subsequently stimuli in the same "dimension", but for a different and new learning task. This body of evidence has been thoroughly reviewed by Sutherland and MacKintosh (1971). The term they use for the second of the two processes is "selective attention". It is implied that "dimensions" of stimuli (or responses) can be selected independently of the selection of specific stimuli or responses which are examples of those dimensions. The term "dimension" is somewhat difficult to define rigorously, but in the experiments discussed by Sutherland and MacKintosh, it includes such things as "shape", "brightness", "position" and so on. These dimensions are not the stimuli/responses themselves, but rather are variables from which a range of specific stimuli/responses could be chosen. These authors present a strong case that in many learned responses, stimuli do not unconditionally dictate responses, but do so only in so far as the "analyzer" for the appropriate dimension has been brought to bear on the learning task.

This conclusion is again reminiscent of Kant's *a priori* concepts, Piaget's "schemata", or, more generally "contexts". Again, the authors do not cite Kant's *Critique*. Nor has this strand of the evidence been integrated with that initiated by Tolman. The reason for the latter shortcoming is that Tolman's emphasis was on *spatial* contexts, as though they were the only example of their class, whereas Sutherland and MacKintosh deal with "dimensions" of many varieties, with no particular emphasis on the spatial dimension. In summary, we have a nice historical parallel between the early post-Renaissance philosophy and early twentieth century learning theory. In both cases, the first endeavours suggested that the appropriate inputs, and the appropriate means of linkage to the output, were sufficient to guarantee the input-output relationship unconditionally; and in both cases, closer scrutiny of the problem led to a revision of this view, such that input dictated output only under the condition that the appropriate context be also in use.

1.5 The Hippocampus and Contextual Representation

Sometime around the year 1970, the idea emerged that the hippocampus played an important role in behaviour involving specific responses to places in the environment.

The origin of this idea lay first of all in the study of behaviour (especially in learning tasks) in animals with hippocampal lesions. Such animals seemed to be particularly deficient in their ability to respond in specific places (in maze learning for instance; see Kimble 1970). Impressed by this fact, O'Keefe and Dostrovsky (1971), most perceptively, looked for (and found) neuronal response in the hippocampus of freely moving rats, which were specific to the animal's location in its environment. Another important strand to the development of ideas at this time was a review by Hirsh (1974). Hirsh makes the very pertinent comment that animals with lesions of the hippocampus or fornix corresponded quite well to the notions of the pure stimulus-response theorists. The full range of context-dependent behaviour relied, however, on the integrity of the hipocampus. More specifically, the widely cited monograph of O'Keefe and Nadel (1978) presented a case that the hippocampus contains an internal representation of space (or place in the environment). This was based on many types of evidence, but especially on O'Keefe's own experiments showing the existence of place units, and a thorough review of the literature available at that time on the behavioural effects of hippocampal lesions. This important monograph does make the historic link between the "cognitive mapping" variety of learning theory, and Kant's requirement for context representations, as pre-requisites for many varieties of information processing. As such, this monograph is a landmark.

Three important features of this synthesis must however be made clear. First, O'Keefe and Nadel (1978) envisage that Euclidian space is the one context of overriding significance (all other contexts, for instance in human use of language, being derived from the spatial context, by virtue of some formal similarity to space). Second, environmental space is topographically represented in the spatial dimensions of the hippocampus. Third, they assume, following Kant, that this representation of space is innate, not acquired. It is the contention of the present author that, on all three of these points, O'Keefe and Nadel's synthesis requires amendment. Thus, the chapters below will advance the following ideas: (1) That there can be a great variety of contexts controlled by the hippocampus. (2) That their representation in the hippocampus is in a far more versatile manner than that of topographical mapping of environmental space against cerebral space. (3) That context representations are acquired, albeit more slowly than in many more peripheral aspects of learning. Of these three issues, the third is the most fundamental, because, if contexts can be acquired, there can be a co-existence of many context representations, for use on different occasions (issue 1), and their manner of representation must therefore be in a more versatile form than simple spatial mapping (issue 2).

The idea that context representations are acquired, rather than innate can be derived from several sources: (1) Sutherland and MacKintosh's (1971) monograph makes it quite clear that there can be a variety of "dimensional representations" in use by animals, and that they are acquired, rather than innate. (2) Piaget, though realizing his debt to Kant, did not actually believe that the schemata he envisaged as underlying developing human cognition were really genetically determined, e.g., a product purely of maturational processes in the brain. Rather he saw them as acquired by interaction with the environment, at critical stages of development, these stages themselves being the result of accumulated experience assimilated by innate mechanisms (Piaget 1955). In a variety of innocuous but ingenious experiments he was able to precipitate and observe the actual moment at which a new schema was actually acquired. (3) There is a small but relevant body of evidence from man, and the beginnings of similar evidence in subhuman primates, on the behavioural effects of damage to the hippocampus, or structures in its efferent pathways. In contrast to small mammals, the deficit is not focussed on ability to handle spatial situations. Rather there is a singular form of amnesia. The impairment does not apply to very old memories acquired before the brain damage. Even for material acquired after the damage, the deficit in retrieval is not absolute, but may be restored towards normal by appropriate use of cues to the class of information to be retrieved. It is thus arguable that this form of amnesia consists of an inability to properly structure memories as they are acquired, and that this inability is, in fact, a failure to "locate" individual items of memoranda within the appropriate context. The amnesia is thus the result of failure to re-establish a specifying context of neural activity with which to narrow down the range of possible items required for retrieval. If this inference from the amnesia literature in primates is sustainable, it inevitably implies that memories of all kinds can be context-linked each to its own appropriate context, of which there can be a great variety established at once, for use on different occasions, but one at a time. (4) In addition to all this fairly technical evidence, there is a variety of more casual evidence coming from common acts of introspection by humans about the way their minds operate, which suggests that contexts are acquired rather than innate. The distinction made by Kant between individual items of information and the background beliefs which underlie mental manipulation of such ideas is, in introspective terms, a valid one. We are all familiar with the fact that we can more easily deal with a certain set of ideas if the framework of ideas in which the ideas are "located" has been already established in our minds. (This is captured in a number of common usages of language, such as: "You're not on my wavelength", etc. Much of our sense of humour employs a similar principle, since it depends on forcing someone to view an item of information out of its usual context). Furthermore our introspection tells us that a particular frame of reference ("context") which we employ, may, in specifying the set of all possible relationships to be dealt with under its aegis, use complex and specific structures of relationship that have no formal similarity to Euclidian space.

Thus, the present work accepts one of the aspects noted above of Kant's Critique (the need for context representations), but rejects the other (their being an innate endowment of an animal). There is perhaps some confusion over the centuries in the use of the term "*a priori*". To Kant, the analyst of reasoning, "*a priori*" may have meant "given prior to a particular instance of inference". Nowadays, we are perhaps too ready to imbue the term with the implication "inborn, part of our genetic endowment". According to Kant a "context" could not be acquired by reasoning. Since this was the only "psychological model" he had by which the environment could lead to acquisition of knowledge, context knowledge must be "*a priori*". The possibilities of other models of acquisition of knowledge, that have been developed more recently under the guise of learning theory, might have allowed him to retain the idea of contexts as "given prior to an instance of inference", as well as the notion that they were still acquired in the earlier experience of the organism.

2 Designs for a Prototype Cerebral Cortex

2.1 Introduction

The most impressive feature of the microanatomy of central nervous tissue is the prodigious number of connections it contains. The soma of each neuron and its dendrites may receive many thousands of input synapses from other neurons, and likewise, the terminal arborization of its axon may give many thousands of synapses to other neurons. We are all used to the idea that the central nervous system encodes information as patterns of connections when those patterns are activated by action potentials. However, with such a proliferation of connections the obvious question is begged: *How can action potentials in specific patterns of connections represent anything specific if their origin upstream can come from so many possible sources, and their influence downstream is spread over so many further diverging and converging connections?* There would seem to be an inherent *ambiguity* in the representation of information in nervous tissue, an ambiguity which appears unresolvable because it arises as a direct consequence of the structure of the neuronal building units of that tissue.

In the present chapter, an attempt will be made to analyze this ambiguity in some detail. We will set out to design a cerebral cortex, first of all in naïvely simple terms, and later with increasing degrees of realism; and at each stage the manner of representation of information will be discussed. In this exercise, there are three important constraints to be reconciled in the design. Firstly, the cortical network should represent patterns of information *specifically* (i.e., *un*ambiguously). Secondly, it should do so *economically*, that is, making effective use of the vast total number of connections for mapping information. Thirdly, *it should not demand that each single neuron has connections with an implausibly large number of other neurons in the cortex.* In reconciling these three requirements, we will find a means whereby specificity and economy can be reconciled, but only in very small blocks of nervous tissue. In larger blocks (including the cerebral cortex of even the smallest-brained mammals), our third constraint prevents the first two from being adequately reconciled. The rest of this book then deals with a hypothetical mechanism by which all three constraints can be reconciled in brains of realistic size.

2.2 Rule-Dependent Synaptic Change

Before describing various prototype cortices, it is necessary to introduce the idea of rule-dependent modification of synapses, as a basis for learning. This concept has been latent from the early years of the century, when the neuron theory first became established. Its explicit formulation is generally attributed to Hebb (1949), who proposed the following rule for strengthening a synapse:

> *When an axon of cell A is near enough to excite a cell B and repeatedly or persistently takes part in firing it, some growth process or metabolic change*

takes place in one or both cells, such that A's efficiency as one of the cells firing B is increased.

In this quotation it is clear that increase in synaptic efficacy is envisaged to occur as a result of contiguous activation of pre- and post-synaptic elements. Activation of the postsynaptic element will occur more readily if several presynaptic elements are active at once than if just one is active. Hence, Hebb's rule implies that groups of synapses converging on a single neuron and having the tendency to fire together will become strengthened as a group. This is the principle of *cooperativity*. This principle has a form similar to the idea of "learning on the basis of signal contiguity" as in classical conditioning or pattern learning, except that it applies at the single neuron level rather than the level of the behaving intact animal.

The techniques for investigation of Hebbian synaptic change were not available when Hebb was writing. In recent years, however, Hebb's rule has attracted much interest from experimentalists. It is inappropriate to review this literature here. Suffice it to say that Hebb-like processes of synaptic change have been described in a number of forebrain regions (hippocampus: Goddard 1980; Wigstrom and Gustafsson 1985; neocortex: Lee 1982; Bienenstock *et al.* 1983; Fregnac *et al.* 1988; thalamus: Green and Wienberger, 1983). Hebb-like processes are also a plausible basis for many of the examples of neuronal plasticity observed in the visual and other sensory cortices (Blakemore and Mitchell 1973; Von der Malsburg 1973; Spinelli and Jensen 1979; Merzenich *et al.* 1983a,b; Linsker 1986 a,b,c). In addition, there is now available a great deal of information about the mechanism of such synaptic change, especially for synapses in the hippocampal formation, where experimentalists have concentrated their efforts (see review by Wickens 1988). The mechanisms now emerging amount, initially at least, to a cascade of rapid biochemical actions, rather than discernible anatomical growth.

Quite apart from consideration of mechanisms, there have been more recent formulations of rule-dependent synaptic modification using rules other than that proposed by Hebb. For instance, Stent (1973) advanced a scheme wherein synapses were weakened if the firing on pre- and post-synaptic sides of a junction were in anti-correlation, a scheme which might be described as "anti-Hebbian". This rule is compatible with some of the evidence for synaptic change induced in the developing visual cortex by abnormal sensory environments, although there is also evidence that complementary Hebbian processes also occur there. Reward- and punishment-mediated learning have also been brought into the realm of rule-dependent modification of synapses (Miller 1981, 1988).

There is no need to review all these matters here. It is merely necessary to emphasize that rule-dependent synaptic modification is an underlying premise to much of the theoretical arguments presented in this monograph. Specifically, synapses modifiable by Hebbian or quasi-Hebbian rules are assumed to be found in excitatory pathways throughout the cortex (including the hippocampus), for both long- and short-axon excitatory links. The complementary "Stent" rule for synaptic weakening is probably also required, but since this paradigm has been less well studied, it will be referred to less frequently here. Synaptic modification dependent on reward and punishment is seen here as a property of the striatal regions. Although these regions are not the focus of attention in the present work, they are sometimes referred to.

2.3 A Cortex to Suit Stimulus-Response Learning Theorists

2.3.1 Sensory and Motor Cortices Alone

Our first prototype cortex (Fig. 1) consists of just two components, a single region of sensory cortex (S) and a single region of motor cortex (M). Between the two there is envisaged to be a set of cortico-cortical connections, such that each projection neuron in the sensory cortex has independent anatomical links with every neuron in the motor cortex. Such a network has been called elsewhere an "omniconnected network" (Miller 1981).

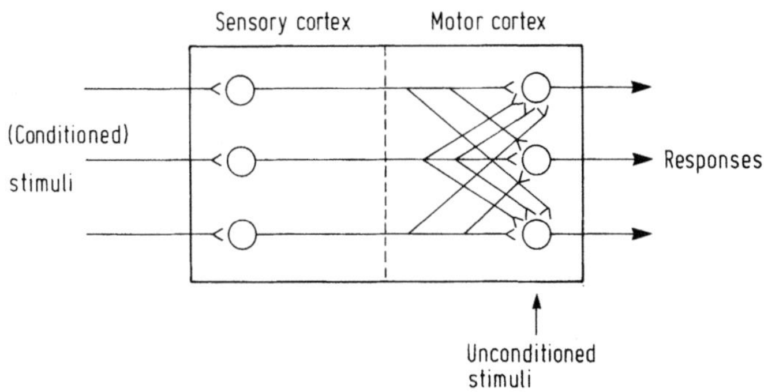

Fig. 1. Schematic diagram of a cortex consisting of just a sensory and a motor area, with connections suitable as a substrate for stimulus-response versions of learning theory, for instance in classical conditioning

If each neuron in the sensory cortex represents a different stimulus, and each neuron in the motor cortex represents a different response, we thus have a cortex where any detectable stimulus has a potential link with any performable response. In other words, we would have an anatomical basis for the ideas of the early learning theorists in the "stimulus-response" tradition. In specific examples of classical conditioning, it would be envisaged that one of these potential (i.e., anatomical) stimulus-response links would be strengthened, to become an actual functional link, allowing the relevant stimulus to control the appropriate response in the unconditional manner which these theorists postulated. The paradigm by which selected connections can be so strengthened is likely to be something like that proposed by Hebb (see above), for instance in classical conditioning.

2.3.2 Sensory Cortex, Motor Cortex and Striatum

For instrumental conditioning, a somewhat more complex relationship is envisaged, as described elsewhere (Miller 1981, 1988). The striatum is seen as the primary site for acquisition of reward- or punishment-mediated learning. The stimulus-response links which become established through that structure represent those behaviours which produce favourable effects or avoid or escape unfavourable ones. Activity representing such responses is relayed from the striatum to the motor cortex and adjacent regions. This input, although dealing with motor information, can be treated by the cerebral cortex in the same way as its sensory inputs. Thus, the Hebbian rule can be used

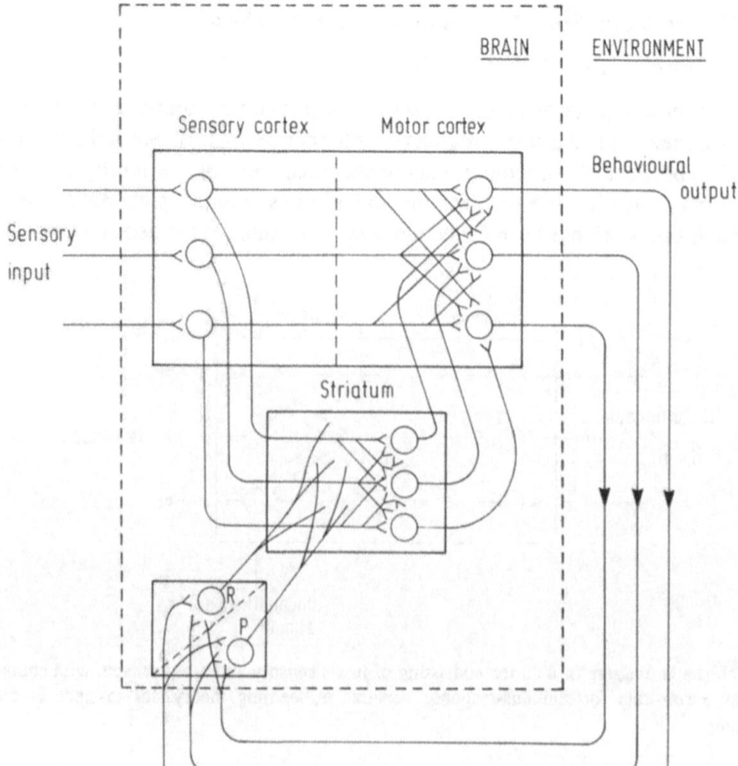

Fig. 2. Schematic diagram of a cortex consisting of a sensory and a motor area, with in addition a loop through the striatum and related structures. This model is seen as a substrate for stimulus-response versions of learning theory, including both classical and instrumental conditioning, and some higher order combinations of the two. (See Miller 1988)

to encode the contiguities between the signals which the striatum sends to the motor cortex and other signals within the cortex (e.g., those traversing direct pathways from the sensory cortex). Various more complex properties may be envisaged to arise from the relation between cortex and striatum than either of them are capable of alone. These emergent properties have been discussed elsewhere (Miller 1981, 1988), so it is not necessary to develop this idea in detail here. However, the implication is that for instrumental learning Fig. 1 should be replaced by Fig. 2, which includes the striatal link.

This relation between striatum and cortex was initially conceived in a framework of stimulus-response theory. However, the relationship should be equally valid for the prototype cortices described later in this chapter and based on more advanced psychological models. In most of what follows this relation between cortex and striatum is implicit. However, most diagrams are simplified by omission of the striatal link. The point of introducing this idea here is to make clear that the input-output links represented in the cortex include those established by reward and punishment (the "effect" of a response), as well as those established by the simpler associative processes dependent only on contiguity of locally converging signals.

2.3.3 Objections to the First Prototype Cortex

There are a number of features of this prototype cortex against which objections may be raised. Amongst these objections two may be mentioned here:

1. One of the premises in the Section 2.3.1, that there can be "omniconnected networks" is implausible except for blocks of neural tissue of quite small size. True omniconnection could apply only where the number of motor cortical recipient neurons was equal to or less than the average number of synapses made by the axonal arborization of a single sensory cortical neuron. Single neurons receive and give approximately 10^4 connections from and to other neurons. This means that the above model is limited to blocks of sensory or motor cortex containing this number of neurons or less, actually a very tiny block of cortex.

2. There are no direct connections from areas of sensory cortex to those of motor cortex, at least for the distance senses (vision and hearing). This has been established in several species (rat: Miller and Vogt 1984; cat: Kawamura 1973; monkey: Kuypers *et al.* 1965; Pandya and Kuypers 1969; Pandya *et al.* 1969; Jones and Powell 1970). Instead, the most direct pathways there are include an additional relay in an intervening area of association cortex (e.g., areas 5 and 7). At this intermediate step of the pathway there is ample opportunity for divergence and convergence. This means that any pathways from neurons in sensory cortex to those in motor cortex cannot be independent of each other: they must interact and overlap.

2.3.4 A Tripartite Cortex

For the time being we will still think in stimulus-response terms. However, without solving all the design problems, we can at least amend our model cortex to one closer to anatomical reality. In Fig. 3, we have a tripartite cortex, with an area of association cortex (A) between sensory and motor cortex. At each stage (i.e., between S and A, and between A and M) we again have an omniconnected set of links.

As mentioned above, this inevitably means that, in anatomical terms, massive divergence and convergence of signals can occur in the association cortex. This would seem to produce a cortex in which signals are so widely mixed that total confusion results. However, it may be inappropriate to think in anatomical terms. After all, we are nowadays accustomed to the idea that many connections which exist in anatomical terms may be "silent", unable to transmit messages across the synapses very effectively. We now believe that the effective connections are a small subset of the anatomical ones, selected by the Hebbian paradigm (or related rules) for synaptic modification. Given this, there are, in principle, two possibilities of the sort of pathways between sensory and motor areas which might be selected in our tripartite cortex:

The *first* possibility, illustrated below (Fig. 4) is that the lines of connection from sensory to motor cortex, via the association cortex, remain quite independent of each other, a principle similar to that in our first prototype cortex.

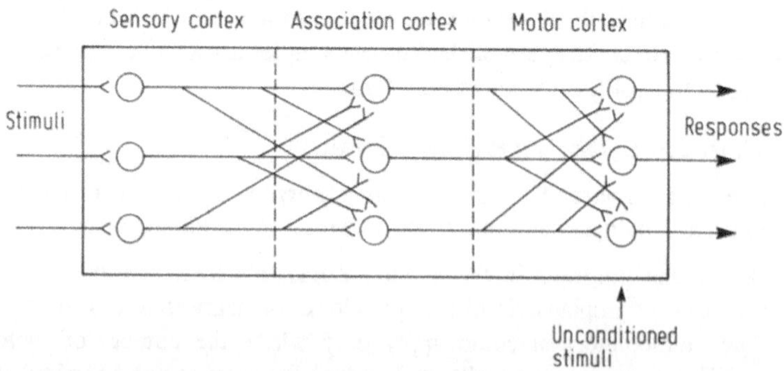

Fig. 3. Schematic diagram of a cortex consisting of a sensory and a motor area, with an intervening area of association cortex. Striatal connections as in Fig. 2 may be assumed

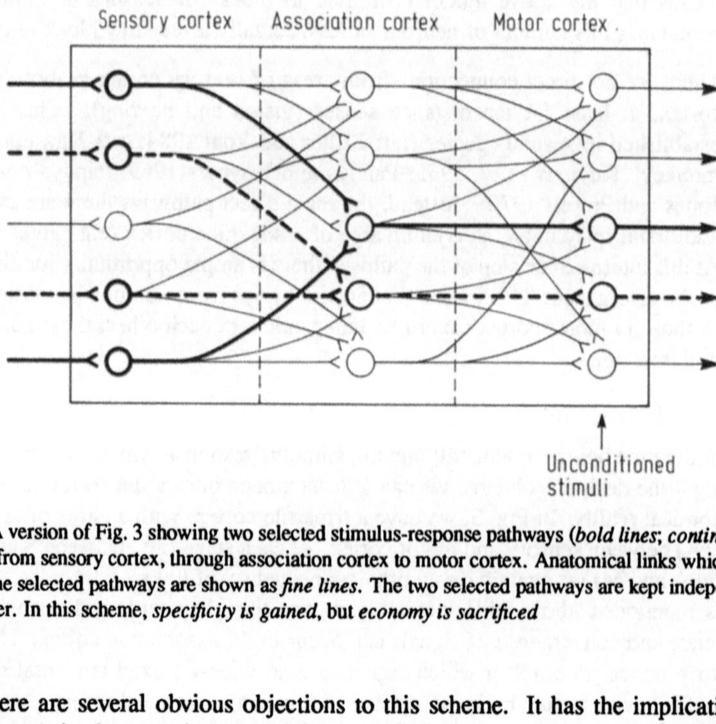

Fig. 4. A version of Fig. 3 showing two selected stimulus-response pathways (*bold lines; continuous* and *broken*) from sensory cortex, through association cortex to motor cortex. Anatomical links which are not part of the selected pathways are shown as *fine lines*. The two selected pathways are kept independent of each other. In this scheme, *specificity is gained*, but *economy is sacrificed*

There are several obvious objections to this scheme. It has the implication that each neuron in the association cortex can participate in only one stimulus-response link. However, this is almost certainly incorrect: a very uneconomical representation of information is implied in which the maximum number of links which our cortex can represent is the same as the number of neurons in the association cortex. The vast repertoire of synaptic connections on each neuron, which each has the potential to act as an independent site of information storage, would be seriously under-utilized. In addition, it may be asked "by what rules of synaptic modification could the formation of such independent links through the assocation cortex be established?" The Hebbian rule would not suffice, since a single stimulus neuron in the sensory cortex, which was

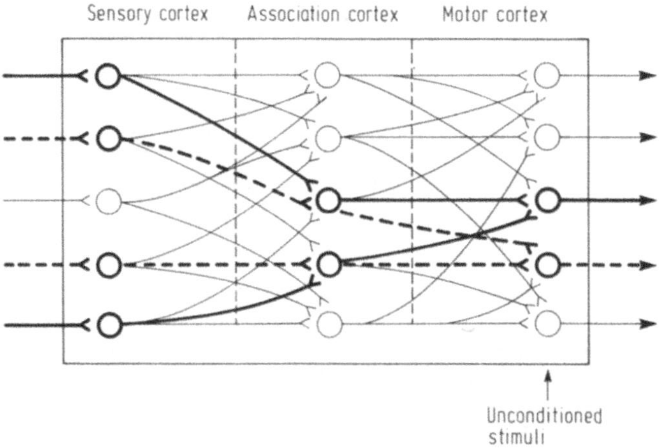

Fig. 5. Another version of Fig. 3, again showing two selected stimulus-response pathways (*bold lines*; *continuous* and *broken*) from sensory cortex, through association cortex to motor cortex. In this case, the two pathways overlap in the association cortex. In this scheme, *economy is gained*, but *specificity is sacrificed*

required to be linked with more than one response neuron in the motor cortex, could as easily form the link via a common neuron in the association cortex as via two different neurons.

The *alternative possibility*, illustrated in Fig. 5, is that each association cortex neuron could participate in several different stimulus-response links.

Once again the objection to this scheme is obvious: the link between a single stimulus and a single response would no longer be inevitable and unconditional, as implied by early learning theorists. The relationship would be irresolvably ambiguous. This is true even though the Hebbian mechanism has been envisaged to operate, severely pruning the anatomical repertoire of connections to produce the functionally effective ones. Nevertheless, if this problem could somehow be overcome, one would have a more economical storage of information, which utilized the number of available synapses, rather than the number of available neurons. (In Fig. 5 two association cortex neurons are depicted as participating in both S-R links; but either one of them singly would presumably be as good as both).

In short, the alternative ways of representation of input-output relations in these two figures represent the horns of a dilemma: *Either specificity is gained and economy sacrificed (as in Fig. 4), or economy is gained, and specificity sacrificed (as in Fig. 5).* How can these two be realistically reconciled?

2.4 Distributed Memory

2.4.1 The "Combinatorial Matrix"

An approach to resolving this dilemma has been suggested by several authors in the last 15 years (Marr 1969, 1970, 1971; Kohonen 1977; Palm 1980, 1982; McNaughton and Morris 1987), and is probably used widely as a design principle in various parts of central nervous systems. This uses a simple mathematical principle, that for a

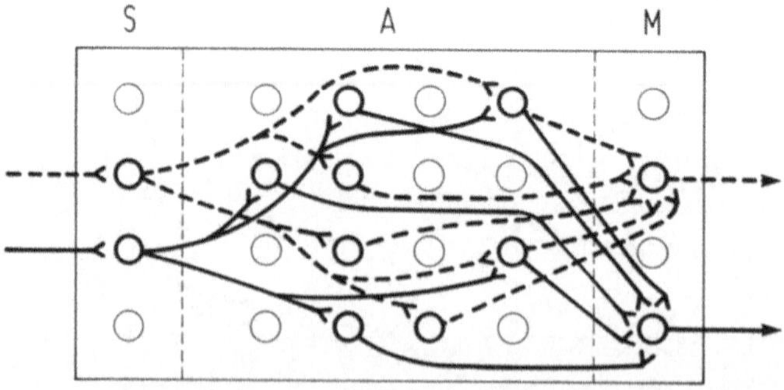

Fig. 6. A version of the tripartite cortex of Fig. 3 in which stimulus-response links are made by *combinations* of neurons in the association cortex, rather than by single neurons

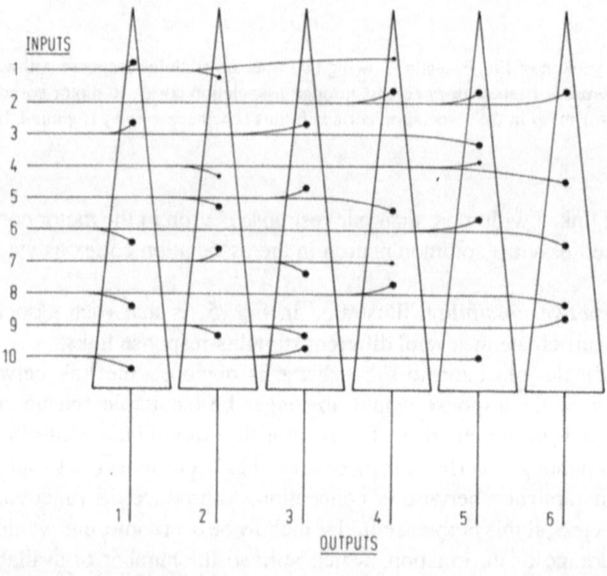

Fig. 7. Operation of the Hebbian principle to allow distributed representation of information, by combinations of neurons in a cortical block

set of N neurons, the number of unique subsets of N is vastly greater than N itself[2]. So, one could imagine (Fig. 6) that each link from a specific stimulus neuron to a specific response neuron involves not a single neuron in the association cortex, but a specific combination of them. The selection of neurons for each combination is envisaged to use the Hebbian principle for synaptic modification. Details of the possible operation of this design principle are illustrated schematically in Fig. 7. In this scheme

[2]In fact, for subsets of size r, the number of unique subsets is:

$$\frac{N!(N-r)!}{r!}$$

In the figure an array of six output neurons and ten input fibres is depicted. Connections are arranged randomly so that each input fibre gives synapses to three neurons, and each neuron receives five input synapses ($3 \times 10 = 5 \times 6$).

Registration of a memory. The Hebbian principle for synaptic strengthening is envisaged to require that *three* input synapses be activated simultaneously. Then input combinations, presented singly, will strengthen connections to outputs according to the following table:

	Input combination	Output combination
a)	2,5,7,9	2,3,4
b)	1,3,7,10	1,3,
c)	1,4,6,8	1,5
d)	2,3,6,10	1,6
e)	2,6,9,10	5,6,

Retrieval of a memory. Assume that the first four of the above patterns (a–d) are stored simultaneously in the network (The corresponding synapses are shown in Fig. 7 as larger "blobs"). For retrieval, these enhanced synapses are more effective than the remainder. Assume therefore that co-activation of only *two* such enhanced inputs is required to fire a neuron. In the above case, correct retrieval is possible with incomplete input patterns, as follows:

	Input combination	Output combination
a)	2,5,7	2,3,4
b)	1,3,7	1,3
c)	1,4,6	1,6
d)	2,3,6	1,5

In other words, the four input patterns may be recognized accurately, even when data about them is incomplete. However, if all five patterns are stored simultaneously, the relations between typical input patterns and corresponding output patterns would be as follows:

	Input combination	Output combination
a)	2,5,7	2,3,4
b)	1,3,7	1,3
c)	1,4,6	1,5,6
d)	2,3,6	1,5,6
e)	2,6,9	5,6

In other words, the specificity of recognition breaks down (in the above case because inputs c, d and e do not produce distinguishable outputs) different stimulus-response links would use different combinations of neurons in the association cortex, but any one of those neurons could participate in several such unique combinations. Memory would then be distributed throughout the association cortex. The idea therefore accords with the consensus from experimental work, that objects in the environment are not represented by single cortical neurons ("grandmother cells"), but probably by groups of neurons. The requirements for both specificity and economy would be satisfied. We can call this configuration a "combinatorial matrix".

For such a principle to be useful, one would, of course, require a mechanism, downstream from the association cortex, which could recognize the specific combinations in the association cortex which joined each stimulus-response link. This could be so if two additional features in the link from association to motor cortex were present: first there should be a sufficiency of connections between association and motor cortex to form representations of any likely combination. Second, these connections should be modifiable by rules which permit groups of co-active synapses converging on a single motor cortical neuron to become strengthened as groups. In other words, these synapses should follow Hebbian or similar rules for synaptic modification. Given these two very plausible premises, the motor cortical neurons, whose representation of the response is defined from other sources, will come to be addressed by just those combinations of association cortex neurons which the corresponding stimulus activates.

2.4.2 *The Cell Assembly*

Tissue of the cerebral cortex contains many intrinsic (local) connections, which find their synaptic targets without leaving the cortical grey matter. It is very likely that these connections also are subject to modification by Hebb-like rules. Given this, it follows that the neurons comprising each specific combination of cells in association cortex, whose concurrent activity links sensory to motor cortical neurons, will become functionally linked by strengthening of mutual local interconnections. This is illustrated in Fig. 8. When this happens, the combination of cells has become a "cell assembly", a term coined by Hebb (1949), implying that when some of the neurons constituting such

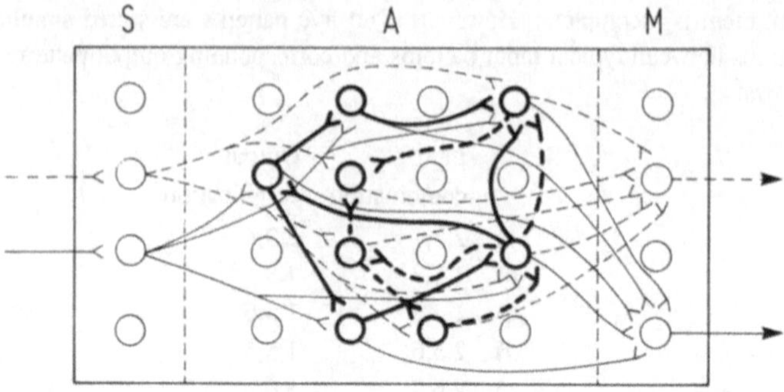

Fig. 8. A version of Fig. 6 in which each combination of cells in the association cortex is coordinated to form a "cell assembly". This is achieved by the strengthening of local connections between members of each combination. Extrinsic connections of the two "cell assemblies" are shown in *thin lines* (respectively as *thin continuous and broken lines*). Intrinsic conections within each "cell assembly" are shown as *thick continuous and broken lines*

an assembly become active, activity will spread throughout the assembly. For reasons that will soon become apparent, such an assembly is designated here as a "local cell assembly".

2.4.3 An Advantage of the Combinatorial Matrix: Economy of Storage of Information

Palm (1980, 1982) has estimated mathematically the maximum information capacity of such a combinatorial matrix. Although there are the order of $N!$ unique subsets of neurons in a network with N neurons (see footnote above), the actual number of utilizable combinations turns out to be considerably less than this. The number of bits of information which can be stored as modified synapses cannot be greater than the number of synapses. In a pathway from a block of N sensory cortex neurons to another block of N association cortex neurons, omniconnected, as in Fig. 3, then the number of synapses is N^2.

There are, however, additional limits on the maximum infomation capacity, set by the fact that if too many combinations are represented in the same matrix, their specificity is once again lost. (An example of this loss of definition during retrieval is given in Fig. 7, as the number of patterns stored simultaneously in the array increases). Making ad hoc assumptions about highly schematic matrices, Palm computed that in a well-filled matrix the number of synapses which could be utilized without undue loss of specificity approached $0.7 \times N^2$.

This value depends on the exact assumptions, for instance concerning the size of the matrix, the number of active inputs which contribute to activating the local cell assembly, and the required fidelity of retrieval of stored information. Clearly there is a "trade-off" between the fidelity of information retrieval and the information storage capacity. The value given above is clearly much less than would be predicted from the number of unique subsets drawn from N neurons. Nevertheless, it is a considerable saving if N is large. Palm (1982) gives the following specific results: For a matrix consisting of 100 neurons with 100 synapses on each neuron, inputs consisting of 6 out of the 100 inputs being active in a typical pattern, and with a criterion of retrieval of 90% (on average) of the stored information, 417 such patterns can be stored. For a 1000×1000 matrix, and input patterns consisting of 9 out of 1000 input lines being active in a typical pattern, 34000 such patterns could be stored with 90% fidelity. If the required fidelity of retrieval becomes less than 90%, the number of patterns which can be stored rises dramatically. In any case the improved storage capacity resulting from use of the combinatorial matrix is very significant, and becomes increasingly large as the number of nerve cells in each matrix increases.

An estimate can also made of the number of distinct cell assemblies which can coexist in a block of (association) cortex. Assume that the size of a cell assembly is defined by the average number of cells in the assembly whose firing is triggered when one other cell in the assembly fires. Let this number be n. Then correspondingly, the number of synapses from sensory to association cortex used in connecting up this assembly will increase by a factor n. Thus the average storage capacity declines when information is represented in cell assemblies by a factor n. Even so, in blocks of cortex of realistic size, considerable economies of storage of information are still likely to accrue from use of combinatorial matrices.

2.5 A Cortex for Gestalt Psychologists

All the above arguments, whether they depend on representation of object and events by single neurons or by cell assemblies, are based on stimulus-response theory. In other words it has been assumed hitherto that stimuli and responses are the categories whose cerebral representation is enough to define behaviour. However, organisms seldom respond to individual, invariant stimuli. More typically they respond to "objects". Each object can present many different stimuli ("images") to the organism, each with its own stimulus configuration. For instance, an object such as a cube will produce quite different sensory messages according to its orientation and distance from the eye. Nevertheless, we can respond to the cube as a whole, although we never see it as a whole. This is the idea underlying the school of psychology which flourished between the two world wars and was generally referred to as "Gestalt psychology". Its central idea was that we can represent "wholes which are greater than the sum of their parts. (i.e., Gestalts)" (Kohler 1938).

Likewise we do not emit unitary, invariant responses. Most of our responses are guided and continually adjusted by concurrent stimuli. Just as we need to have representations of Gestalts which are abstractions from the many sensory images of a single object, so we need to have representations of "motor Gestalts" ("response intentions"), which are abstractions or generalizations from many individual forms of a particular response. In other words, we need to be able to represent the "essence" of a response, incorporating the fulfillment of the immediate intentions of that response, whatever the details of execution dependent on exigencies of sensory guidance.

Thus, the key variables controlling input/output relations are not stimuli and responses, but higher-order derivatives of them. It is reasonable to suppose that primary sensory and motor cortices (such as in our initial prototype) represent stimuli and response programmes, but it would seem that some intermediate stage(s) are required if objects and response intentions are to be represented. This may be the design factor which necessitates polysynaptic connectivity between sensory and motor cortex, via intervening areas of association cortex (see Sect. 2.3.3).

It is relatively easy to specify the conditions under which the formation of Gestalt representations (i.e., representations of whole objects or response intentions) can come about. Suppose an object can present three different images to the nervous system (i1, i2 and i3, in Fig. 9). Each one of these images will come to control a small subset of neurons in the association cortex. These may each be called "image cell assemblies". The different subsets will have some degree of overlap. Initially, if the three images activate quite different sets of neurons in the sensory cortex, this overlap will be determined just by the statistics of the anatomical connectivity between sensory and association cortex. The overlap will therefore be no greater than that between images of different objects. However, as mentioned above, all areas of cortex contain prolific intrinsic connections (i.e., confined to the same area). If these local connections are subject to Hebb-like modification processes, as are the long cortico-cortical connections, the initial state of haphazard overlap between the cell groups representing different images of an object is not likely to last long. The different images of a single object will inevitably tend to occur together in close temporal juxtaposition, or in transitional forms. This high degree of correlation can be registered in the neural network by strengthening of local

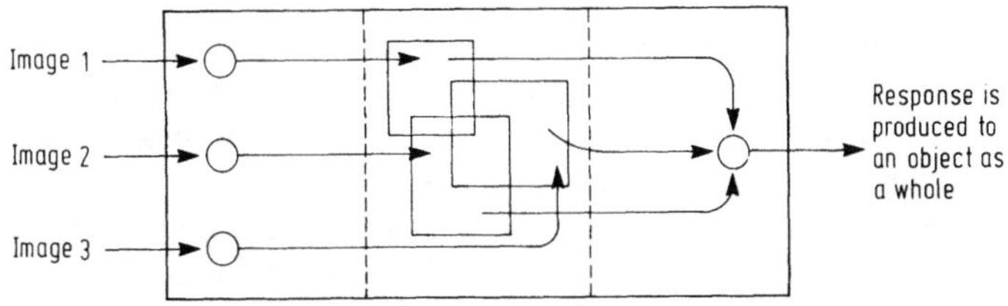

Fig. 9. An object can present three different images to the organism. Cell assemblies for each individual image can form in the association cortex (*three squares*). These can then become coordinated as a single, larger cell assembly. This represents the object as a whole, although the object is never seen as a whole. *Alternatively*, the three squares in the association cortex can be regarded as representations of individual response programmes, fulfilling the same intention. When they become coordinated, the generalized intention is represented. Taken together, a generalised intention may be carried out to an object as a whole, although the object is never seen as a whole, and the generalized intention is only an abstraction

connections in the association cortex, between neurons representing the different images of an object. The different image cell assemblies will thus gradually coalesce into a larger cell assembly representing the object as a whole, an abstraction, since the object is never seen as a whole. (Both the image cell assemblies, and their resultant, an "object cell assembly", would still fall under the heading of "local cell assemblies").

Likewise, whenever an intentional act is performed, it will involve a small group of association cortex neurons. Since the same intention will be served by responses which differ in detail, there will come into being different collections of neurons which each represent a different version of the same response intention. These would inevitably have large elements in common (many neurons common to several versions of the response), and different versions would be used in tight temporal sequence. There would thus emerge a cell assembly (in association cortex rather than motor cortex) representing the motor Gestalt as a generalization (see also Fig. 9).

Let us summarise the argument so far: in the early prototype cortices (Sect. 2.3, Figs. 1, 2 and 3) there was a one-to-one mapping of stimuli or responses against nerve cells, but the conflicting demands of specificity and economy could not be met. Later, with distributed mapping of information, and the formation of cell assemblies (Sect. 2.4), these two design constraints could be reconciled, but the categories represented were still stimuli and responses. The prototype cortex for the Gestalt psychologist has an advantage over all of these prototypes based on stimulus-response theory. It contains a more useful representation of the environment. The complex relations of several images of an object, or of several response programmes which all fulfill the same intention can be mapped in a way which is formally similar to their actual relations. However, there is still a serious weakness.

2.5.1 Ambiguity in the Environment

So far in this chapter we have considered just the ambiguity inherent in cortical neural networks, by virtue of their manner of interconnection. We have accepted a central tenet of stimulus-response theory and have not questioned it when discussing Gestalt psychological notions. This tenet is that a discrete input (whether a stimulus or an object) provides all the information required by a nervous system (of whatever configuration) to specify the appropriate output (whether a response or a response intention). In other words, we have assumed that the environment itself is never ambiguous. However, this tenet is clearly wrong. It is clear that ambiguity arises for representation of both stimuli/objects and responses/response intentions.

A single image (stimulus), taken in isolation, may be an aspect of several different objects. Take for instance the design shown in Fig. 10A. This might be interpreted as the corner of a cube, or as an inverted letter "Y"; but consider it in the guises in which it appears in Fig. 10B or C. Clearly this simple design has quite different meanings in these different settings. Thus the information structure of the sensory environment is such that small-scale patterns of sensory information may, in themselves, be quite ambiguous.

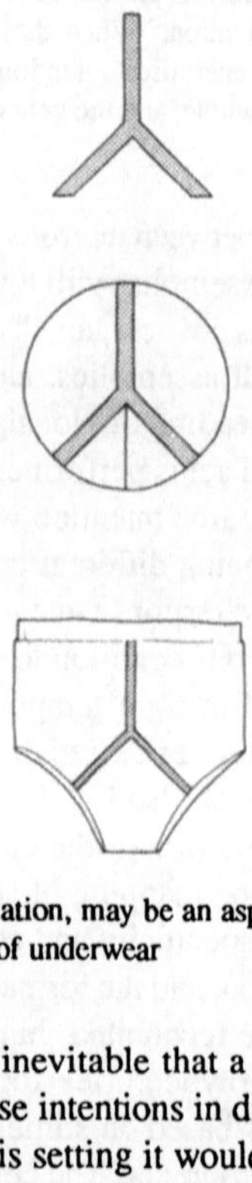

Fig. 10. A single image, taken in isolation, may be an aspect of different objects. A: An inverted letter "Y". B: A peace symbol. C: An item of underwear

On the response side, it is inevitable that a single object can acquire the capacity to control quite different response intentions in different settings. Take, for example the rose depicted in Fig. 11A. In this setting it would be quite appropriate to pick the rose, and put it in a vase on one's dinner table; but a very similar rose, seen in a different setting (Fig. 11B) would generally inhibit one from carrying out such a response. Thus, even if one can unambiguously identify the object to which a response should be made, this object has an ambiguous relation to response intentions it might elicit.

In terms of neuronal representation, an image which belonged to more than one object (as in our inverted Y example above) would have a representation which comes to be part of more than one object cell assembly. (As discussed above, the object cell assemblies thus overlap). If several images belonging to the same object occurred close together, the ambiguity inherent in any one of these would be resolved, because their combination would uniquely specify the object cell assembly; but when a single image occurs in isolation, there may an ambiguity in the representation of that image (Fig. 12).

Fig. 11. A single object acquires the capacity to control different response intensions in different settings. A: A rose in back garden. B: A rose in a public park (see text)

Likewise, the ambiguity of representation on the output side can be thought of in terms of cell assemblies. Suppose that a particular object, represented in the association cortex, is a necessary trigger for two incompatible response intentions. One must imagine that in one situation one particular group of cortical neurons has a heightened tendency to discharge, as a result of instrumental or classical contingencies. In a different situation a different group of cortical neurons is brought close to the discharge point

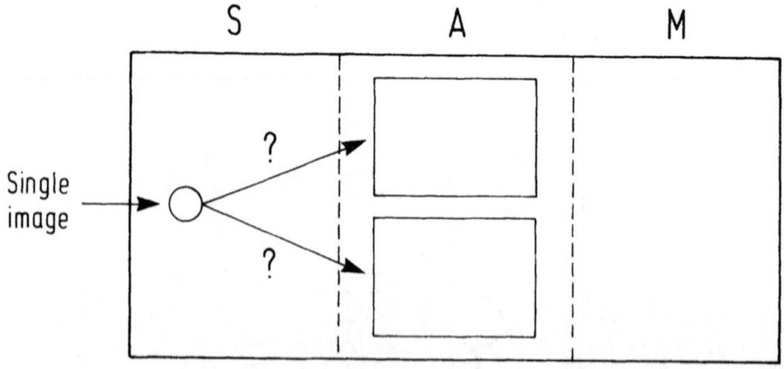

Fig. 12. A single image representation may not be able to uniquely address a specific object cell assembly

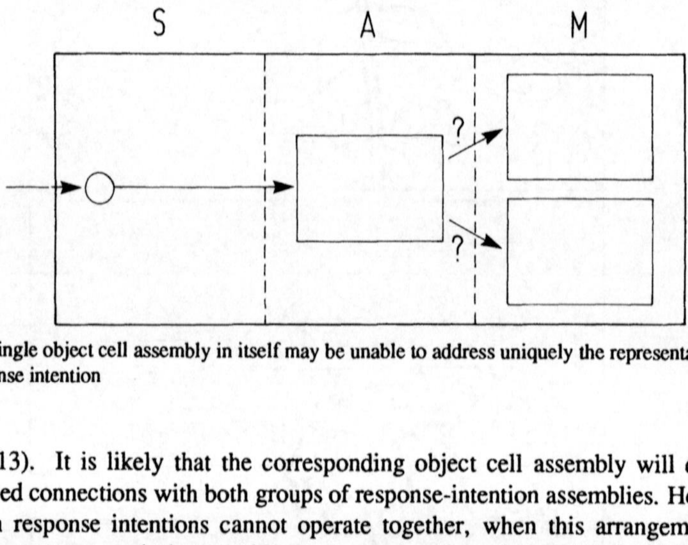

Fig. 13. A single object cell assembly in itself may be unable to address uniquely the representation of a single response intention

(see Fig. 13). It is likely that the corresponding object cell assembly will develop strengthened connections with both groups of response-intention assemblies. However, since both response intentions cannot operate together, when this arrangement has been set up we once again have ambiguity in the representation of the object-response relationship.

Thus, despite the great advantages of a distributed memory system, representing Gestalts in overlapping cell assemblies, it cannot resolve the ambiguities inherent when environmental information is considered in small chunks. In one sense it faithfully codes the contiguities in the environment (on a small scale) both in simple circumstances when these leave no ambiguity, and in more complex circumstances when these are highly ambiguous. Of course, the advantages of distributed mapping of memory amount to nothing if we cannot unambiguously retrieve information from this network.

In view of this, we cannot develop our prototype cortex further, unless we move to another level of realism for the representation of information in the brain. In principle, the ambiguity in both cases (Figs. 12 and 13) can be resolved, but this involves not the properties of the discrete image (or object), but the configuration of the environment as a whole. Given the totality of information from the whole environment, the conditions can be set up so that a single image can address the correct object cell assembly, or a single multipurpose object cell assembly can address the appropriate response intention. However, for this to be achieved one must depart from the molecular approach of stimulus-response theory and (to a lesser extent) Gestalt theory, and seek a more molar representation of information, typical of the cognitive theorist.

2.6 Towards a Cortex for the Cognitive Psychologist

2.6.1 Resolution of Ambiguity by the Environment as a Whole

In principle, the types of ambiguity inherent in images and objects could commonly be resolved if one could take account of the environment as a whole in which these various images and objects present themselves. However, the easy illustrations of the use of the environment as a whole to resolve the ambiguity of a small part of the environment (Figs. 10 and 11) should not betray us into thinking we have solved our problem. *We still need to enquire into the mechanism by which this can occur*. Here we run into surprising difficulties.

At first sight there appears to be a quite obvious mechanism. In the discussion above it was argued that the ambiguity inherent in a single image of an object could be resolved if the representation of several images of the same object could be coordinated. The necessary anatomy and physiology in which this could occur were easily defined. By analogy, it would seem simple to utilize cues from the environment as a whole to establish a "global cell assembly", which incorporates the single image of our object. Provided this global cell asembly has been activated by the environment as a whole, then it will be clear to which object the image belongs. The global cell assembly will incorporate the representation of only those objects which have occurred consistently in that environment. Therefore the single image cell assembly will be interpreted as belonging specifically to one of those objects, rather than to an object which only occurs in a different environment.

There is, however, a catch. In the above schema, it is necessary to represent the environment as a whole. It is therefore necessary to have a much more diverse cortex. It will be required to have sensory areas for all senses—visual, auditory, somatosensory, olfactory. Each of these would need its own areas of association cortex. In addition, it would probably be necessary to have several separate regions of association cortex where information from each of these "first-order association cortices" could be combined further (e.g., auditory with visual, visual with somatic). In short, one would need a very much bigger cortex altogether (see Fig. 14).

2.6.2 The Effect of the Size of the Cortex

Does this put any constraints on the principle of the "combinatorial matrix", which would prevent the formation of a "global cell assembly"? In fact it probably does, for the reason that formation of cell assemblies requires a sufficiency of connections. For instance, to form any local cell assemblies in the association cortex (Fig. 9), one needs sufficient connections such that there are a significant number of association cortex neurons where inputs from sensory cortex converge to contribute to representing the image or object. To coordinate the different image cell assemblies to form an object cell assembly (Fig. 9), one needs sufficient connections intrinsic to the association cortex so that there are at least a few neurons mutually connected between the different image cell assemblies. To form a specific link between an object cell assembly and the assemblies representing response intentions, one needs sufficient connections between the relevant

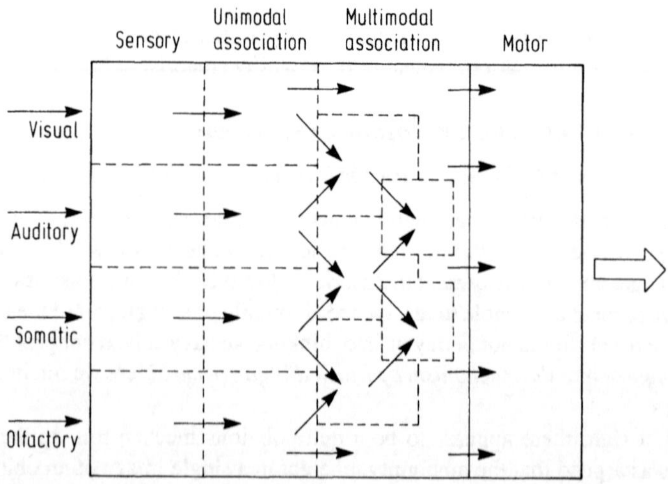

Fig. 14. A more realistic plan for a cortex which can represent the environment as a whole. Compare this with the prototype cortices in Figs. 1, 2, 3, etc.

areas of association cortex, so that at least a few cortical neurons in the response intention assembly can be activated.

These statements require a somewhat quantitative approach. The first thing which we need to estimate is: "How many active axonal inputs must converge on a single neuron to raise its membrane potential from the resting level to the threshold for firing?" If one knew this, one could calculate the chances that a single neuron would receive sufficient connections to allow the recognition of the combination of neurons which form its source of input. Although this question has, as yet, no firm answer, there are experimental results which provide probable lower limits to the number of converging afferents required. Creutzfeldt and Ito (1968) and Creutzfeldt *et al.* (1969) produced evidence, by intracellular recording from cortical neurons, which suggested that a short burst of impulses in a single fibre afferent to the impaled neuron could raise membrane potential from the resting level a substantial way towards the threshold for firing. Specifically, the single EPSP's had amplitudes of between 0.5 mV and 3 mV. This may be compared with a figure of between 5 and 10 mV, which represents the depolarization from resting levels required to reach the threshold for neuronal firing. If their inference is accepted, it is clear that threshold may be reached if only a very few active afferent axons converge on a single neuron. The actual number required will be smaller, the larger the number of closely spaced impulses occurring in each axon. In the paragraphs below, the somewhat arbitrary, but plausible assumption is made that convergent activation in at least *three* axons is required to produce postsynaptic impulse discharge. (A minimum rather than an average figure is chosen because it makes the conclusion of the following argument more secure. The circumstances in which this minumum figure apply are presumably for synapses which are already strengthened by the Hebbian mechanism).

Imagine two connected regions of cortical tissue each of dimensions about 1 mm^3. These would each contain of the order of 10^5 neurons. Each one of these neurons would receive (and give) about 10^4 connections to other blocks of neurons. So the probability of a neuron of the afferent region making contact with one in the recipient region would, on average, be about 0.1. If the convergence of three coactive afferents was sufficient to fire a postsynaptic neuron, then, assuming random distribution of axons, the probability of recipient neurons having such anatomical convergence would be roughly $0.1 \times 0.1 \times 0.1$, i.e., one neuron in a thousand. Thus in the 1 mm^3 cortical block, approximately 100 neurons would be capable of representing the converging combination of three coactive afferent neurons. This might be sufficient to allow the block of recipient neurons to effectively recognize the combination.

Suppose, however, that the dimensions of these regions were not 1 mm^3, but more realistically in the mammalian brain, 1 cm^3. There would then be 10^8 neurons in this volume, but each neuron would still give (and receive) only about 10^4 connections. In this case, the probability that a single afferent neuron would directly contact a single recipient neuron would be 0.0001. The number of neurons in which the necessary triple convergence could occur would be 0.0001^3, i.e., about 1 in 10^{12}. In other words, as the size of the brain regions increases, not only does the *proportion* of neurons with the required connections decrease, but in fact there is a decline to vanishingly small *absolute numbers*[3].

These estimates are based on the idea that individual projecting and receiving neurons encode information, whereas the discussion earlier in the chapter indicated that cell assemblies, combinations of neurons which tend to fire together, are probably the meaningful entities. Bearing this in mind, the number of appropriately connected neurons is certainly much larger than the above argument suggests. In principle one could calculate cell assembly sizes on the basis of convergence and connectivity ratios. However, the real determinants of the size of cell assemblies are likely to be factors governing cell excitability, for example after-hyperpolarization (Alger and Nicoll 1980; Hotson and Prince 1980; Schwartzkroin and Stafstrom 1980), inhibitory interneurons, or global control systems originating in brainstem and other nuclei. It is impossible to express these factors quantitatively.

However, the size of the cell assemblies is probably irrelevant to our argument in any case. If a cell assembly is large, its information-containing capacity (per cell) is correspondingly small. For cell assemblies of 1000 neurons, giving 10 million synapses, approximately 10000 bits of information could be stored, i.e., 10 bits per neuron. For cell assemblies of only 100 neurons, giving 1 million synapses, approximately 10000 bits of information could also be stored, this time at a ratio of 100 bits per neuron. The real force of the argument is thus unchanged by considerations of the size of the cell assembly. It is this: In a large cortex one may, on the one hand, have small cell assemblies (in the limit a single cell) and a problem with inadequate numbers of appropriately connected cells; or one may, on the other hand, have large cell assemblies in which case there is an equally severe problem with inadequate economy of storage of information. *In the former case, a global cell assembly representing the whole environment simply could*

[3]Braitenberg (1978) has suggested that there may be a tenfold increase in the number of connections possessed by a cerebral cortical pyramidal cell in the human cortex, compared with that in the mouse cortex. However, even this difference is quite small compared to that required if the connectivity ratio of a single neuron was to match the number of nerve cells with which that neuron should be connected in the above calculation.

not form. In the latter case it could form but would not contain sufficient information to be much use in resolving ambiguities of small chunks of information.

The assumptions of the above calculation were chosen so as to understate the case. For instance, it was assumed that all the connections of the projecting neurons were distributed to a single recipient block, whereas in fact only 10% might be a more realistic proportion. Nevertheless, the conclusion is clear. *In a block of nervous tissue that is anything like as large as real mammalian cortices, it is highly improbable that "global cell assemblies" can organize themselves in any useful manner. Either there are simply insufficient cortico-cortical connections to establish cell assemblies (except in rather small regions of cortex), or the global cell assemblies contain insufficient information to resolve the ambiguity of the local cell assembly for an image or object.*

This chapter started with cortical models based on stimulus-response theory, and proceded via progressively more realistic ones. Yet we still have not resolved the basic ambiguity in representation of information. This ambiguity can be regarded as a biological counterpart of the indeterminacy we noted in Chapter 1, with respect to philosphical and psychological issues. Thus we have a threefold analogy: (1) In deductive inference, premises do not dictate conclusions unconditionally. (2) In learning theory, stimuli do not dictate responses unconditionally. (3) From consideration of the microanatomy of nervous tissue, representations of input apparently cannot uniquely address the appropriate representations of output. There thus seems to be a severe paradox. We know that, given the right conditions, premises can dictate conclusions, and stimuli can dictate responses. Therefore there must in fact be some way in which the apparent ambiguity of representation in neuronal tissue can be resolved. If global cell assemblies could be formed, the resolution of the ambiguity (inherent both in cortical connectivity and in the structure of small scale information structure of the environment) could be accomplished. *One therefore needs a scheme of anatomical or physiological organization other than the combinatorial matrix* to achieve this, although the latter scheme is still likely to be an important contributory principle. To discover this manner of resolving ambiguity of representation will therefore be tantamount to explaining the conditions under which premises dictate conclusions, or stimuli dictate responses. This is very closely related to the question put at the start of the first chapter ("What is a context?"), since it is some function of the environment as a whole (generally speaking, the "context") which provides these conditions. But what is this function?

Actually, the stumbling block we have met with in considering the use of global cell assemblies has in a sense already been solved, in the conventional digital computer. This can store large amounts of information, quite economically, and without any ambiguity. However, compared with real brains, conventional computers are absurdly slow in information retrieval, and, unlike real brains, their speed of retrieval depends on the amount of information stored. This is because in computer systems information is usually "listed", whereas in brains (and in all the prototype brains considered above) it is "mapped". The latter characteristic means that it is not necessary to sort through the whole of a memory bank, item by item, to retrieve a memory. Thus we should not use computer design to solve our problem of design in the real brain. Mapping (as opposed to listing) of information must continue to be a feature of any scheme used to resolve our design problem.

2.7 Discussion

Without global cell assemblies many learning tasks could still occur in the same manner as shown by intact mammals. These are tasks where the contingencies relating stimuli, responses and reinforcers occur in a highly controlled, constant environment, and always in the same relation to each other, in other words in the sort of experiment beloved of the stimulus-response theorists. However, in an ecologically more realistic environment, such as those employed by the cognitive theorists, where the relation between stimuli, responses and reinforcements is conditional upon the global configuration of the environment (in other words, in context-dependent learning situations, such as discussed in Chapt. 1), the animal is likely to be severely impaired unless this design problem can be overcome.

We started Chapter 1 by asking why it was necessary to subdivide information patterns into "items" and "contexts". The arguments presented in this chapter give us the start of an answer to this question. Pattern recognition for single stimuli or objects (involving the formation of a "local cell assemblies") can apparently be accomplished in real neural networks by the use of the connections that are available, and the Hebbian premise for synaptic modification. Using such mechanisms alone, pattern recognition for larger-scale information groupings (formation of a "global cell assembly") fails due an insufficiency of connections. A radically different, and more complex method of formation of a cell assembly appears to be necessary, to coordinate the widely distributed neurons in such a global cell assembly. It is the information content of this coordinating mechanism which we can designate as a "context".

In brief, a context representation is required to resolve the ambiguity in the information structure of the environment, when that ambiguity can be resolved in no other way. This is the initial answer to the question posed at the start of Chapter 1. The answer will become clearer as the argument proceeds in later chapters.

2.8 Hints that the Hippocampus Might be Involved in Forming Global Cell Assemblies

To form such global cell assemblies (i.e., ones which represent the whole environment as a Gestalt) it seems necessary for there to be some region of the forebrain where connections from all parts of the cortex converge. Geschwind (1965a,b) has pointed out that in the typical mammalian cortex, it is only in the limbic areas where inputs from all other parts of the cortex intermingle. Elsewhere in the cortex there are, for instance, no regions where visual, auditory and somatic sensory information converge. This generalization does not hold for the higher primates, where areas specialized for language (or the evolutionary precursors of such areas) also show such multimodal convergence. However, this exception need not obscure the significance of the limbic lobe of the cortex for the present discussion. As will be detailed in the next chapter, the hippocampus has connections with the neocortex, and these are strongest for the entorhinal cortex, and still strong, though less direct, for limbic areas such as the cingulate cortex, and for the prefrontal cortex. Thus, the process of multimodal convergence starts in the connections of the latter areas, continues more powerfully through the entorhinal cortex, and reaches its culmination in the hippocampus.

We cannot immediately conclude from this that it is the hippocampus which coordinates global cell assemblies. With such massive convergence it is, at first sight, difficult to see how the specificity of a global Gestalt representation could be maintained. Moreover, the pathways from the areas of association cortex such as 5 and 7 (which deal directly with sensorimotor information) to the hippocampus involve several synaptic relays, before they can influence the hippocampus, each of them allowing convergence and divergence. There are therefore further difficulties to be resolved before the role of the hippocampus in representing global Gestalts is clarified. Nevertheless, it is an indication which helps us to narrow down possibilities of the regions where we should look for global Gestalt representation. Thus the argument based on reconciling our design constraints (economy, and specificity of information coding, and realistic connectivity ratios of single neurons) seems to point to the same structure as that based on other evidence alluded to at the end of Chapter 1. *The hippocampus, appears to be a crucial region for representation of contexts, or for formation of global cell assemblies.*

The details of this role are the subject of the rest of this monograph. The central idea, (enunciated in detail in Part 3), is that global cell assemblies can be selected during learning, and subsequently used during performance if they establish resonant circuits entrained to the rhythm of a neural oscillator situated outside the main part of the cerebral cortex. This neural oscillator is identified as the hippocampus, and the neural oscillations are the hippocampal theta rhythm. Before this idea can be presented in convincing manner however, a great deal of experimental evidence must be discussed. This is accomplished in the chapters in Part 2. Some readers may look in advance for indications of where the argument is leading, in order to focus their mind on the somewhat difficult review material of Chaps. 3–8. They are directed to Chap. 11 (especially Sect. 11.1), which gives a synposis of the whole argument. This section is designed to be read either before or after tackling the detail of chapters 3–10, but will be more accessible to readers with some prior familiarity with neuroscience literature on the forebrain and related psychological issues.

A topic which will not be dealt with in this monograph is the behavioural effects of lesions of the hippocampus (and fornix). It is admitted that this subject is highly relevant to the case being argued. However, the literature on this is extremely large, and growing year by year. To cover it adequately would double the size of this book (at least) and possibly obscure its main thrust in a welter of detail. The interested reader is referred to extensive reviews by O'Keefe and Nadel (1978), Gray and McNaughton (1983) and Chozick (1983).

Part II

Reviews Of The Experimental Evidence Relevant To The Formulation Of The Theory Of Cortico-Hippocampal Interaction

3 Anatomy of the Hippocampal Complex and Related Regions

3.1 Introduction

Any attempt to construct theories about the brain, or about parts of it must recognize the constraints on the theoretical possibilities set by anatomy. The first part of this chapter deals with the overall topography of the hippocampal formation and related structures. This information has been available for a long time, but is summarized here for the benefit of those readers without this anatomical background. However, the main objective of this chapter is to describe the connectional anatomy of these regions, that is, the detailed origin and distribution of some of the connections to and from each of the recognizable components of the hippocampal complex. Our knowledge of this has undergone major revision in the last 10 years, following the introduction of modern techniques for tracing neural pathways using anterograde or retrograde transport methods. Not all of the connectional anatomy of the hippocampus is reviewed in this chapter. However, those connections which are clearly of relevance to understanding the theta rhythm, and the interplay between hippocampus and cortex are described in some detail.

3.2 Topographical Anatomy of the Hippocampal Formation

The hippocampus, like all limbic cortical structures, is a medial continuation of the mantle of cortical grey matter (see Fig. 15). It is therefore connected indirectly to fairly typical regions of cortex ("isocortex"), which are a layer of grey matter usually described in terms of six laminae. The hippocampus itself, though still a laminated region of cortex, has a much simpler pattern of lamination than the isocortex, having lost the equivalent of the superficial four layers. It is thus designated "allocortex".

Between the rhinal fissure, in the ventrolateral aspect of the hemisphere, and the hippocampus in the ventromedial aspect, is a transitional region. This includes some regions of cortex which are quite similar to typical isocortex: following round the base of the hemisphere from the rhinal fissure laterally to the hippocampus medially, one first encounters the entorhinal cortex, and then its medial relations in the cortical layer, the parasubiculum and presubiculum (terminology of Blackstad 1956). The pattern of lamination of these regions is similar to that of the entorhinal cortex. Further medially one encounters the subiculum and prosubiculum (Blackstad 1956). These two are also laminated, but their pattern of lamination is similar to that of the hippocampus itself, with only the deeper layers of cells being represented. Thus, between presubiculum and subiculum there is a sudden reduction in the thickness of the cortex (the "subicular cliff"). Subiculum and prosubiculum consist of a rather dispersed layer, consisting mainly of pyramidal cells in register with laminae V and VI of the presubiculum and

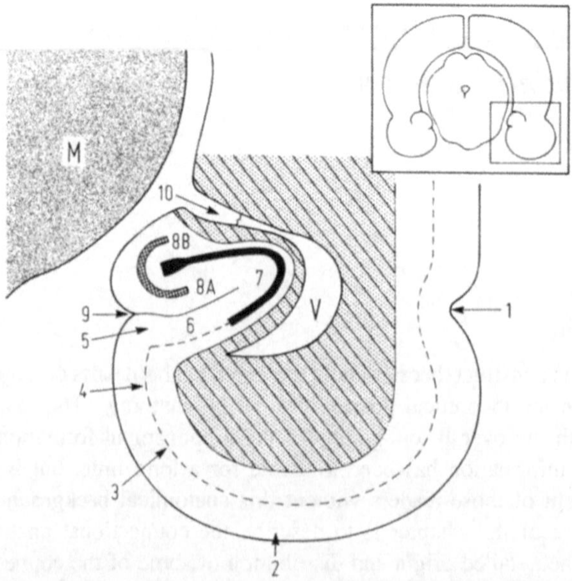

Fig. 15. A diagrammatic section, roughly in the coronal plane, through the ventral quadrant of the forebrain at the antero-posterior level of the parahippocampal gyrus, showing the various cytoarchitectural fields, other than those in Ammon's horn. 1 Rhinal sulcus; 2 entorhinal cortex; 3 parasubiculum; 4 presubiculum; 5 subicular "cliff"; 6 subiculum; 7 Ammon's horn; 8A dentate gyrus (buried blade); 8B dentate gyrus (exposed blade); 9 hippocampal fissure; 10 choroid fissure; V lateral ventricle; M midbrain. *Oblique lines* indicate white matter. The *dashed line* indicates the approximate position of lamina V of the cerebral cortex, continuing into the pyramidal cell layer of Ammon's Horn

entorhinal cortex (see Shipley 1975, plate I). Detailed cytoarchitectonic descritions of all the above regions have been given, but are not relevant here.

The hippocampus itself (otherwise known as Ammon's Horn, Cornu Ammonis, or CA) is distinguished from these regions because, in most regions, its pyramidal cells are grouped into a dense band, rather than loosely scattered. Nevertheless, the pyramidal cell layers of Ammon's horn and the subiculum are basically in register, as a continuous sheet. Ammon's horn is subdivided into a number of cytoarchitectural fields. In coronal section (Fig. 16), one encounters, in order, fields CA1, CA2, CA3 and CA4. Fields CA2 and CA4 are small and not well defined, so the most important CA fields are CA1 (otherwise known as "regio superior") and CA3 ("regio inferior"). There are detailed cytoarchitectural differences between the various CA fields, which need not be described here, recognizable in Nissl-stained sections, and also striking differences in their connectional patterns (see below). In field CA4 the dense packing of pyramidal cells gives way to a more diffuse arrangement, with a much wider variety of cell types being observed here (in Golgi preparations) than in the other CA fields (the "polymorph zone").

The CA fields as a whole are folded so as to be U-shaped when seen in coronal section. The hippocampal fissure, mainly obliterated in the adult, separates the two halves of this U-shaped fold. The "free edge" of Ammon's horn, that is the limit of

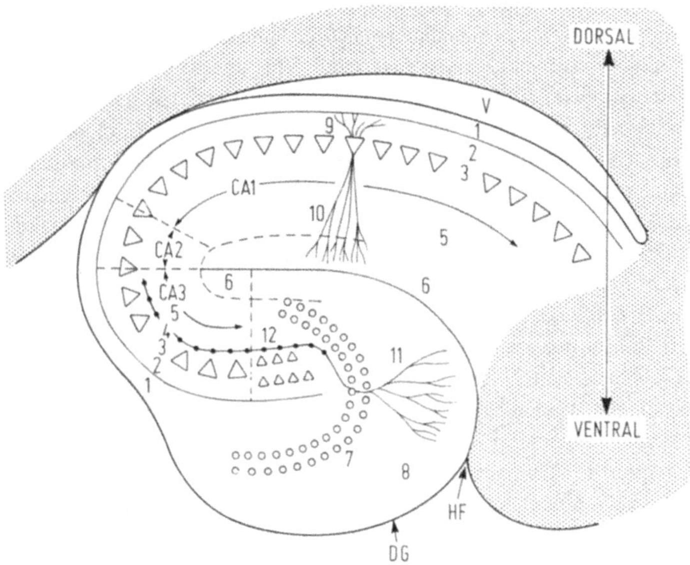

Fig. 16. Schematic coronal section of the hippocampus as in the rat brain, showing cytoarchitectonic fields and the laminae in the dentate gyrus and Ammon's Horn. (Note that this figure represents the dorsal hippocampus, and therefore is inverted compared with Fig. 15). Typical dendritic trees of principal neurons are also shown. 1 Alveus; 2 stratum oriens; 3 stratum pyramidale; 4 stratum lucidum (field CA3 only); 5 stratum radiatum; 6 stratum lacunosum/moleculare; 7 stratum granulosum; 8 molecular layer of dentate gyrus; 9 basal dendrites of a pyramidal cell; 10 apical dendrites of a pyramidal cell; 11 dendritic tree of a granule cell; 12 mossy fibre; V lateral ventricle; DG dentate gyrus; HF hippocampal fissure

field CA4, is enclosed by a "cap" of grey matter, the dentate gyrus. In coronal section this also is folded to appear U-shaped, interlocking with Ammon's horn. There are thus two blades to the dentate gyrus, a buried blade, enclosed within the fold of the CA fields, and an exposed blade, lying on the medial surface of the hemisphere (viz. of the temporal lobe in primates). Field CA4 is included in the CA nomenclature, but there is no consensus on whether it should be viewed as part of Ammon's horn or as a "hilus" of the dentate gyrus. The two terms will be used synonymously here.

The pyramidal cells of the CA fields have two sets of dendritic arborizations. The more extensive ones extend towards the "centre" of the hippocampus, (i.e., the hippocampal fissure) as the apical dendrites. The smaller trees, the basal dendrites, extend towards the "outer" surface of the hippocampus, where it bulges into the lateral ventricle. The dentate gyrus consists of many small, very densely packed neurons, the granule cells. Their dendrites extend in the "outer" region of the dentate gyrus (i.e., that furthest from field CA4). In the buried blade of the dentate gyrus these dendrites approach the hippocampal fissure.

There is a widely used nomenclature for the different laminae of the dentate gyrus and Ammon's horn, in which there are three laminae in the dentate gyrus, and five (or six) in the CA fields (see Fig. 16). In the dentate gyrus, the principle cells, the granule cells occupy the *stratum granulosum*, their dendrites fill the *stratum moleculare*, and beneath the granule cells is the hilus or polymorph cell region ("CA4"). In *regio superior* of Ammon's horn the lamina immediately adjacent to the ventricle is termed the *alveus*, containing mainly efferent axons of pyramidal cells, and some afferent axons. Deep to this is the *stratum oriens*, containing the same mix of axons, plus the basal dendrites of the pyramidal cells, and the cell bodies of some interneurons (including a distinctive class of cells called "basket cells"). Deep to this is the *stratum pyramidale*, where the somata of the pyramidal cells are located. The thickest lamina is deep to this, the *stratum radiatum*, which comprises mainly the proximal parts of the apical dendrites of pyramidal cells, running in parallel, and with relatively few branches. Deepest of all is the *stratum lacunosum/moleculare*, which borders on the hippocampal fissure, and contains the distal dendrites of pyramidal cells, more highly branched than in stratum radiatum. In this lamina are also some unusual interneurons, mentioned by Cajal (1911), and recently subjected to electrophysiological and electroanatomical study (Lacaille and Schwartzkroin, 1988 a,b). They appear to have axonal and dendritic ramifications which cross the hippocampal fissure into the dentate gyrus, and will be mentioned again in Chapter 8. The laminae are almost the same in *regio inferior*, except that a thin layer on the apical side of the pyramidal cell layer is distinguished from most of the stratum radiatum, because of a prominent band of transversely oriented axons. This band is the *stratum lucidum*, and the transverse axons are the mossy fibres (see below).

This description emphasizes the coronal or transverse appearance of the hippocampus. However, Ammon's horn and the associated dentate gyrus are elongated. In most mammals one end lies under the corpus callosum and is close to the septal nuclei. This is the septal or anterior end. The other end bends round the curvature of the lateral ventricle to the medial ventral aspect of the hemisphere (the temporal end). In primates the whole of the hippocampus is shifted ventrally, around the curve of the lateral ventricle, into the temporal lobe, so that its septal end is posterior to its temporal end. The longitudinal dimension of the hippocampus is often ignored, because some of the most striking connections of the hippocampus are arranged in the transverse plane. Nevertheless, longitudinal associational connections in the hippocampus are prolific. Thin transverse slice preparations sever most of these connections. Figure 17 attempts to convey the three-dimensional appearance of the hipocampus as a whole, as seen in animals such as the rabbit. It will be noticed that an electrode track passing vertically downwards from the dorsal surface of the hemisphere encounters the lateral ventricle, then field CA1, and beneath this either the dentate gyrus or field CA3 (depending on medio-lateral position of the track).

A few words are also appropriate here on the layout of the septal region, because it is so closely linked to the hippocampus. Its position is well in front of the hippocampus, situated on either side of the midline, between the anterior horns of the lateral ventricle. It is not a laminated cortical structure, but consists instead of a number of distinct nuclear groups. The largest of these, forming the medial wall of the anterior horn, are the lateral septal nucleus (dorsally), and the nucleus accumbens (around the ventral limit of the anterior horn). Close to the midline are smaller nuclei, more important in the present context. These are the medial septal nucleus and its ventral continuation, the nucleus of the diagonal band (of Broca). There is in addition a basal nucleus (of Meynart) ventral to both these components. The posterior part of this region is penetrated by two important

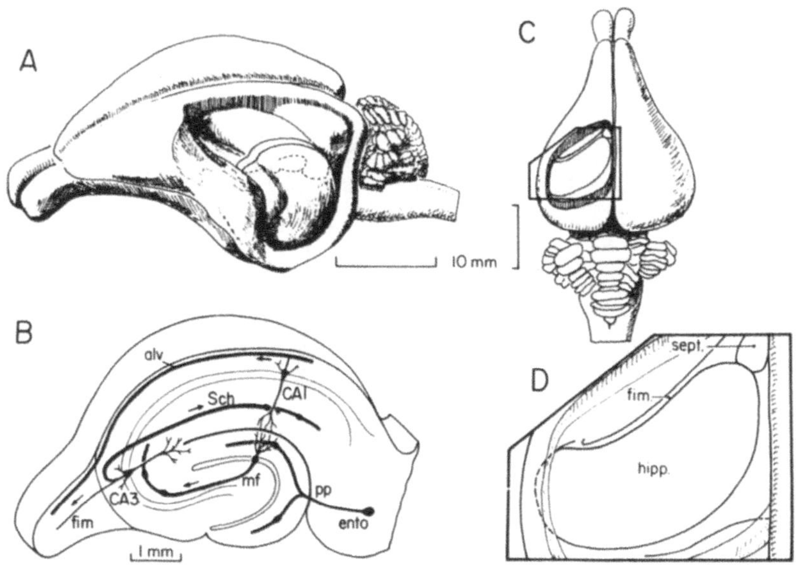

Fig. 17. A. Schematic three-dimensional view of the left hippocampus as it might be seen in a rabbit, from above and posterolaterally, with the superficial part of the hemisphere removed. B. The characteristic pattern of Ammon's Horn is seen in a section through the middle region of the hippocampus. The continuity between the pyramidal cell layer of Ammon's Horn and lamina V of the isocortex is also indicated. Alv, Alveus; CA1, CA3, fields of Cornu Ammonis; ento, entorhinal cortex; fim, fimbria; mf, mossy fibres; pp, perforant pathway; Sch, Schaffer collaterals. C. View of forebrain from above, showing orientation of hippocampus, revealed by removal of part of neocortex. D. Hippocampus as in C (detail). sept, septal region; hipp, hippocampus. (Andersen *et al.* 1971) (From *Experimental Brain Research* Vol 13 pp. 222–238, 1971)

fibre tracts, the fornix, and the anterior commissure. The layout of nuclei in this region is depicted in Fig. 18.

3.3 Connections of the Hippocampal Complex

For the dentate gyrus and each field of Ammon's Horn the following account will describe the *extrinsic* afferents and efferents (those connecting to regions outside the hippocampal complex) and the *intrinsic* connections (those entirely within the hippocampal complex).

3.3.1 The Dentate Gyrus

3.3.1.1 Extrinsic Afferents

An important projection of fibres originates in the entorhinal cortex, perforates the fold of tissue including the subiculum, then crosses the obliterated hippocampal fissure, so approaching the dentate gyrus. This is the *perforant pathway*. The origin of these fibres appears to be mainly in lamina II of the entorhinal cortex (rats: Ruth *et al.* 1988; cats: Habets *et al.* 1981; Witter and Groenewegen 1984), though a few neurons in deeper layers have also been seen to project to the dentate gyrus in rats (Kohler 1984). There is an interesting relation between the region of origin of these fibres in the entorhinal

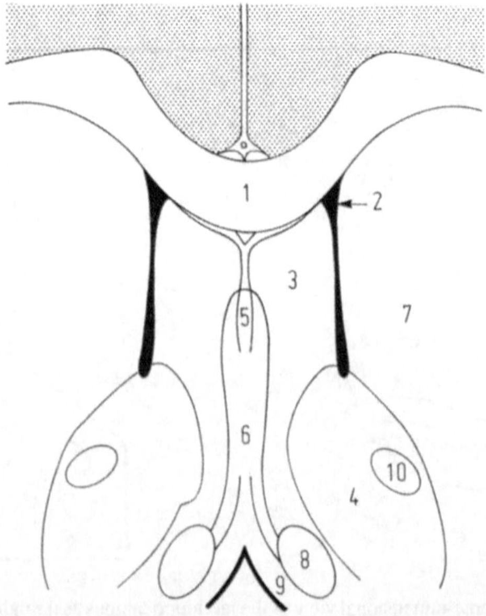

Fig. 18. Coronal section diagram of septal region showing main nuclear groups. 1 Corpus callosum; 2 lateral ventricle; 3 lateral septal nucleus; 4 nucleus accumbens; 5 medial septal nucleus; 6 nucleus of the diagonal band (Broca); 7 caudate nucleus; 8 basal nucleus (Meynart); 9 diagonal band. (After Konig and Klippel 1967 Fig. 14a)

cortex and the region of termination on the granule cell dendrites: fibres originating in the medial entorhinal cortex terminate on the middle third of these dendrites, while those that originate in the lateral entorhinal cortex terminate more distally along the dendrites. This pattern has been demonstrated in rat (Hjorth-Simonsen 1972; Steward 1976; Wyss 1981; Ruth *et al.* 1988), mouse (Stanfield *et al.* 1979), cat (Habets *et al.* 1981) and monkey (Van Hoesen and Pandya 1975b). It has been suggested, on the basis of electrophysiological evidence (Andersen *et al.* 1969; Lomo 1971; Rawlins and Green, 1977), that this projection is organized on a lamellar basis, that is, perforant path axons pass transversely with little divergence of connections in the septo-temporal axis (see Fig. 19B). However, recent anatomical studies of this issue (Wyss 1981; Ruth *et al.* 1988) have shown that this is true only for the projection from the medial entorhinal cortex. The projection from the lateral entorhinal cortex to the dentate gyrus shows considerable spread in the septo-temporal axis. This has been confirmed by Ruth *et al.* (1988).

Another extrinsic pathway afferent to the dentate gyrus consists of fibres arising in the *medial septal nucleus* and the adjacent *nucleus of the diagonal band.* These fibres are, in part at least, cholinergic. Since this is the most important pathway relaying the signals to the hippocampus which initiate rhythmic theta activity in it, it is appropriate to describe it in some detail. Early studies using correlated silver-stained degeneration and acetylcholinesterase (AChE) staining suggested that cholinergic afferents from the medial septum terminated in the molecular layer of the dentate gyrus (Mosko *et al.* 1973;

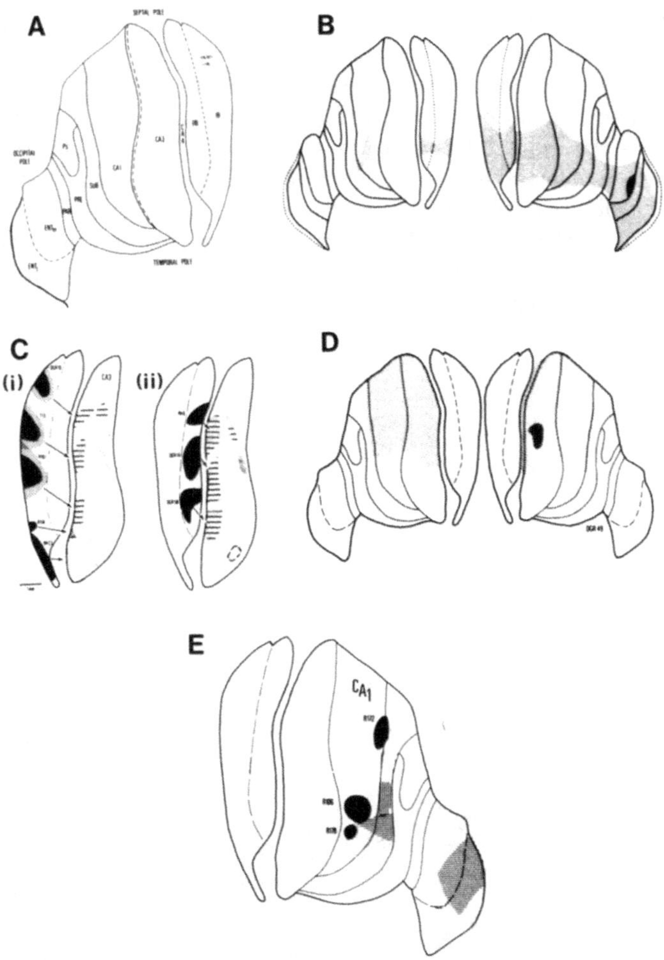

Fig. 19. A composite diagram from various sources showing longitudinal spread of perforant path (lateral and medial), mossy fibres, CA3 association fibres, recursive fibres from CA3 to CA4 to DG, and Schaffer collaterals. A (Swanson *et al.* 1978, Fig. 1, III). Unfolded representation of the entorhinal cortex (*ENT^l*: lateral entorhinal cortex; *ENT^m*: medial entorhinal cortex); the dentate gyrus (*OB*: exposed blade; *IB*: buried blade), the CA fields, and the subicular regions (*SUB*: subiculum; *PS*: postsubiculum; *PRE*: presubiculum; *PAR*: parasubiculum). This diagram serves as a template for the other parts of this figure. B (Wyss 1981 Fig. 3B) The spread of projections from a small injection site (*black*) of the medial entorhinal cortex within the dentate gyrus and the CA fields. C (Swanson *et al.* 1978, Fig. 6A, B). The spread of projections from small injection sites (*black*) in the exposed (*i*) and buried (*ii*) blades of the dentate gyrus to field CA3 (slightly modified from original). D (Swanson *et al.* 1978, Fig. 14B). Spread of connection from a small injection site in CA3 to large parts of fields CA3, CA4 and CA1 on both sides. E (Swanson *et al.* 1978, Fig. 19). Spread of projections from small areas of field CA1 within the subiculum and the entorhinal cortex. Note that B uses a slightly different template from that depicted in A, with the entorhinal cortex divided into three, rather than two subfields. (From *Journal of Comparative Neurology*)

Mellgren and Srebro, 1973) as well as (more densely) in the dentate hilus. (A figure from Milner *et al.* (1983), showing the laminar distribution of acetylcholinesterase in the hippocampal formation is reproduced here as Fig. 20.) Rose *et al.* (1976) gave this finding quantitative support, by showing that, after injection of tritiated leucine into the septum (mainly but not exclusively the medial septum), autoradiographic grain counts were elevated in both the dentate granule cell layer and in the molecular layer. Later work from this laboratory (Lynch *et al.* 1978; Swanson and Cowan 1979), using several techniques emphasised the septal input to the hilar region only. However, the uncertainty whether this pathway terminates upon granule cells themselves or on local circuit neurons in the hilus of the dentate gyrus is now being clarified with more sophisticated techniques.

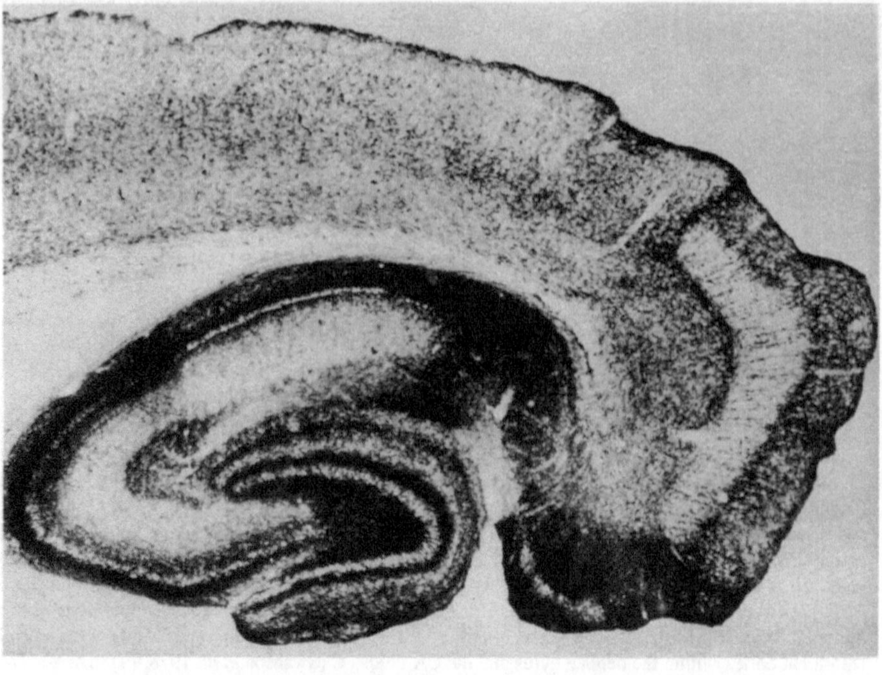

Fig. 20. A transverse section through the hippocampal complex stained for acetylcholinesterase. (Milner *et al.* 1983) (From *Developmental Brain Research* Vol 8 pp. 343–371, 1983)

Crutcher *et al.* (1981) and Chandler and Crutcher (1983), using anterograde transport from the septal region identified not only a dense projection to the dentate hilus, but also a modest input to the dentate molecular layer. With the electron microscope they detected inputs to hilar cell dendrites and somata, and to granule cell somata. Nyakas *et al.* (1987) studied hippocampal terminals from the septum/diagonal band complex using a sensitive method involving orthograde transport of a lectin. In the dentate gyrus, thick axons, with few boutons were found to terminate in the hilar region, while finer axons, with many *en passant* terminals, were found to end in the supragranular and molecular layers, most densely in the middle third of the molecular layer. The latter fine fibres were preferentially labelled by injections confined to the medial septum and vertical limb of the diagonal band where cholinergic cell bodies are located. Moreover, their terminal zones correlated well with that described by others for cholinergic markers (Storm-Mathisen, 1977; Houser *et al.* 1983). Thus, with some exceptions, these studies show a substantial septal input to the hilus, and a weaker one to the molecular layer of the dentate gyrus. The latter may be very fine unmyelinated fibres which escape detection with some techniques.

AChE staining, used in early studies, is not an ideal marker of cholinergic elements. A better marker is the enzyme choline acetyl transferase (ChAT). Results using stains for this enzyme confirm the early results: Houser *et al.* (1983) and Matthews *et al.* (1987) using antibodies against ChAT, and the light microscope, confirmed the presence of cholinergic release sites in the supragranular layers as well as in the part of the hilus nearest the granule cell layer. These bands of staining were almost entirely abolished by electrolytic lesions of the medial septum/diagonal band complex. Frotscher *et al.* (1986) used the electron microscope and Golgi impregnation in combination with ChAT staining. All parts of the dentate granule cells received ChAT-positive terminals. There were symmetric contacts on the dendritic shafts and cell bodies, while there were symmetric labelled terminals on the heads of small dendritic spines, and asymmetric ones on the stalks of more complex spines. These findings support the conclusion of the previous paragraph, providing definite evidence that some of the septo-hippocampal fibres which contact granule cell dendrites are cholinergic.

One problem of interpretation here is that there appear to be a few cholinergic somata in the hippocampus (Houser *et al.* 1983; Wainer *et al.* 1985; Frotscher *et al.* 1986; Blaker *et al.* 1988). Thus there is a possibility that not all ChAT-positive puncta are terminals of medial septal input fibres. However, lesioning of the medial septum/diagonal band shows that the vast preponderance of the cholinergic elements of the neuropil appear to originate there rather than in intrinsic neurons (Matthews *et al.* 1987; Blaker *et al.* 1988). Despite the existence of these cholinergic cell bodies intrinsic to the hippocampus, the conclusion of the previous paragraphs still stands.

There is, however, another the issue to be dealt with: are all the septo-hippocampal fibres cholinergic? This appears not to be so. According to Baisden *et al.* (1984), the AChE stains alone may fail to detect all septo-hippocampal neurons, because over 50% of such neurons, labelled by retrograde transport from the hippocampal formation, fail to react to AChE stain. Rye *et al.* (1984) and Wainer *et al.* (1985) confirm this, using an antibody to ChAT, in addition to retrograde labelling from the hippocampus. Amaral and Kurz (1985) give further details on distribution of cholinergic and non-cholinergic projection cells in the medial septal region, and on the topography of their projection along the length of the hippocampus. Separate clusters of cholinergic and non-cholinergic cells projecting from the medial septal region to the hippocampus were identified. A septo-temporal gradient of cholinergic innervation was found in the hilus of the dentate gyrus (see also Matthews *et al.* 1987).

It is likely that gamma-amino-butyric acid (GABA) is one of the transmitters used by the remaining neurons in these regions. A proportion of the large cell bodies in the medial septum and diagonal band are immuno-reactive with antibodies to glutamic acid decarboxylase (GAD) (Panula *et al.* 1984). These cell bodies are distributed amongst the population of hippocampally-projecting cells, thought to include cholinergic neurons. According to Kohler *et al.* (1984) 30% of cells labelled by retrograde tracers injected into the hippocampus or entorhinal cortex are GAD-positive. Aware of the probable heterogeneity of septo-hippocampal fibres with regard to transmitters, Nyakas *et al.* (1987) suggested that the fine fibres which they found innervating the supragranular and molecular layers of the dentate gyrus were cholinergic afferents to the dentate gyrus, while the coarser fibres found in the hilus may be both cholinergic and non-cholinergic. However, there is, as yet, no direct proof that GABAergic axons from septal inputs synapse with hilar cells, although Bilkey and Goddard (1985, 1987) have provided indirect electrophysiological evidence for this.

In addition to the perforant path, and the septo-dentate pathway, there are extrinsic fibres to the dentate gyrus from the opposite hippocampus. However, since these commissural fibres have a similar distribution to the ipsilateral associational fibres, they will be considered below along with other intrinsic connections.

3.3.1.2 Intrinsic Connections

The sole intrinsic efferent pathway of the dentate gyrus granule cells consists of the *mossy fibres*. The name "mossy fibre" derives from the very large boutons of these fibres [up to 6 microns in diameter, according to West *et al.* (1982a)] which make synapses with pyramidal cell dendrites. The mossy fibres pass transversely into Ammon's Horn, with a rather exact lamellar organization (i.e., very little spread in the septo-temporal axis) (see Fig. 19C). Anatomical evidence for this statemant comes from Blackstad *et al.* (1970), Lynch *et al.* (1973) and West *et al.* (1982a). Lynch *et al.* (1973) report that in rats and mice the labelled axons occupied a thin transverse slice of tissue no wider than the injection site (i.e., 250–300 microns). Corresponding electrophysiological evidence is provided by Rawlins and Green (1977). The mossy fibres make such synapses with dendrites of pyramidal cells in CA3. They also synapse with basket cell interneurons in this field (Frotscher 1985) and (via short collaterals) with neurons in the hilus of the dentate gyrus (CA4) (Claiborne *et al.* 1986) where they make contact with a very heterogenous collection of neurons which use a variety of transmitter substances. They never extend anywhere beyond the ipsilateral hippocampus, and never extend as far as CA2/1 (Blackstad *et al.* 1970; West *et al.* 1981, 1982a) (see: Fig. 21). The synaptic contacts with pyramidal cells are found in two bands, that is on the proximal regions of both the apical and the basal dendrites. The main (suprapyramidal) layer of mossy fibres makes synaptic contact with the proximal apical dendrites of pyramidal cells (in the so-called stratum lucidum). West *et al.* (1982a) also report that, after running transversely to their limit in fields CA3a (most remote from the dentate gyrus), they turn and run temporally for several hundred microns. Histochemically, a prominent feature of the mossy fibres is that they contain large quantities of zinc (Haug 1967).

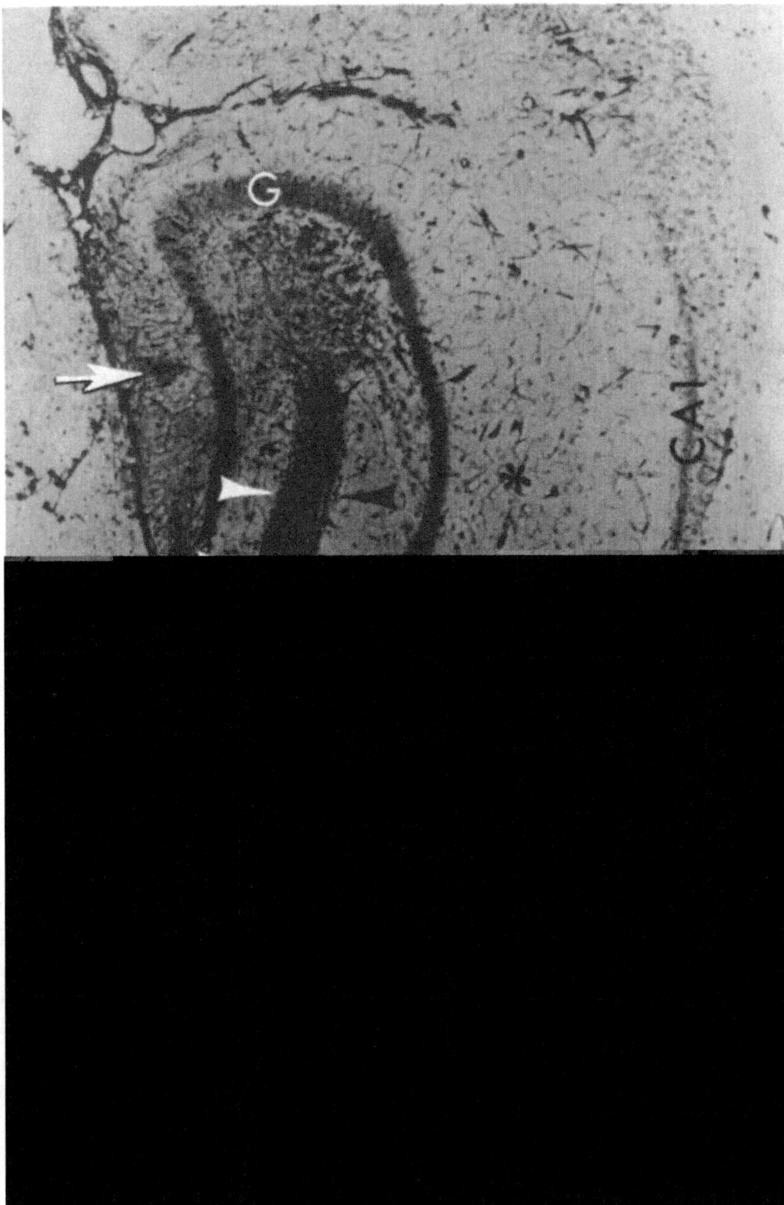

Fig. 21. Trajectory of mossy fibres, visualized after injection of horseradish peroxidase into the buried blade of the dentate gyrus. G, granule cell layer; *, hippocampal fissure; white arrow heads, infrapyramidal layer of mossy fibres; black arrow head, suprapyramidal mossy fibres; white arrow, tissue damage from micropipette (West *et al.*, 1982a) (From *Brain Research Bulletin* Vol 6 pp. 119–124)

3.3.2 The CA Fields

3.3.2.1 Extrinsic Afferents of Field CA3

Field CA3, like the dentate gyrus, receives an input from the entorhinal cortex, though it is sparser than the pathway to the dentate gyrus (Swanson and Cowan, 1977). The projection to CA3 is similar to that to the dentate gyrus in several respects: in its distribution along the length of the hippocampus (Wyss 1981), and along the length of the apical dendrites of the pyramidal cells (mouse: Stanfield, *et al.* 1979; rat: Hjorth-Simonsen 1972; cat: Steward 1976 and Habets *et al.* 1981; monkey: Van Hoesen and Pandya, 1975b). The pathway appears to originate in the superficial laminae of the entorhinal cortex (Ruth *et al.* 1982; Witter and Groenewegen 1984). Whether the cells of origin are the same as those which project to the dentate gyrus is not known.

Field CA3 also has an input from the medial septal nucleus in common with that to the dentate gyrus. After septal lesions, degenerating fibres are found in the infrapyramidal layer as a major pathway, while a minor pathway traverses the suprapyramidal layers (Mosko *et al.* 1973; Mellgren and Srebro 1973; Swanson and Cowan 1979). This distribution matches that of AChE staining (see Fig. 20). These results have been confirmed by Crutcher *et al.* (1981) using anterograde transport from the septal region. The distibution of cholinergic elements has been confirmed by Houser *et al.* (1983) and Frotscher *et al.* (1986) using anti-ChAT-antibody. An early result of Lynch *et al.* (1978) suggested that the cholinergic fibres project to interneurons in stratum oriens of CA3. However, there is, as yet, no study using ChAT staining in combination with electron-microscopic and tracing methods, so the cellular target of the medial septal cholinergic fibres is uncertain. However, Frotscher *et al.* showed, using EM and ChAT immunohistochemistry, that there *are* cholinergic synapses of the asymmetric type on the cell bodies and smooth proximal dendrites of pyramidal cells in the CA fields.

One other electron microscopic study should be mentioned. Nyakas *et al.* (1987) found two types of degenerating axons in CA3 labelled by injections into the septal region. Coarse fibres with few synaptic contacts were found in all layers of CA3, most densely in stratum oriens. Fine fibres with many *en passant* terminals were found in the pyramidal cell layer, but, in contrast to CA1 (see below) were scanty in stratum lacunosum-moleculare. The congruence of the distribution (throughout the hippocampal complex) of such fine fibres with that of cholinergic markers suggested that they were cholinergic axons, undetectable with classical methods. There also appear to be GABAergic septo-hippocampal projections to CA3: Kohler *et al.* (1984) found that 30% of cells labelled by retrograde tracers injected into the hippocampus or entorhinal cortex were GAD-positive, suggesting that there is a substantial GABAergic projection from septum to hippocampus. Little detail of the distribution of these fibres in CA3 is available, though they may correspond to Nyakas *et al.*'s coarse fibres.

3.3.2.2 Intrinsic Efferents of CA3

Field CA3 has very widespread intrinsic connections in the hippocampus. Since their distribution in the ipsilateral hippocampus is very similar to that in the contralateral hippocampus, the associational and commissural pathways will be dealt with together. In addition, CA3 is the principal source of intrinsic connections to the CA4 (hilar) region. These will be dealt with first.

The distribution of recurrent connections, passing from CA3 "against the main flow" to the hilar region and the dentate gyrus has been the source of some controversy. Gottlieb and Cowan (1973) and Laurberg (1979) concluded that CA3c (certainly) and polymorph cells of the hilus (less certainly) projected to the inner third of the dentate molecular layer on both sides. However, this was not confirmed by Swanson *et al.* (1978). Injections restricted to CA3 or its subfields labelled fibres in CA4, but never those in the dentate gyrus. Only if the injection site included CA4 was there labelling in the dentate gyrus. The situation was clarified by the retrograde tracing experiments of Swanson *et al.* (1981). Dye injections limited to the dentate gyrus labelled a high proportion (80%) of neurons in the CA4 region, on both sides. It thus seems likely that there is a sequential recursive pathway from CA3 to CA4 to the dentate molecular layer. Seress and Ribak (1983) using immunocytochemical methods, showed that a high proportion of hilar neurons (60%) are GAD-positive. This indicates that the direct link from the hilus to both the ipsilateral and contralateral dentate gyrus may be largely or entirely inhibitory. On the other hand electrophysiological results of Douglas *et al.* (1983) suggest that the inhibitory effect produced in granule cells by electrical stimulation of the contralateral hilus is mediated by the commissural axons (of presumed hilar origin) *exciting* inhibitory interneurons in the opposite hilus.

The associational and commissural projections of field CA3 are directed also at fields CA3 and CA1, on both sides. The associational connections from CA3 to CA1 are well known as the Schaffer collaterals (Schaffer 1892). Gottlieb and Cowan (1973) showed that the CA3-to-CA3 connections terminate on basal dendrites (stratum oriens) and the middle portion of the apical dendrites (stratum radiatum). Those to CA1, both ipsi- and contralaterally terminate on the basal dendrites (stratum oriens) and proximal 3/4 of the apical dendrites (stratum radiatum) of CA1 pyramidal cells.

The spread (in the longitudinal axis of the hippocampus) of connections from CA3 to all its target fields (CA4, CA3 and CA1) appears to be considerable (see Fig. 20). This also appears to be true for the projection from CA4 to the dentate gyrus. There is a distinct longitudinal band of fibres running in a temporal to septal direction in subfield CA3c (the subfield nearest the dentate gyrus) (Swanson *et al.* 1978). Swanson *et al.* (1978) and Laurberg (1979) find that lesions in CA3 produce greater degeneration in CA1 on the septal side of the lesion than on the temporal side of it. Laurberg (1979) adds that the connections projecting septally from a lesion site terminate more proximally on the CA1 pyramidal cell dendrites than do those projecting in the temporal direction.

3.3.2.3 Extrinsic Afferents of Field CA1

There is a projection from the entorhinal cortex to field CA1, but the details are a matter for some debate, with the possibility that there are species differences. Nafstad (1967) said that there was a relatively modest input from the entorhinal cortex, if any. Steward (1976) found evidence in rats for a more substantial projection. The mediolateral axis of the entorhinal cortex was mapped transversly by connections terminating in CA1, from the the CA1/CA2 boundary to the subicular border. They argued that this manner of termination had led Nafstad (who used EM methods, and therefore only studied small regions of CA1, which might not be the part receiving fibres from the lesioned zone) to report very sparse connectivity. Witter *et al.* (1988), using rats, reported a substantial projection to CA1 from the entorhinal cortex, but this originated in the rostral and lateral parts of the entorhinal cortex, whereas the projection to CA3 and the dentate gyrus originates in medial and caudal parts. In cats, Habets *et al.* (1981) found a projection from entorhinal cortex to CA1, and, as was found in CA3,(but in contrast to the results of Steward 1976) the medio-lateral axis of the region of origin was mapped along the CA1 dendrites in proximal to distal direction. Witter and Groenewegen (1984) found that this projection arises from lamina III of the entorhinal cortex. In rhesus monkeys, Van Hoesen and Pandya (1975b) found no projection from entorhinal cortex to field CA1, except in the CA1/CA3 zone of overlap.

The possibility of an extrinsic input to field CA1 from the medial septal nucleus is controversial. In the older literature Raisman (1966) finds that the medial septum projects only to the dentate gyrus and CA3/4 but not to CA1/2. This view is supported by Rose *et al.* (1976), Swanson (1978) and Swanson and Cowan (1979). However, other studies have seen a septal projection to field CA1, albeit a sparse one. Powell (1963) and Ibata *et al.* (1971) find degenerating fibres from the medial septum in fields CA1, as well as in CA2 and CA3. Mellgren and Srebro (1973) correlated the degeneration and the disappearance of AChE staining in various parts of the hippocampal formation after lesions of the medial septal nucleus. They concluded that CA1 did receive cholinergic septal afferents, but they were less prolific than those to CA3. Mosko *et al.* (1973) find that degeneration after medial septal lesions is very sparse in CA1, and could be accounted for as fibres *en route* to the subiculum. Lynch *et al.* (1978) found a very thin band lying between stratum moleculare and stratum radiatum which was labelled by both AChE stain and tracers injected into the medial septum. Crutcher *et al.* (1981) describe a sparse projection from septum to stratum oriens and stratum lacunosum-moleculare of CA1, most of the labelled axons apparently passing through CA1 to more distant regions. Houser *et al.* (1983) describe the same laminar pattern of distribution of ChAT-positive puncta in CA1 as in other CA fields. The predominant view from this evidence is thus that the cholinergic septo-hippocampal projection to field CA1 exists but is sparse and possibly limited to axons passing through CA1.

Three papers stand against this view. In two of them (Sakanaka *et al.* 1980; Monmaur and Thompson 1983) horseradish peroxidase was injected into CA1 and succeeded in labelling neurons in the medial septum or diagonal band. Monmaur and Thompson (1983) argue that this was unlikely to have been due to retrograde transport from axons passing though CA1 without bearing terminals, because the septal neurons known to pass through the alveus there (which was included in the injection site) were not labelled. Thus a definite projection from septum to CA1 has been established, but it appears to be sparse. In the third paper Nyakas *et al.* (1987) used a sensitive orthograde transport method and labelled coarse fibres with few terminals originating widely in the medial septum/diagonal band region. These were located in all laminae of CA1, but most densely throughout the strata oriens and lacunosum/moleculare. More delicate fibres with many *en passant* terminals had more restricted laminae of termination, viz. most richly in the pyramidal cell layer, and less densely in the stratum lacunosum/moleculare. The latter fibres were labelled only by injections in the medial septal nucleus and vertical limb of the diagonal band, where cholinergic cell bodies are found. Moreover, the distribution of these fine fibres correlated well with that of cholinergic markers described by others (Storm-Mathisen, 1977; Houser *et al.* 1983; see also Fig. 20). It thus seems likely that such fine fibres are the cholinergic axons ramifying within the hippocampus.

Since axons of this type were found to be much more dense in the dentate gyrus and CA1 than in CA3 (see above) the interesting possibility is raised that they are the connectional basis of the fact that the hippocampal theta rhythm (in urethanized preparations at least) is present in the dentate gyrus and CA1, but apparently not in CA3 (see Chap. 7). Despite the fine calibre of these fibres, they have many terminals (presumably on pyramidal cell dendrites) and so may be expected to have an influence out of proportion to their prominence using most neuroanatomical tracing techniques. Nyakas *et al.* also raise the suggestion that the coarser fibres are non-cholinergic, as in the dentate gyrus. However there is not yet a detailed description of the termination in the CA fields of fibres using transmitters other than acetylcholine.

At this stage, the comment can be made that, apart from the septal input, *the predominant direction of fibre trajectories in the hippocampus is clearly from entorhinal cortex to dentate gyrus to CA3 to CA1*. As explained below this pathway continues from CA1 to subiculum and thence back to the entorhinal cortex and to various neocortical regions. There is however a little evidence that the subiculum projects back to the CA fields, more directly than by the well-known route from entorhinal cortex via the dentate gyrus and CA3 field (Sikes *et al.* 1977; Berger *et al.* 1980a). However Meibach and Siegel (1977b), using rats, were unable to confirm the existence of this "against-the-flow" pathway. Evidence of inputs to CA fields, from further afield, in the isocortex itself, will be considered below (Sect. 3.6).

3.3.2.4 Extrinsic Efferents of CA3 and CA1

The best-known extrinsic efferent pathway from the hippocampus is that via the fornix passing to various subcortical regions in telencephalon, diencephalon and brainstem. However, when the extent of this connection was examined in rats by Swanson and Cowan (1977), using modern methods, the extent of the projection from the hippocampus proper was shown to be rather small. There was a projection from CA1 and CA3 pyramidal cells via the pre-commissural fornix to the lateral septal nucleus, the projection from field CA3 being bilateral. Only a small proportion of pyramidal cells in these fields was involved, the remaining majority projecting to structures in the parahippocampal gyrus other than the CA fields, and to regions of limbic cortex elsewhere (see below). The bulk of the fornix, and all the post-commissural fornical projections to diencephalon and brainstem originated not in Ammon's Horn, but in these other parts of the parahippocampal gyrus.

It has generally been assumed that pyramidal cells were the only ones to send projections from the CA fields to regions outside the hippocampus. However this appears now to be incorrect, at least for the projection to the septal region (Chronister and DeFrance 1979). According to Alonso and Kohler (1982), the projection to the medial septal region appears to be from non-pyramidal cells, that to the lateral septal region from pyramidal cells. Cells in the hilus labelled from tracer injections into the medial septal nucleus were entirely non-pyramidal, those in the CA fields were more commonly non-pyramidal and those in the subiculum were entirely pyramidal. When non-pyramidal cells were labelled they were very heterogeneous morphologically, and were widely distributed across the different laminae. Non-pyramidal cells have also been reported to project to the contralateral hippocampus (Schwertfeger and Buhl, 1986), the nucleus accumbens (Totterdell and Hayes, 1987) and the supramammillary region (Ino *et al.* 1988).

The major projection from fields CA3 and CA1 to the parahippocampal gyrus has been described in detail in the rat in a number of published studies (Swanson and Cowan, 1977; Swanson *et al.* 1978; Swanson *et al.* 1981). Areas of terminal distribution were identified as the subiculum, and to lesser extents to pre- and para-subiculum, the entorhinal cortex (especially lamina IV). The projections to the subiculum arose in CA3, in the same cell bodies as the projection to lateral septum, and the associational projection to CA1 and CA3 both ipsi- and contralaterally. The projection to the entorhinal cortex originated in CA1, from cell bodies which also projected to the lateral septum. Using cats, Irle and Markowitsch (1982a) showed a projection to the subiculum from all CA fields, the smallest number of these originating in CA1. Rosene and Van Hoesen (1977) and Saunders and Van Hoesen (1988), using rhesus monkey, found a projection from CA1/prosubiculum to the deeper layers (mainly lamina V, but, in some regions, lamina III) of the entorhinal cortex. The full projection from the parahippocampal gyrus to the remainder of the cortex will be described below. In the main it does not actually arise from Ammon's Horn.

3.4 The Subiculum, Pre- and Para-subiculum

3.4.1 Extrinsic Afferents

Afferents to the subicular regions from the medial septal/diagonal band complex have been seen in a number of experiments in rats. Mosko *et al.* (1973) and Mellgren and Srebro (1973) saw a conjunction of AChE staining and degeneration of fibres of septal origin in the subiculum, and pre-, and parasubiculum. Meibach and Siegel (1977b) confirmed the projection to the subiculum with both anterograde and retrograde techniques. Swanson (1978) states that the septal input to the parasubiculum was amongst the most dense areas of termination of this pathway, the subiculum itself receiving a low level of connections (see Fig. 20). Swanson and Cowan (1979) found labelled fibres of medial septal origin in all parts of the subiculum, pre- and parasubiculum, the densest projection being to the deep part of the molecular layer of the subiculum.

3.4.2 Extrinsic Efferents

As noted above, the chief hippocampal region giving rise to efferents to the diencephalon is not Ammon's horn, but the subiculum and adjoining regions. This was documented in rat by Swanson and Cowan (1977), who found a projection directly to the anterior thalamus from pre- and/or parasubiculum and a projection to the mammillary body from dorsal subiculum, as well as pre- and/or parasubiculum. Similar results have been obtained in other species (cat: Irle and Markowitsch, 1982a; squirrel monkey: Krayniak *et al.* (1979).

The connections from subicular regions to the mammillary bodies and the anterior thalamus are part of the circuitry which used to be called Papez circuit (included in Fig. 22). This concept comprised connections from the hippocampus (now known to originate in subicular regions) via mammillary bodies, the mammillo-thalamic tract, and thence to the cingulate gyrus (see Papez 1937; Brodal 1969). Apart from the detail of its origin, the existence of this circuit of connections has not been challenged by modern research [see papers by Watanabe and Kawana (1980) and Seki and Zyo (1984) in the rat, and by Berger *et al.* (1980b) in the rabbit].

3.4.3 Intrinsic Connections

Some of these have already been considered as afferents or efferents of the CA fields. The subiculum also appears to send projections to the presubiculum and entorhinal cortex (rat: Kohler *et al.* 1978; Sorensen and Shipley 1979; cat: Irle and Markowitsch 1982b; rhesus monkey: Rosene and Van Hoesen 1977). These projections appear to be only ipsilateral, and their termination in the entorhinal cortex is confined to the deeper layers. There are also projections from pre- and para-subiculum to the entorhinal cortex (Shipley 1975; Segal 1977; Kohler *et al.* 1978; Rosene and Van Hoesen 1977).

3.5 Septal Afferents to the Entorhinal Cortex

Like the hippocampal complex, the entorhinal cortex receives an input from the medial septum/diagonal band complex. Mellgren and Srebro (1973), using rats, detected falls in AChE staining, and degenerating silver-stained fibres throughout the entorhinal cortex after lesions of the medial septal nucleus. This connection was confirmed by Segal (1977), Meibach and Siegal (1977b), Swanson (1978) and Swanson and Cowan (1979). [Rose *et al.* (1976), on the other hand showed that in the entorhinal cortex, grain counts after [^{3}H]-leucine injections into the medial septum were little higher than the background, assessed in the caudate]. Insausti *et al.* (1987b) succeeded in labelling neurons in medial septum and the nucleus of the diagonal band after WGA/HRP injections into the entorhinal cortex. (A few neurons in the basal nucleus of Meynart were also labelled.) From these studies it seems clear that there is a projection from medial septum to the entorhinal cortex, at least part of which may be cholinergic. More details of its distribution come from two further studies. Milner *et al.* (1983) showed AChE staining in laminae II, IV and VI of the entorhinal cortex. Alonso and Kohler (1984) used an anterograde tracer method in conjunction with AChE staining and found a septal projection to laminae II and (more densely) to lamina IV, which was partly but not entirely cholinergic, and overall a less dense septal projection than to the hippocampus proper. These findings are compatible with the earlier results of Mosko *et al.* (1973) (see also Fig. 20).

3.6 Connections Between the Isocortex and Hippocampal Formation

The reciprocal connections between the hippocampal formation and the isocortex are of particular importance in the present work, because they are the anatomical substrate for interplay between cortex and hippocampus. This interplay is envisaged to serve some important aspects of information processing in the forebrain, because it forms the global cell assemblies (see Chap. 2) which can resolve the ambiguity of representation of small-scale patterns of information held within the cortex. The circuit of Papez has already been mentioned. In recent years it has become clear that there are also more direct reciprocal links between hippocampus and isocortex. The following section reviews this evidence.

3.6.1 *Afferents to the Hippocampal Formation*

Afferents connections from the isocortex to Ammon's horn are summarized in Table 2A.

Afferents to the fields of *Ammon's horn* from isocortical regions have been detected so far only in primates. They arise in prefrontal, parietal, temporal, cingulate and peri- and ento-rhinal cortices, and terminate in CA1 and occasionally in CA3, but not in CA4 or the dentate gyrus.

Afferents to the *subicular* regions from other cortical regions have been reported in cats as well as primates. These originate in both medial and dorsolateral prefrontal cortex, cingulate cortex (especially posteriorly), temporal, and parts of parietal and occipital cortices (but not from primary visual or somatosensory cortices, or from the primary motor cortex of cats). They also arise from peri- and entorhinal cortices. Terminals are distributed in subiculum, and pre- and parasubiculum.

Afferents to the *ento- and perirhinal cortices* from a variety of cortical regions have been reported. So far, direct isocortical afferents to the entorhinal cortex have been seen only in primates, though there are also connections from perirhinal to entorhinal cortex. [Room and Lohman (1984) using cats found inputs from the neocortex to the perirhinal cortex but failed to find any direct connection from the neocortex to the entorhinal cortex]. Regions of origin include orbitofrontal and dorsolateral prefrontal cortices, prepiriform cortex, insula, cingulate cortex, and many parts of the temporal lobe. Insausti *et al.* (1987a) comment that all the cortical areas projecting to the entorhinal cortex could be regarded as polysensory association cortex. Amaral *et al.* (1987) define a cytoarchitectonic subdivision of the entorhinal cortex in macaques which alone receives direct input from the olfactory bulb.

3.6.2 *Efferents from the Hippocampal Formation*

Efferents from the hippocampal formation to the isocortex are summarized in Table 2B.

Of the *CA fields*, at least CA1 appears to send some connections to the isocortex. This has been shown in both rats and monkeys, the target of these efferents including perirhinal cortex, retrosplenial cortex and various parts of frontal and temporal cortices.

The *subicular components*, as well as receiving from various cortical regions, also project back to the isocortex, a result based on experiments in primates, cats, guinea pigs and rats. These projections terminate in posterior cingulate cortex, dorsal prefrontal cortex, temporal, parietal and prepiriform cortex. In some experiments the projection from the convexity of the parietal cortex has been shown to be sparse or absent.

Efferents from the *entorhinal cortex* to other cortical areas have been described in cats and monkeys. The target of these projections includes the medial surface and convexity of the prefrontal cortex.

3.7 Connections Between the Isocortical Areas Directly Linked with the Hippocampal Formation, and Other Isocortical Areas

In the above section it was apparent that the isocortical regions most directly linked to the hippocampal formation were the limbic (e.g., cingulate) and prefrontal areas, though other isocortical regions received less abundant connections. The cingulate and

Table 2. Connections Between Hippocampus and Isocortex

A Afferents to Hippocampal Formation from Isocortex

From: Frontal[a]	Parietal[a]	Temporal[a]	Occipital[a]	Cingulate "lobe"
1. To Ammon's Horn				
Squirrel monkeys: medial prefrontal to CA1-3: (1)	/	/	/	/
Marmosets: areas 4,6,7 10,11 to CA1-3 (2)	(2)	Areas 20,21, 22 to CA1(2)	/	To CA1-3(2)
2. To Subicular regions				
Monkeys: medial and lateral PFC (5,9)	(7)	TE,TF,TG, TH (8[b])	/	(6)
Marmosets: /	(2)	20,22(2)	/	/
Cat: /	(4[b])	(4)	(4[b])	(3,4)
3. To Entorhinal cortex				
Squirrel monkeys: Orbito-frontal (1)	/	/	/	/
Rhesus monkeys: Orbito-frontal (11,13), dorso-lat. PFC (9) (12), TA, TE (13)	/	TE,TF,TG TH, pre pirifor ctx (11, 27,49,51	/	/
Macaques: Orbito-frontal, convexity, (15) (14,15)	/	Sup.temporal gyrus, ventral insula, temporal operc.	/	(15)
Marmosets: Orbito-frontal (10)	/	/	/	/

Table 2. (continued)

B Efferents from Hippocampal Formation to Isocortex					
To:	Frontal[a]	Parietal[a]	Temporal[a]	Occipital[a]	Cingulate "lobe"
1. From Ammon's horn					
Monkey:	From CA 1 to areas 9 and 10 (2) CA1 to area TE (24)	/	From CA1 to areas 20,2 (2); from	/	/
Rats:	From CA1 (17,18)	/	/	/	From CA1 (16)
2. From Subicular regions					
Rhesus monkeys:	medial PFC (23), convexity of PFC (9)	/	/	/	/
Marmosets:	areas 9,10 (2)	/	20,21 (2)	/	/
Cat:	medial PFC, weakly from convexity of PFC (21,22)	Weak (21)	Weak (21)	/	Cingulate, retrospenial (21)
Guinea pig:	/	/	/ (20)	/	Retrosplenial
Rats:	/	/	/ cingulate (19)	/	Post.
3. From entorhinal cortex					
Cats:	medial PFC convexity of PFC (22)	/	/	/	/
Rhesus monkeys:	convexity of PFC (9)	/	/	/	/

Table 2. (continued)

[a] or equivalent regions in non-primate species.
[b] excluding primary sensory and motor areas.
The symbol "/" indicates that no report of such a connection was made. It does not imply that such a connection was excluded. Connections tabulated above usually involve only parts of the lobes concerned. PFC prefrontal cortex.

Sources:

(1) Leichnetz and Astruc (1975a, 1976).
(2) Schwertfeger (1979)
(3) Cragg (1965)
(4) Irle and Markowitsch (1982a)
(5) Nauta (1964)
(6) Pandya *et al.* (1979)
(7) Seltzer and Van Hoesen (1979)
(8) Van Hoesen and Rosene (1979)
(9) Goldman-Rakic *et al.* (1984)
(10) Leichnetz and Astruc (1975b)
(11) Van Hoesen *et al.* (1972)
(12) Van Hoesen and Pandya (1975a)
(13) Van Hoesen and Pandya (1975b)
(14) Amaral *et al.* (1983)
(15) Insausti *et al.* (1987a)
(16) Swanson and Cowan (1979)
(17) Swanson (1981)
(18) Ferino *et al.* (1987)
(19) Meibach and Siegel (1977a)
(20) Sorensen (1980)
(21) Irle and Markowitsch (1982b)
(22) Cavada et al.(1983)
(23) Rosene and Van Hoesen (1977)
(24) Iwai and Yukie (1988)

prefrontal cortices send and receive many connections to the rest of the cortex. These are relevant in the hypothesis of cortico-hippocampal loops which is advanced later in this book, and are described briefly below.

In small-brained mammals, such as the rat and rabbit, the cingulate cortex has reciprocal connections with certain visual regions and frontal areas (Vogt and Miller 1983; Vogt 1985). In monkey, more information is available. A larger proportion of the cortex is association cortex (non-primary receiving areas) and the connections of the cingulate cortex are, predictably, directed more to these asociation areas: *efferents* from the posterior cingulate cortex (area 23) have been seen projecting to the inferior and posterior parietal cortex, dorsal and lateral prefrontal cortex, orbito-frontal cortex, parieto-temporal cortex, medial and lateral temporal cortex (Mesulam *et al.* 1977; Pandya *et al.* 1981; Baleydier and Maugiere 1980). Efferents of area 24 (an anterior subdivision of the cingulate cortex) are more limited but have been seen projecting to premotor areas 6 and 8, orbitofrontal area 12, the rostral part of the inferior parietal lobule, and the anterior insular cortex (Pandya *et al.* 1981). Returning *afferents* to the cingulate cortex have been seen arising in various cortical areas to area 23, from several frontal and temporal areas and a few parietal areas (Vogt and Pandya 1987; Baleydier and Maugiere 1980), and to area 24 from many areas of the frontal and temporal cortex, from some parts of the parietal and insular cortex (Baleydier and Maugiere 1980; Petrides and Pandya 1984; Vogt and Pandya 1987). Pandya and Yeterian (1985) describe afferents to the cingulate arising from both auditory and somatosensory association areas. In addition, afferents have been detected passing from area Tpt (which forms the planum temporale—a part of Wernicke's area—in humans), to area TH (a region just behind the entorhinal area) (Tranel *et al.* 1988).

Cortico-cortical connections of the prefrontal cortex have been extensively studied in primates and have recently been reviewed by Fuster (1981: Chap. 2). In the monkey, the pattern is that the prefrontal cortex receives connections from the association areas associated with all the major sensory systems. Usually these connections are at least the third in succession from the primary receiving areas. For instance, somatic sensory pathways pass from the primary areas (1, 2 and 3) to association areas 5 and/or 7 before they impinge on the appropriate area of prefrontal cortex (in this case area 46). According to a fairly general rule, cortical regions that project to an area of prefrontal cortex also receive connections back from the same area. Thus, the various regions of unimodal association cortex both send to and receive from a portion of the prefrontal cortex. Within individual areas of the prefrontal cortex, some convergence of information from different sensory systems seems to be possible. However, this is usually an intermingling of just two types of information, not a thorough multimodal mixing of all sensory systems. In sub-primate mammals, these same principles of connectivity apparently also seem to hold (e.g., see Heath and Jones 1971 in cat), though less data is available. The number of sequential steps involved *en route* to the prefrontal cortex is sometimes in dispute, and there may be some exceptions to the rule of reciprocal connectivity. It is also possible that bimodal convergence may occur earlier in a sequential pathway in subprimates than in primates.

3.8 Summary of Connections, and Comment

It is appropriate to summarize the complex connectional data just reviewed, and to point out some singular features, which bear on the theory of cortico-hippocampal relationships to be developed later in this book. Fig. 22 summarizes the main connections described above, both within the hippocampal formation, and between it and the isocortex.

Superficial layers of the entorhinal cortex project strongly to the dentate gyrus, and less densely to CA3 and CA1. In the first two of these projections the medio-lateral dimension of the entorhinal cortex is mapped along the proximo/distal length of the dendrites. In CA1 it is not clear whether the mapping is along the dendritic length or the CA2-to-subiculum dimension orthogonal to the dendrites. Via mossy fibres the dentate granule cells project to hilar cells and CA3 pyramidal cells, terminating close to pyramidal cell bodies. CA3 pyramidal cells project via Schaffer collaterals to CA1 pyramidal cells, terminating in the mid-region of their apical dendrites. From here CA1 pyramidal cells project to subicular regions, entorhinal cortex, and more sparsely to more distant regions of cortex. This sequence of transverse connections appears to be the main route through the hippocampal complex. There is however a recurrent path from CA3 to hilar neurons, and from there back to dentate granule cells. The net effect of this

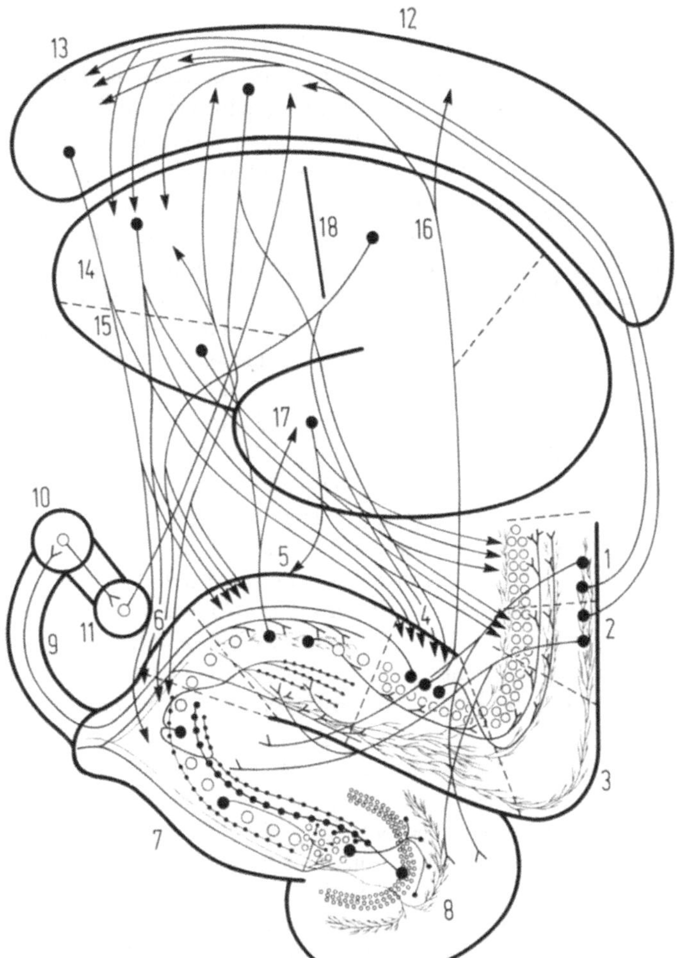

Fig. 22. Summary of connections within the hippocampus and between it and the isocortex. The section *below* is meant to represent the hippocampal formation, including the parahippocampal gyrus on the *right*. A schematic representation of the circuit of Papez is added to the *left*. *Above* is shown a schematic lateral view of the left cerebral hemisphere, with its medial surface represented above it. For the sake of clarity, a primate hemisphere is depicted, but the connections between hippocampal formation and the hemisphere are generalized across species. The *numerals* represent the following structures: *1* entorhinal cortex (lateral); *2* entorhinal cortex (medial); *3* pre- and parasubiculum; *4* subiculum; *5* CA1; *6* CA2; *7* CA3; *8* dentate gyrus, with CA4 (dentate hilus) enclosed between exposed and buried blades; *9* fornix; *10* mammillary body; *11* anterior thalamus; *12* cingulate gyrus; *13* medial prefrontal cortex; *14* lateral surface of prefrontal cortex; *15* orbital surface of prefrontal cortex; *16* parietal cortex; *17* temporal lobe; *18* central sulcus. The various connections shown summarize those described in the text. Those within the hippocampal formation attempt to depict the field of distribution and the laminae of cell bodies and fibre termination. Those between hippocampal formation and isocortex depict only areal distribution

pathway is inhibitory, though the sign of the individual parts of it is not yet resolved. In addition there is a little evidence for another "against the flow" loop, from subiculum to the CA fields. The perforant path and the mossy fibres are quite lamellar in distribution, while projections from CA3 (whether to CA3, CA1, or recurrent to the hilus and dentate gyrus) show much greater divergence along the longitudinal axis of the hippocampus.

The medial septum and vertical limb of the diagonal band send fibres to all parts of the hippocampal formation. Some of these are cholinergic, while others use another transmitter (GABA for at least a proportion of the non-cholinergic fibres). There are suggestions that the cholinergic components consists of very fine calibre axons. In the dentate gyrus they terminate both on hilar cells and various parts of the granule cells, including somata and dendrites. Although requiring confirmation, results of Nyakas *et al.* (1987) provide interesting hints of the exact laminae of termination of the cholinergic fibres. In the dentate gyrus, the hilus and the middle region of the molecular layer seem likely to be a major region of termination of cholinergic fibres. In the field CA1, cholinergic fibres also are probably distributed throughout the strata oriens and pyramidale, and also in the stratum lacunosum/moleculare. On anatomical grounds these fine, putatively cholinergic fibres probably have an influence out of proportion to their prominence in most neuroanatomical staining methods. GABAergic septo-hippocampal fibres are recognized to exist but their exact terminal distribution is not understood. Septo-hippocampal fibres to various subicular regions are well documented, are quite prolific in some regions, and are partly cholinergic. No evidence is available about non-cholinergic septo-subicular fibres. Septo-hippocampal fibres also innervate the entorhinal cortex, in part via cholinergic axons, but probably less densely than to most of the hippocampal complex. Laminae II and IV appear to be be the main destination of these fibres.

There are a few reports of afferents from CA fields to isocortical regions, and many more of projections from subicular regions and the ento- and peri-rhinal cortices to the isocortex. The most consistent target of these projections is to various parts of the prefrontal cortex. There are, however, projections to most of the temporal cortex (in primates) and to the cingulate cortex, as well as more sparse projections to the rest of the convexity of the hemisphere. Various regions of the isocortex project back to the hippocampus, directly or indirectly. Connections direct to CA fields are scanty, with origin from a variety of cortical regions, prefrontal, limbic and other. There is, however, a richer projection from cortex to subicular regions and to ento- and perirhinal cortices, originating especially in prefrontal cortex, and (in primates) in the temporal cortex. Cingulate and (sparsely) parietal cortex also contribute to these hippocampal afferents. The areas of origin of cortico-subicular fibres include limbic and association areas, but (with the possible exception of the temporal lobe) not the primary sensory or motor areas. Some of the isocortical afferents to the entorhinal cortex are indirect via the perirhinal cortex.

In conclusion, some general comments are appropriate about the overall configuration of these pathways. For the cortico-cortical connections within the isocortex, it is generally true that connections are reciprocal: a region which is afferent to another region usually (but not always) receives efferents from that region as well. *In the hippocampus there are important exceptions to this rule. Most notably, the perforant path from entorhinal cortex to dentate gyrus and CA fields is unidirectional.* The highly lamellar mossy fibres and the Schaffer collaterals, and the projections from CA fields to the subiculum are also predominantly unidirectional. (Although there are pathways against the main direction, they are either very sparse, or distributed much more widely in the longitudinal axis of the hippocampus, so that they can hardly be called reciprocal). Nevertheless, the overall distribution of pathways between hippocampal formation and the isocortex are generally reciprocal, with many isocortical areas both giving and receiving connections from one or other fields of the hippocampal formation. Thus, both within the hippocampal formation, and between it and the isocortex there are many possibilities for recursive loops of connections. Within the hippocampal complex there are potential loops between dentate gyrus and CA3 via the hilus. Somewhat larger loops are possible between hippocampal complex and entorhinal cortex. Very complex polysynaptic loops are possible between hippocampus and the limbic/polysensory association areas of the isocortex. Since the latter areas have wide connections throughout the cortex, the hippocampus has the potential to form loops with almost any region of the isocortex, except perhaps the primary sensory and motor areas (especially in primates) which seem to be the most remote from the hippocampus. However, since the connections within the hippocampal complex, unlike most of the connections between isocortical areas, are highly polarized (because the perforant pathway is *not* reciprocal), *any loops which involve the hippocampal circuitry can transmit signals only one way through that circuitry—from entorhinal cortex, to dentate gyrus, to CA fields and then to the subiculum—but not in the reverse direction.* This striking anatomical fact has an important significance for the hypothesis of self-organizing loops, resonating with the hippocampal theta rhythm, which will be developed in later chapters.

One further point is also relevant to that hypothesis: the area of isocortex which is most strongly and consistently connected with the hippocampus is the prefrontal cortex. This is the part of the cortex which is spatially most remote from the hippocampus. If there be delays due to axonal conduction in the pathways between hippocampus and isocortex, they are likely to be greatest for just such connections. The same point can be made about the other pathway by which the hippocampus can influence the cortex, namely the Papez circuit: it seems to be remarkably circuitous, as if trying to maximize rather than minimize conduction delays. This point is amplified in Part 3 of this book (Chap. 9).

4 Discovery and General Behavioural Correlates of the Hippocampal Theta Rhythm

4.1 Introduction

The central idea of the present book is that rhythmic neural signals passing between hippocampus and isocortex can set up resonance between the two structures; and when they do so important types of information processing become possible. The rhythm around which the hypothetical resonance is envisaged to be entrained is the hippocampal theta rhythm, the most regular electrographic rhythm in the mammalian brain. As an introduction to the theta rhythm its most obvious behavioural correlates in various species will be described in the present chapter. The approach adopted in the present chapter is somewhat behaviourist in tone, though not in philosophical commitment, for reasons described below.

4.2 Discovery and early studies

The very regular EEG rhythm, of slow frequency, which may be generated by the hippocampus under various conditions was first described by Jung and Kornmuller (1938). They recorded a regular rhythm of 5–6 Hz in the hippocampi of unanaesthetized rabbits, elicited by strong noxious stimuli, or occasionally by auditory stimuli. In 1954 Green and Arduini (1954) published a more detailed study of this phenomenon. They recorded hippocampal and cortical EEGs, in several preparations: rabbits, cats and monkeys, in either unanaesthetized, curarized state or with brainstem transections, and in a smaller number of rabbits and monkeys with chronic electrodes in place. A variety of stimuli (olfactory, tactile, visual or auditory) could elicit a regular 5–7 Hz electrical rhythm from the hippocampus. In the acute rabbit experiments, olfactory or tactile stimuli were more effective at eliciting this rhythm than were distance stimuli (visual and auditory) though in acute cats auditory stimuli were also very effective. There was some suggestion that motivationally significant stimuli (e.g., for a rabbit: a cat roaming around the experimental room) were particularly effective in inducing these EEG rhythms. In chronically prepared rabbits visual stimuli were also quite effective. With repeated presentation of a stimulus, the electrographic rhythmic response habituated within a minute or so. In monkeys the hippocampal EEG findings were more complex and it was not possible to evoke the regular slow rhythm very predictably. When it occurred it was less well sustained, tending to consist of short series of slow waves overlaid and interrupted by faster activity. Often slow waves appeared with the first presentation of a stimulus, but not with subsequent presentations.

In all species the stimuli which elicited hippocampal slow rhythms also commonly produced low voltage fast activity (LVFA) in the cortical leads, though the two manifestations were not inevitably linked. The hippocampal slow rhythms survived large lesions of the overlying cortex. The combination of hippocampal slow rhythm and cortical LVFA could also be produced by high frequency (100 Hz) electrical stimulation of various sites in the brainstem or diencephalon, and in the septal region of the telencephalon. However, stimulation of the fornix or the mammillary bodies was without effect. The hippocampal slow rhythm was abolished, temporarily by deep barbiturate anaesthesia, or permanently by lesions of the septal region or fornix. Thus Green and Arduini's findings clearly show a strong relation between the septal region and the hippocampal slow rhythms, while the relation of hippocampal slow rhythmic activity and cortical LVFA was found to be less inevitable. The hippocampal slow rhythm has been termed the theta rhythm in consistency with terminology used generally for EEG rhythms. It is also sometimes called rhythmic slow activity (RSA). The two terms will be used interchangeably here.

In the second half of the 1950's, several further studies amplified the findings of Green and Arduini, in all cases using rabbits. Brucke *et al.* (1959a) extended the observations of Green and Arduini concerning rhythmic hippocampal activity evoked by septal stimulation (see below Sect. 5.5.1). The same authors (Brucke *et al.* 1959b) showed that microinjection of the local anaesthetic procaine into the septal region temporarily abolished the theta rhythm. The earliest pharmacological studies of the theta rhythm also date from this period: Mayer and Stumpf (1958) and Stumpf (1959) showed that hippocampal theta activity could be activated with the drugs eserine, apomorphine, methamphetamine or nicotine. In all such cases, the theta activity was abolished by septal lesions, just as was the theta rhythm induced by other means in Green and Arduini's study.

From about this time, the study of the theta rhythm expanded greatly. Some of the most detailed and elegant work dates from shortly after this time. However, it is preferred in the following chapters to pursue the hippocampal theta rhythm via a conceptual organization of information, rather than in historical sequence.

4.3 General Behavioural Correlates of Hippocampal Theta Activity in Several Species

It has become clear that the behavioural circumstances in which theta activity occurs vary from one species to another (for reviews see Winson 1972; Robinson 1980). This need not lead us to conclude that there is no functional role for theta rhythms common amongst species. After all, different species have different behavioural repertoires, so it is no cause for surprise that behavioural correlates of the theta rhythm also show species variation. However, it might be anticipated that the different behaviours, which, in different species, are associated with theta activity, have some more general significance for the ecology or psychology of each species, which may allow generalization across species.

In the subsections below, available data will be reviewed on the general behavioural correlates of theta activity in commonly used experimental animal species: small rodent species (mouse, rat, gerbil, guinea pig, and hamster) will be dealt with together. Separate descriptions will also be given for rabbit, cat, dog, monkey and man. By "general behavioural correlate", it is intended to concentrate on correlations with features of behaviour, observable at a single point in time, without taking into account the position in a longer sequence of events. Many of the experimental results to be dealt with have been obtained incidentally in the course of learning experiments, or performance of previously learned tasks. However, for the time being specific discussion of the relation between hippocampal RSA and learning will be excluded. This is because correlations of EEG with learning is categorically different from the more simple-minded EEG/behaviour correlations upon which we concentrate below. Learning is not something which can be directly observed, but is rather to be inferred after collation of a sequence of events. In addition the precise nature of the inference that learning has occurred is closely linked with theory development about learning. Even if one attempted to present "the facts" only, leaving their interpretation until later, there are many pretheoretical biasses which could determine the manner of presentation of information about learning. It is thought preferable to present the evidence on the relation between theta activity and learning in close association with the theoretical interpretation which this evidence will be given (see Chap. 9 and 10). This inevitably requires a deeper grounding in the physical mechanisms of theta rhythm generation than is available at the start of this chapter.

At the end of this chapter an attempt will be made to establish general principles which might apply across species to these general behavioural correlates of theta activity. These principles are, however, only ecological generalizations, rather than a detailed exposition of the psychological significance of theta activity. As such they give some clues to what the function of theta activity might be. The *details* of how the function of theta activity can be generalized across species is left until a later chapter.

4.3.1 Rat

4.3.1.1 Behavioural Correlations of Vanderwolf and Co-Workers

Since the rat is the species in which study of the hippocampal EEG has been most intensive, it will be described first. The work of Vanderwolf's group is dominant, and so it will be taken as the point of departure and comparison with other workers' results. Three broad correlations between hippocampal EEG and behaviour (all illustrated in Fig. 23) emerge from Vanderwolf's work:

1. Vanderwolf (1969) found that when rats engage in any movement involving the whole body (e.g., walking, running, rearing or jumping) hippocampal theta at frequencies from 6–12 Hz is an invariable accompaniment. Whishaw and Vanderwolf (1971) add swimming to this list. The frequency is related to the vigour of movement, and during sudden vigorous movements it might occasionally reach the maximum frequency of 12 Hz for brief instants (Kramis *et al.* 1975).

2. During small body movements from a fixed base (isolated movement of head or of a single limb during immobility, grooming or feeding) theta activity of lower frequency (4–7 Hz) is found in the hippocampal EEG. Theta activity observed in these circumstances is also of much lower amplitude than during more vigorous whole body movements (Whishaw and Vanderwolf 1973).

3. During behavioural immobility in the alert state, and during "automatic" behaviour patterns (such as blinking, scratching, washing face, licking or biting fur, chewing food or lapping water), the hippocampal activity is usually of irregular

rather than rhythmic type. Feder and Ranck (1973) add to this list teeth chattering, yawning and vomiting.

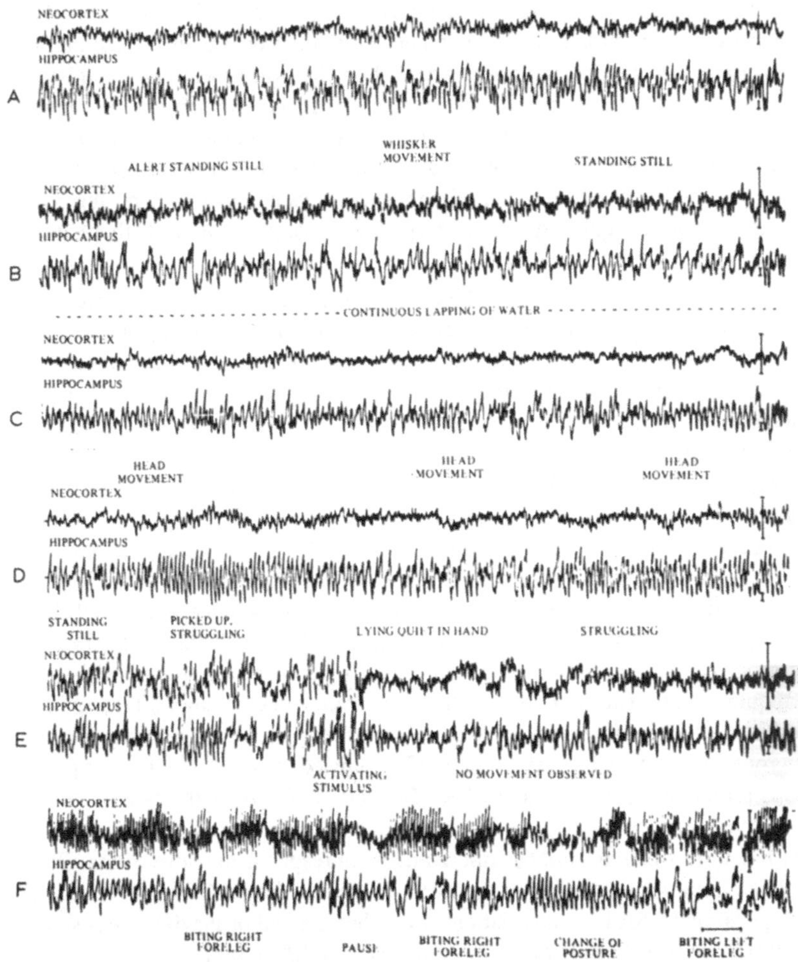

Fig. 23. Rhythmic slow activity recorded from the hippocampus of a single rat, during specified behaviours of various types. (Vanderwolf 1969: Fig. 2). In each pair of traces the *upper trace* is recorded from occipital neocortex, the *lower* from the CA1 pyramidal cell layer. Calibrations: 100 μV and 1 s (From *Electroencephalography and Clinical Neurophysiology* Vol 26 pp. 407–418, 1969)

These correlations between hippocampal EEG and behavioural activity appear to apply without modification to the mouse (Frederickson *et al.* 1982) and the Mongolian Gerbil (Kramis and Routtenberg, 1969; Whishaw, 1972).

Vanderwolf initially referred to the first and second categories of behaviour as "voluntary" and the third as "automatic". Later (Kramis *et al.* 1975; Vanderwolf *et al.* 1978) he preferred more neutral terms ("type 1" and "type 2" behaviours respectively).

4.3.1.2 Corroboration and Extension of Vanderwolf's Correlations

Whatever terms are used, Vanderwolf and co-workers provided a very simple correlate of hippocampal RSA: certain specific types of bodily movement. However, if these are to be regarded as the fundamental circumstances under which hippocampal theta activity is generated, it is necessary to show that when these particular behavioural items occur in a wide variety of other overall behavioural settings, the presence or absence of theta activity correlates with the specific items of behaviour listed by Vanderwolf and colleagues, rather than with the other concomitant behaviour. In fact there is a considerable collection of published evidence of this sort, summarized in Table 3, though with a more quantitative approach, many exceptions to these correlations are also on record (see below).

Some of the circumstances in which rats show theta activity deserve special mention. The correlation of high frequency theta activity and movement appears to apply even during sleep. During slow-wave sleep, the hippocampus shows irregular activity (Morales *et al.* 1971; Vanderwolf *et al.* 1978). During REM sleep, low-frequency RSA is shown most of the time during deep muscular relaxation (Brugge, 1965; Morales *et al.* 1971; Irmis *et al.* 1971; Vanderwolf *et al.* 1978). During the phasic bursts of eye and bodily movement RSA at high frequency is shown (Vanderwolf *et al.* 1978). Sano *et al.* (1973) found close association between eye movements and hippocampal RSA during REM sleep, the RSA starting regularly 0.5 s before each eye movement and increasing in frequency from about 7 Hz to 8–9 Hz as the eye movements began. Black (1975) used rats which were operantly conditioned to produce either high-frequency RSA (up to 10.5 Hz, mean 7.5 Hz) or low-frequency RSA (down to 4.6 Hz, mean 6.5 Hz). The groups were also subdivided so that half of each group was trained to produce the required EEG pattern under conditions of behavioural immobility, and half under conditions of free movement. Regardless of the behavioural training requirements, the groups producing high-frequency RSA showed greater activity ratings than the groups producing low-frequency RSA. Bennett (1975) trained rats on a "differential reinforcement of low rates" schedule (DRL). In this schedule, a time-out period (20 s) occurred, after which reward became available in response to a lever press. The end of the time-out period was either unsignalled, or signalled by light onset. The occurrence of hippocampal RSA bore no relation to whether the end of the delay was signalled or not. Instead *it correlated better with performance of the lever press.*

Three papers have investigated whether the absence of hippocampal RSA when rats are licking or chewing also applies to the circumstance of these responses being used as trained operants rather than as stereotyped behaviour patterns (Table 3). Black (1975) trained rats in an avoidance task in which shocks from the grid floor were avoided either by lever pressing, or by operant licking. In accordance with Vanderwolf's correlations, avoidance by lever pressing was regularly accompanied by 7 Hz RSA, while that accomplished by licking was associated with irregular hippocampal activity. Similar results were obtained by Frederickson and Whishaw (1977); and by Young (1976) in experiments where rats were trained to lever press in response to one discriminative stimulus, and to operantly lick in response to another, with either brain stimulus or repeated (Sidman) avoidance as the reward. Thus if the term "voluntary" in Vanderwolf's original categorization is taken to mean "conditionable", these two studies do not support the voluntary/automatic distinction. They relate the occurrence of theta activity to the pattern of response *per se*, rather than its method of generation.

Much of the above evidence suggests a relationship between the occurrence or frequency of hippocampal RSA and the *vigour* of certain types of movement. It has led several investigators to attempt to *quantify* the relation between vigour of movement and RSA frequency. Bland and Vanderwolf (1972) and Whishaw and Vanderwolf (1973) showed that in rats trained to make a jump as an avoidance response, the frequency of hippocampal RSA rose from 7–8 Hz on approach to the jump and to 10–12 Hz just before take-off. Vanderwolf and Cooley (1974) studied hippocampal EEG in a similar task in which rats repeated a jump to exhaustion. The same results were obtained, but when the rats were fatigued, failure to jump correlated, on a trial-by-trial basis, with either absence of RSA altogether, or failure of RSA frequency to rise just before take-off. McFarland *et al.* (1975) studied the relation between running speed on a treadmill, and incidence and frequency of hippocampal RSA. As runway speed increased from 1–22 cm/s, the incidence of 6–8 Hz theta increased monotonically. There was also some increase in 10–12 Hz theta at the highest running speeds. Amongst the observations of Whishaw and Vanderwolf (1973) are that the frequency rise at the time of launch is greater, if the jump distance is greater. Morris *et al.* (unpublished, cited in O'Keefe and Nadel 1978, pp. 180–182) and Morris and Hagan (1983) studied these relationships more systematically, varying the height of the jump, and the loading of the animal, and measuring force exerted on the ground at take-off. There was a systematic increase in frequency (decrease in period) for the higher jumps at take-off, and during the flight phase (which lasted one to three cycles of RSA). This was related to take-off velocity, distance or acceleration, rather than force generated. In the later study the wave period also decreased as a function of increasing jump-height for the waves immediately before after take-off.

There are a few reports for rats of correlation between hippocampal EEG and the specific coordinated response of head, ears and body called *orienting*. Unfortunately, this term is sometimes used interchangeably with "exploration". For instance Brugge (1965) found that RSA occurs during "orienting", but fails to separate this correlation from that with exploratory locomotion. This point is of some importance, since some writers regard exploration as the real sign of orienting in the rat (Knoll, 1956). Feder and Ranck (1973) distinguished orienting from exploring and *inter alia* saw RSA during orienting. Buzsaki *et al.* (1979a) provide a specific scale for assessing orienting in rats, including locomotion as the highest rating. Highest orienting scores occurred at the time when RSA scores were also highest, but did not reach the intensity at which locomotion was predominant. It is likely from these data that RSA can occur as an accompaniement to orienting responses that do not include locomotion. Such correlations can still be included under Vanderwolf's heading of "small body movements from a fixed base".

There have been a number of reports of hippocampal EEG correlates of vibrissal twitching, and other simple rhythmically repeated acts. Vanderwolf (1969) and Whishaw and Vanderwolf (1971) stated that theta bursts during vibrissal twitching only occurred when there were other concomitant movements of body or head. However, Gray (1971) describes and illustrates theta bursts closely time-linked to the duration of vibrissal twitching in an otherwise immobile rat. Komisaruk (1970) also found a consistent phase relation between rhythmic vibrissal twitching/sniffing and the concomitant

Table 3. Corroboration of Vanderwolf's RSA correlations in rat

Reference	Circumstances of experiment	Finding
Brugge *et al.* (1965)	Arousing stimuli (sound or air-puff)	RSA correlated with exploration and orienting*
Routtenberg (1968)	Habituation to novel enclosure (1 h daily sessions)	RSA habituated pari passu with exploration
Irmis *et al.* (1970)	Sequences of acoustic stimuli	RSA duration and frequency tended to correlate with exploratory activity
Gray and Ball (1970) Kimsey *et al.* (1974)	Runway task	RSA frequency increased from start box to runway
Morales *et al.* (1971)	Novel cage, or introduction of novel object	RSA occurred in association with exploration, but not after habituation of exploration
Albino and Caiger (1971)	Classical conditioning	RSA greater during climbing or exploring than during immobility, or movement from fixed base
Schwartzbaum *et al.* (1973)	Repeated flashes	Decrease in 7–8 Hz RSA associated with decrease in locomotion
Schwartzbaum and Kreinick (1973)	Repeated flashes predicting food	Increase in 7–9 Hz RSA as food came nearer in time, associated with increase in locomotion
Kurtz and Adler (1973) Kurtz (1975)	Sexual behaviour in either sex	RSA associated with locomotion
Feder and Ranck (1973)	Approaching food	RSA occurs during locomotion
Black (1975)	Operant conditioning of high frequency RSA	Locomotion greater than for operant conditioning of low frequency RSA
Bennett (1975)	DRL schedule	RSA correlated with lever presses
Black (1975)	Avoidance by lever pressing or operant licking	RSA present during lever pressing but not during licking
Young (1976)	Lever pressing or operant licking, to obtain brain stimulus reward, or avoidance of shock	More RSA during lever pressing than licking
Frederickson and Whishaw (1977)	Bar pressing or chewing/licking to obtain food	More RSA during bar-pressing than during chewing/licking.
O'Keefe and Nadel (1978)	Jumping task	RSA correlated with movement

* see text

theta cycle. This has been investigated in more detail by Macrides, *et al.* (1982) who observed a consistent time lag of 120–180 ms (mean 150 ms of the sniff cycle with respect to the theta cycle. This delay was the same regardless of changes in the theta frequency. Related findings were made by Semba and Komisaruk (1978) with respect to repetitive lever pressing. The phase of hippocampal RSA was consistently related to the moment of lever pressing and of lever release, opposite phase relations being observed for these two phases of lever pressing. The authors comment that motor acts such as lever pressing appear to be more easily elicited at certain phases of RSA than at others. Macrides (1975) also studied exploratory sniffing in hamsters, and found that RSA always occurred during bouts of sniffing. This was not always at exactly the same frequency as the sniffing cycle, and when entrainment occurred the phase relation between RSA and sniffing varied widely between subjects. The suggestion was made that RSA exerts an influence over the sniffing rhythm, but does not directly drive it. Eichenbaum *et al.* (1987) also suggest that, for investigatory sniffing, the correlation with theta activity, though quite common is not obligatory. It was observed that there was often a consistent time-locking of sniff cycles with theta waves, but this could be accounted for more plausibly as a *sensory-processing* aspect of the sniff cycle than as a precursor of the motor aspect. (Other evidence, discussed in Sect. 10.2, also shows that theta activity may be initiated in correlation with processing of important or novel sensory stimuli).

To summarize so far, *there is a considerable body of information of relatively simple correlates between large whole body movements and occurrence of high-frequency RSA in rats; and between smaller movements from a fixed base and lower-frequency RSA.* However, against this background we can now describe a variety of *exceptions* to the above correlations, and there are also correlates of hippocampal theta activity with aspects of behaviour other than movements of whole body or segments of the body.

4.3.1.3 Apparent Exceptions to Vanderwolf's Correlations

The correlations of RSA defined by Vanderwolf and colleagues have received such strong support that they can hardly be questioned; but they are only correlations, not explanations. It is possible that these correlations could coexist with a variety of exceptions to the general rules. In the present subsection exceptional data obtained for the rat are discussed in some detail (see Table 4).

The *first* set of exceptions (Table 4A) challenges the earlier observations by showing that under some circumstances theta activity, of whatever frequency, can be produced in an immobile rat. Some of these exceptions are obtained by the use of arousing stimuli in an immobile animal. As Kramis *et al.* (1975) point out, rats respond to most stimuli with bodily movement, so it is not easy to discover the effect of arousing stimuli in an immobile animal. Hence the early observation that theta activity does not occur during immobility may be more a biproduct of special aspects of the behaviour and ethology of rats, than a fundamental limitation. However Macadar *et al.* (1970) showed that immobilized (curarized) rats could generate a low-frequency (6 Hz) hippocampal theta rhythm in response to tactile stimuli. Winson (1976a) also attempted to evoke hippocampal RSA by tactile stimuli in curarized rats. This was not a very effective method, since RSA bursts, when they occurred, were of only brief duration. However, it was found that a vestibular (rotatory) stimulus is a highly reliable and effective way of evoking RSA in the immobilized rat[4]. Albino and Caiger (1971)

[4]This evidence is discussed with the implication that curarized animals are immobilized by a peripheral action of the drug, without confounding effects produced by a drug acting on the central nervous system. However this implication may be invalid: Curare in rat (but not rabbit) modifies the laminar theta profile compared with free moving animal (Winson 1974, 1976a,b) (see Chap. 7), and may therefore be having its own central effects.

also provide evidence of theta activity during immobility: in quantitative assessments of the abundance of theta activity during various types of behaviour, there was some theta activity during immobility, albeit much less than during movement. Another exception to the association of theta activity with movement, reported by Vanderwolf (1975) is that in the brief immobile period before a jump-response the hippocampus showed 6–7 Hz RSA. Furthermore, in the phase of REM sleep when there is deep muscle relaxation, low frequency (6 Hz) RSA is shown (see above).

Another, relatively non-physiological, situation in which hippocampal RSA in rat is dissociated from its usual behavioural correlates is when the hippocampal EEG rhythm is driven by electrical stimulation of the medial septal nucleus, or nucleus of the diagonal band: Kramis and Routtenberg (1977) and Kramis and Vanderwolf (1980) found that when such rhythmic activity was elicited it did not disrupt behaviours (eating, drinking, grooming, immobility) normally associated with a desynchronized hippocampal EEG. Likewise, when septal stimuli succeeded in desynchronizing the hippocampal EEG this did not disrupt the head movements, walking or rearing which are usually correlated with hippocampal theta activity. However, this dissociation may not apply when septo-hippocampal rhythmic activity is elicited by brainstem stimulation (see Chap 5.5.2).

Secondly, beyond these exceptions there are some hints from learning experiments that the intensity of hippocampal RSA may correlate with aspects of a learning situation other than the concomitant levels of locomotor activity, the latter correlations actually overriding those demarcated by Vanderwolf *et al.* There are some results (Table 4B) from experiments involving learning where locomotor activity can occur in the absence of theta rhythms. In the experiments of Routtenberg (1968), in which several days of habituation to an initially novel enclosure occurred, the parallel between theta activity and locomotor activity was not precise: On later days there was very little theta activity, but the animal still walked about. Pickenhain and Klingberg (1967) presented repeated shocks to rats, either unsignalled or signalled, with opportunity for them to learn avoidance responses. It was observed in the unsignalled situation that the animal initially made chaotic attempts to escape, which were not accompanied by hippocampal RSA, but when the first successful avoidance responses occurred, so did RSA. Buzsaki *et al.* (1979a) trained rats in an experimental design which was, formally, classical conditioning. However, the tone used as CS was delivered in the opposite side of the apparatus from the water delivered as UCS, so that behaviour directed at CS and UCS could be assessed separately (respectively as the first and second second of the 2 s CS-UCS interval). It was found that hippocampal RSA was concentrated in the first second, related to CS-directed behaviour (i.e., orienting). As training proceeded, UCS-directed behaviour (in the second second) gradually increased, but without any association with hippocampal RSA in this period.

There are some data (Table 4C) showing that the theta rhythm can occur in the absence of locomotor or other whole body movements. Pickenhain and Klingberg (1967) used rats in the situation of signalled avoidance learning (with repeated flash as warning signal). When the avoidance response had been well learned, the signal presented on each trial evoked hippocampal RSA well before any avoidance response occurred and in the absence of other motor behaviour. Similarly, Bennett (1975) illustrates (his Fig. 6) that in DRL experiments the start of RSA preceded lever pressing by some seconds. Schwartzbaum (1975) trained rats to expect food pellets on presentation of a light flash, and then introduced a clicker as a signal of impending shock. This signal caused the rats to come to ignore the food when the clicker was sounding. When this conditioned emotional response (CER) was well trained, hippocampal EEG records were made in epochs before, during and after the clicker. The clicker caused an increase in theta counts, but at the same time the animal "froze", as part of the CER.

Table 4. Exceptions to Vanderwolf's correlations in rats

Reference	Circumstances of experiment	Finding
A Theta activity observed during immobility: non-learning situations		
Macadar *et al.* (1970), Winson (1976a)	Curarized rats, tactile stimuli	Low frequency RSA evoked
Winson (1976a)	Curarized rats, rotatory stimuli	RSA evoked strongly
Vanderwolf (1975)	Immobility just prior to a jump	RSA present
Vanderwolf (1975)	Deep relaxation of REM sleep	Low-frequency RSA present
Kramis and Routtenberg (1977) Kramis and Vanderwolf (1980)	Electrical stimulation of septal area	RSA seen during immobility
B Theta activity observed during immobility: learning situations		
Pickenhain and Klingberg (1967)	Well-learned avoidance response	RSA occurred *before* onset of each response
Bennett (1975)	DRL schedule	RSA started before a bout of lever pressing
Schwartzbaum (1975)	Conditioned emotional response	RSA seen during "freezing"
C Locomotor activity in the absence of theta rhythms		
Pickenhain and Klingberg (1967)	Avoidance conditioning	On early trials vigorous attempts to escape but lack of RSA
Routtenberg (1968)	Habituation to a novel enclosure	On later days, locomotion persisted despite lack of RSA
Kramis and Routtenberg (1977) Kramis and Vanderwolf (1980)	Electrical stimulation of septal region	Locomotion persisted despite desynchronization of hippocampal EEG
Buzsaki *et al.* (1979a)	Well-trained classically conditioned response	Motor responses to UCS not accompanied by RSA
D Failure of quantitative correlation between locomotor activity and RSA		
Black (1975)	Operant conditioning of RSA	Locomotion did not correlate well with RSA
Bennett *et al.* (1978)	Cued and non-cued DRL schedules	RSA in delay period greater in cued than non-cued schedule

Thirdly (Table 4D) there are further results from experiments involving trained responses in which there is a clear failure of quantitative correlation between locomotor activity and theta counts. Black (1975) in experiments mentioned above, trained rats to produce either high- or low-frequency RSA, in combination with either free movement or immobility. In the four groups, it was found that high-frequency RSA counts correlated to some extent with behavioural activity ratings, but nevertheless, animals producing high-frequency RSA under requirements of behavioural immobility, had similar ratings of behavioural activity to those producing low-frequency RSA under conditions of free movement. Incidentally, it was found that it was more difficult to train high -frequency RSA under conditions of immobility than under free-moving conditions. Bennett *et al.* (1978) trained rats in either a cued or a non-cued DRL schedule. Although response requirements in the two situations were similar, there was a significantly greater amount of RSA recorded in the hippocampi in the interresponse intervals of the cued schedule than of the non-cued schedule.

4.3.1.4 Theta Activity in Rat—Summary

In summary there is a strong relation between locomotor activity (and other types of whole body movement) and high frequency theta in rats in many types of behaviour; and small movements from a fixed base *do* generally coincide with lower-frequency theta activity. However, lower frequency RSA can occasionally be seen in the immobile rat, or in one performing small and selective movements such as orienting or vibrissal twitching. In addition there are many exceptions to the correlation between locomotion and theta activity, both in general behaviour and in learning experiments. Admittedly some of these discrepancies could be accounted for by unreported variations in the "small movement from fixed base" category of action. Nevertheless, there are enough *prima facie* exceptions to give us pause before regarding Vanderwolf's correlations as *the* fundamental correlate of hippocampal RSA under all circumstances. In particular, it would be prudent to leave open the idea that RSA may have particular correlations with certain phases of the learning process, or with performance of learned behaviour of some categories.

4.3.2 Guinea Pig

Sainsbury (1970) reported that the hippocampus of the guinea pig generates RSA during activities such as walking, head movement, digging (but not the pauses between bouts of digging) and climbing (but not "hanging" motionless after having reached the objective of climbing), and postural changes. The frequency of the RSA was higher during walking and digging than during head movements. Low amplitude, irregular activity was seen during sitting, freezing, groomimg (unless it involved postural changes), chewing, urination, defaecation and vocalization (unless it involved walking or struggling). Sainsbury also reports on the hippocampal EEG correlates of sexual and social activity. Automatic components (pelvic thrusting, anogenital licking) were accompanied by irregular activity, while walking, chasing, mounting, or fighting between two males (whether attacking or withdrawing) were accompanied by RSA. In a later paper (Sainsbury and Montoya 1984), more attention was devoted to RSA occurring during immobility. Such a correlation was seen in the guinea pig's response to a snake model when it stood immobile facing the snake, when a tape recording of an owl hoot was played, causing "freezing" in the animal, and during some phases of sexual behaviour. The dominant frequency of RSA during immobility was lower (about 7 Hz) than during movement theta (about 8 Hz). Neutral stimuli which initially do not elicit RSA could acquire a capacity to elicit it, if paired with RSA-eliciting stimuli such as sounds of owls, hawks or kestrels. Thus the general picture seems to be similar to that in the rat, though RSA occurs more commonly during immobility than in the rat. The suggestion that RSA has some relation to learning or learned behaviours receives some support.

4.3.3 Rabbit

A number of papers report on behavioural correlations of hippocampal EEG in rabbits. The most detailed of these is the paper of Kramis *et al.* (1975), so this will be described first. Fig. 24 illustrates some of their results. When the animal is moving around, including activities such as hopping, walking and running, RSA is invariably seen. Its frequency is usually between 7.5 and 12 Hz, increasing as activity increases, and occasionally reaching as high as 15 Hz in sudden vigorous movements. During eating, drinking and grooming a low frequency (5–7 Hz) RSA was usually seen. In the immobile animal it is reported that large amplitude, irregular activity is usually seen, even in an apparently alert animal (as judged by alert posture, erect ears and wide open eyes). The exception to this is when a sensory stimulus is presented. Stimuli such as lights, buzzers, and tones presented to an immobile animal elicit an RSA of frequency 6.3–7.5 Hz, often locked to the stimulus duration. For very short stimuli RSA would outlast the stimulus, and for stimuli longer than 3 s, the RSA ceased before the stimulus. When stimuli were repeated the RSA habituated completely, sometimes in as few as 20–30 presentations (but more slowly than the orienting response). On the basis of this paper it is clear that there are correlates of behaviour similar to those found in rat (bodily movement). There are also significant differences, in that it is much easier to produce RSA in the immobile rabbit than the immobile rat, and RSA is more commonly seen during eating and drinking in rabbit than in rat.

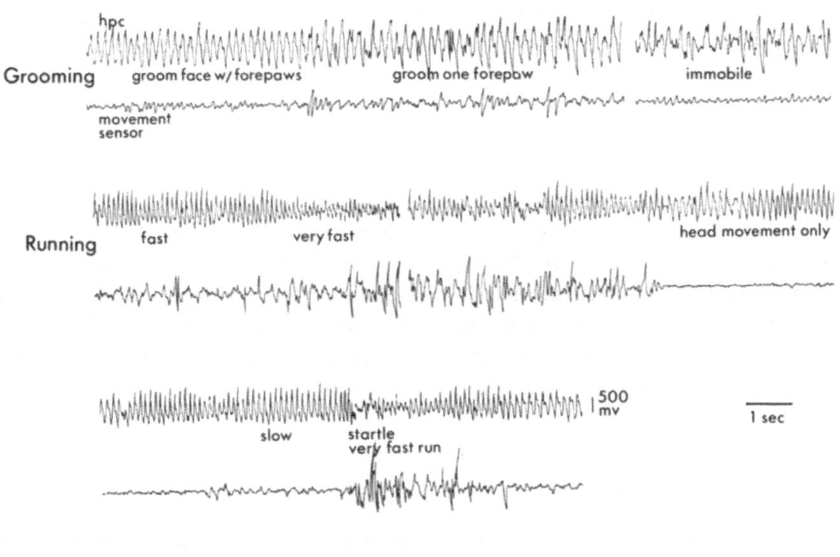

Fig. 24. Rhythmic slow activity recorded from the hippocampus of the rabbit during specified behaviours of various types. (Kramis *et al.* 1975, Fig. 5). In each pair of traces the *upper trace* is recorded from the dorsal hippocampus (unspecified field) and the *lower* from a movement sensor. Calibrations: 500 µV and 1 s. (From *Experimental Neurology* Vol 49 pp. 58–85 1975)

In other papers the correlation between RSA and movement has also been noted (Harper 1971; Klemm 1971; Winson 1976b). Sadowski and Longo (1962) noted high frequency RSA (up to 8–9 Hz) as a rabbit ran to retrieve a piece of cabbage when the conditioning signal sounded. Powell and Joseph (1974) studied a differential classical conditioning task in rabbits and report that both behavioural activity measures and hippocampal theta counts went on increasing through the acquisition period, in a roughly parallel manner. The CS+ evoked both more RSA and more behavioural activity than the CS−. During extinction, there was a decline in both CS-induced RSA and behavioural activity.

The occurrence of low-frequency RSA during eating and drinking is also confirmed by Torii and Sugi (1960) and by Sadowski and Longo (1962). Huston and Brozek (1974) find theta activity (frequency not specified) in all phases of ingestion, though mixed with frequencies below the theta range during chewing.

A few exceptions, comparable to those cited for rats, are reported, where bodily movements do not correlate very well with RSA. Klemm (1971) notes that correlations do not depend on actual movement, e.g they could occur during isometric contraction. Powell and Joseph (1974) found in their differential classical conditioning experiments that gross activity levels tended to rise particularly in the later part of 30 daily acquisition sessions, while hippocampal RSA in response to the CS+ and CS− increased and differentiated more in the earlier phase of acquisition.

That RSA also occurs during immobility in the rabbit is confirmed by a number of papers. Torii and Sugi (1960) report an 8 Hz RSA when the rabbit is "looking for or sniffing" but not moving about, and a lower frequency RSA (5–7 Hz) is seen in the stationary animal, when it stands on hindlegs and licks its forepaws, or rubs its nose. RSA during immobility has also been seen during freezing (showing signs of fear: pupils dilated, ears laid back, head lowered to touch the ground) (Torii and Sugi 1960; Harper 1971), in response to a sudden stimulus (e.g., banging on cage) (Sadowski and Longo 1962; Martin *et al.* 1975)), during and for a short time after a vestibular (rotatory) stimulus (Costin *et al.* 1967) and in the immobile preparatory phase for movement (Harper 1971).

However, not all states of immobility are accompanied by RSA. Torii and Sugi (1960) and Harper (1971) found that in the perfectly still, relaxed animal 2–3 Hz activity is seen rather than theta activity. The same is true of the sleeping[5] rabbit (Torii and Sugi 1960). Klemm (1971) observes that in rabbits under "animal hypnosis" RSA is only occasionally seen. However, Whishaw *et al.* (1982) show that the influences which elicit RSA in a normally immobile rabbit (sensory stimuli, eserine or brain stimulation [see below]) will also do this in a "hypnotized" rabbit.

4.3.4 Cat

4.3.4.1 Similarities to Dominant Correlations in Rats

In cats, the frequency range for RSA is somewhat lower than in smaller mammals, ranging from 3–8.5 Hz. In addition, the cat has behavioural correlates of hippocampal

[5] This datum on sleep probably corresponds to slow wave sleep: Harper (1971) obtains data more in keeping with the rat data on sleep hippocampal EEG, i.e., in REM sleep 7.5–8 Hz RSA is seen, increasing further in frequency during the phasic events of REM sleep.

RSA, which, *prima facie* at least, are rather different from those for most of the smooth-brained animals considered above. There are fewer reports of movement-related RSA and more of vigorous RSA in the immobile animal. Correlations between bodily movement and hippocampal RSA are summarized in Table 5. Three papers (Whishaw and Vanderwolf, 1973; Frederickson *et al.* 1978; Arnolds *et al.* 1984) have specifically observed hippocampal RSA in cats during walking. In the first of these reports RSA was continuous during 8 h walking on a treadmill. RSA was also seen during orienting movements or other movements of the head. However, RSA in cats is typically of higher frequency during locomotion (5–6 Hz) than during skeletal immobility (4–5 Hz). Observations made incidental to other types of behavioural experiment (Table 5) add to this evidence of correlations between locomotion (and other bodily movements) and theta activity, as they do for rat, rabbit, etc. In contrast to this, the hippocampal EEG showed no RSA during lapping milk, chewing food (Whishaw and Vanderwolf 1973), grooming and "reflexive head movements" (Kemp and Kaada 1975). During slow wave sleep no RSA occurs except during phasic body movements (Glotzbach 1975). Thus the times of occurrence of hippocampal RSA show at least some similarities to that found in rats and other small mammals.

Table 5. Examples of correlations between body movement and hippocampal RSA in cats

Reference	Circumstances of experiment	Findings
Adey *et al.* (1960)	Approach learning, extinction and retraining	5–6 Hz RSA correlates with approach response
Holmes and Adey (1960)	Performance and extinction of a delayed response discrimination	Entorhinal theta correlated with bridge walking, rather than with "correctness" of response
Adey *et al.* (1961) Adey *et al.* (1962)	T-maze discrimination Delayed response task	6 Hz RSA correlates with approach response
Porter *et al.* (1964)	Exploration	Maintained 4–5 Hz RSA
Porter *et al.* (1964)	T-maze discrimination	6 Hz RSA during approach
Elazar and Adey (1967a)	Light-dark discrimination	6 Hz RSA during approach; 5 Hz RSA during "escape" responses
Holmes and Beckman (1969)	GO/NO GO task	RSA associated with GO response
Whishaw and Vanderwolf (1973)	8 h continuous walking	Continuous RSA
Whishaw and Vanderwolf (1973) Kemp and Kaada (1975)	Lapping milk, chewing food	Suppression of RSA
Hatfield (cited by Black, 1975)	Cats trained to lever press or remain immobile	RSA occurred more during lever pressing
Frederickson *et al.* (1978)	Walking and immobility	RSA frequency higher during walking
Gralewicz (1981)	Shuttle-box avoidance	RSA associated with walking
Arnolds *et al.* (1984)	Walking	RSA seen

Table 6. Exceptions to the correlation between RSA and movement in cats

Reference	Circumstances of experiment	Findings
A Production of RSA in immobile cats		
Brown and Shryne (1964) Brown (1968) Whishaw and Vanderwolf (1973) Kemp and Kaada (1975)	Visual search, or fixed staring in immobile cats	RSA present
Sakai *et al.* (1973)	REM sleep	In association with eye movements
Bennett and French (1977)	"Gape" response in immobile animal	4–6 Hz RSA
Frederickson *et al.* (1980)	Trained immobility	4 Hz RSA, especially when a cue was used to signal the end of the immobile period
Gralewicz (1981)	"Conflict" trials in shuttle box	High frequency RSA in immobile animal
Buzsaki *et al.* (1981)	Conditioned emotional response; omission of expected food reward	RSA in immobile animal
B Bodily movement in the absence of RSA		
Grastyan *et al.* (1966)	Well-trained active avoidance response	No RSA during response
Elazar and Adey (1967a)	Light/dark disrimination	Little RSA on incorrect approaches, or in overtrained approaches
Bennett (1970)	Well-trained approach response	No RSA during approach
Grastyan and Vereckei (1974)	Well-trained approach response in alleyway	No RSA during approach or T-maze
Frederickson *et al.* (1978)	Alley-running	RSA sometimes absent
Gralewicz (1981)	Shuttle-box avoidance	RSA absent on short-latency responses
Buzsaki *et al.* (1981)	Cat finds empty food cup	Head movement occurs without RSA
C Failure of quantitative correlation between movement and RSA		
Holmes and Adey (1960)	Delayed response discrimination involving choice of two bridges	RSA could occur during delay *before* responding; RSA failed during non-purposeful movements
Porter *et al.* (1964)	Light/dark discrimination in T-maze	RSA seen best when performance is improving; declines when response is automatic
Grastyan *et al.* (1966)	Appetitive or avoidance task	RSA seen *after* completion of response
Elazar and Adey (1967a)	Approach response	RSA less in 2nd half of approach phase, despite maintained walking
Bennett (1970)	Discriminated lever pressing	RSA occurred between incorrect lever presses (animal alert but not exploring)

Table 6. (continued)

Bennett *et al.* (1973)	CRF lever pressing	No RSA seen except during orienting
Bennett *et al.* (1973)	Cued and non-cued DRL task	RSA prominent *between* responses, for cued task only
Coleman and Lindsley (1977)	CRF lever pressing	RSA declined in well-trained task
Coleman and Lindsley (1977)	Alternating periods of reinforcement and non-reinforcement	RSA prominent when responses withheld
Buzsaki *et al.* (1981)	GO/NO GO discrimination	High frequency RSA correlated better with signal function of CS than with responses

4.3.4.2 Discrepant Evidence

The evidence which is discrepant with the movement/RSA correlation is summarized in Table 6. Table 6A lists examples where hippocampal RSA has been seen in immobile cats. Several workers have described hippocampal RSA produced in cats, whose only active behaviour was visual search or fixed staring (Brown and Shryne 1964; Brown 1968; Whishaw and Vanderwolf 1973; Kemp and Kaada 1975). Bennett and French (1977) studied a species-specific response, called the "gape" response, elicited by cats when they encounter a conspecific's urine. The response consists of several recognizable components, including some produced in an immobile animal. RSA at 4–6 Hz occurs throughout this attentive response, including the periods of immobility (though frequency varies somewhat in different parts of the response). RSA has also been reported several times during immobility occurring as part of a conditioning process (Frederickson 1980; Gralewicz 1981; Buzsaki *et al.* 1981).

Conditioning experiments have revealed a number of instances where bodily movement can occur in the absence of hippocampal RSA (see Table 6B). Some of these instances concern the situation of very well trained responses (Grastyan *et al.* 1966; Grastyan and Vereckei 1974; Bennett 1970). Other instances come from the study of individual trials in a learning sequence, where RSA does not occur during movement (Elazar and Adey 1967a; Gralewicz 1981; Buzsaki *et al.* 1981).

There is also a variety of experiments, usually involving conditioning, where there is a failure of detailed correlation between locomotor (and other) movements and hippocampal RSA, although in a less sharp manner than in the above experiments. These results are summarized in Table 6C.

4.3.4.3 RSA Correlating with Orienting

"Orienting" is a coordinated response which appears to indicate alertness, arousal or readiness to process information, and has been mentioned several times in the above paragraphs. Of course, orienting is quite closely linked to learning processes, which will be discussed in Part 3 of this book. Here it is appropriate to collect the evidence about hippocampal EEG correlates of orienting, seen simply as a behavioural pattern, rather than as an indicator of concurrent learning. Some of the evidence was obtained in relation to eye movements, but this is included here because of the close relation

Table 7. Correlations between orienting and RSA in cats

Reference	Circumstances of experiment	Findings
A Spontaneous orienting		
Radulovacki and Adey (1965); Whishaw and Vanderwolf (1973)	Spontaneous orienting	4–5 Hz RSA
Brown (1968) Kemp and Kaada (1975)	Active visual search, and fixed staring	RSA present (frequency higher during active visual search)
Bennett (1970)	Periodic orienting, after exploratory behaviour had habituated	RSA present
Bennett and Gottfried (1970)	Non-cued DRL performance	No RSA except during orienting
Bennett *et al.* (1973)	Continuously reinforced lever pressing	No RSA except during orienting
Kemp and Kaada (1975)	Active visual search	RSA correlated with pupillary dilatation, and degree of eye opening
Coleman and Lindsley (1975)	Alert vs. relaxed posture	RSA present during alertness
Radil-Weiss *et al.* (1976)	Acute pretrigeminal cat	RSA correlated with ocular and EEG signs of orienting
Gralewicz (1981)	Shuttlebox avoidance	RSA during attentive postures
B Orienting acquired during conditioning		
Grastyan et al. (1959); Lissak and Grastyan (1960)	Appetitive and escape tasks	CS came to elicit both RSA and orienting during training. With overtraining both disappear together
Brown and Shryne (1964)	Repeated tones	RSA and orienting develop together
Brown (1968)	Repeated tones	RSA initially absent, despite orienting, but develops later at times of orienting
Bennett (1970)	Clicker signalling food delivery	Orienting and RSA develop together, and disappear together at a later stage of training
Grastyan and Vereckei (1974)	Appetitive training, CS and goal box spatially separate	Orienting and RSA develop together, and disappear together at a later stage of training
Coleman and Lindsley (1977)	Continuously-reinforced lever pressing; or alternating periods of reinforcement and non-reinforcement	RSA develops in association with attentive postures and orienting
Buzsaki *et al.* (1981)	GO/NO GO training	RSA frequency increased as orienting developed
Buzsaki *et al.* (1981)	Unavoidable shock signalled by tone CS	Orienting declined although RSA at CS persisted

between eye movements and the complete orienting response. The data falls into two categories: correlation with spontaneous orienting response, and that with conditioned orienting responses. These data are summarized respectively in Table 7A and 7B.

4.3.4.4 Summary

In summary, hippocampal theta activity does occur in cats in association with locomotion or other body movements, in a manner similar to that seen in rats and other small mammals. However, there are a great many exceptions to these correlations, where theta activity can occur during immobility, or movement can occur without theta activity. Many of these exceptions occur in the course of learning tasks. There are suggestions that theta activity occurs at the time when performance of a learned task is improving most rapidly, and declines as tasks become very well learned. The precise significance of these data must remain enigmatic at the present stage of the argument, though the matter will be discussed in detail in Chapter 10.

Apart from these results, there is a lot of evidence that both spontaneous and conditioned orienting is accompanied by low-frequency theta in cat. There are, however, minor qualifications to these correlations, for instance in the earliest stages of some conditioning tasks, where partial orienting responses occur without RSA, or where orienting declines without decline of RSA.

4.3.5 Dogs

In dogs the frequency of RSA is low (3–8 Hz), as it is in cats. There are several studies reporting clear correlations between such hippocampal activity and movement, as well as other evidence which does not fit so well with this pattern. These data are summarized in Table 8. Reports of movement-related RSA are summarized in Table 8A. As with other species, there are a number of observations available which do not fit into the idea of direct correlation between any movement and hippocampal RSA. There is some indication that RSA at a slightly lower frequency can occur during immobility, or absence of locomotor activity in dogs (Table 8B). In addition, there are some reports where there is a failure of detailed correlation between theta activity and movement (Table 8C). RSA appears to be absent during automatic actions such as eating (with one exceptional result), as summarized in Table 8D.

4.3.6 Monkey

In the monkey the situation is far less clear than in any subprimate species. Crowne and Radcliffe (1975) have provided the most detailed observations. In confirmation of Green and Arduini (1954) they find that RSA detectable by visual inspection of EEG records is infrequent during spontaneous behaviour, and when it occurs, consists of quite short bursts (less than 1 s, at 6–12 Hz). Such rhythmic activity could be elicited by electrical stimuli applied to the septum (e.g., at 10 Hz), but the frequency range over which such theta driving was possible was not reported. High frequency septal stimulation was reported not to desynchronize the hippocampal electrical activity as it does in lower mammals. However, the significance of this finding is unclear, since the site of stimulation in monkey was not precisely known. The only natural way found to evoke such short RSA bursts was during absence of expected reward.

Table 8. Correlations between theta activity and behaviour in dogs

Reference	Circumstances of experiments	Findings
A Movement-related RSA		
Yoshii *et al.* (1967)	Walking, escape reactions. Walking to retrieve food after a delay	6–7 Hz RSA
Storm van Leeuwen *et al.* (1967)	Walking, especially to a goal	RSA present
Dalton and Black (1968)	Discriminated lever-press avoidance; use of the same discriminative stimuli in paralysed dogs	CS generating lever press produced more RSA than that suppressing lever press
Lopes da Silva and Kamp (1969); Kamp *et al.* (1971)	Walking, either to a goal or after retrieving a reward	RSA frequency increases to 6 Hz
Lopes da Silva and Kamp (1969)	Walking away from reward bowl	RSA frequency shifts from 5 Hz to 6 Hz
Black and Young (1972)	Appetitive training to lever-press or to hold still; use of same discriminative stimuli in paralyzed animal	More RSA, and higher frequency during lever press than during holding still
Black *et al.* (1970) Black (1975)	Operant conditioning of RSA or of desynchronized hippocampal EEG	More movement shown by dogs conditioned to produce RSA, even when initial conditioning in paralyzed dogs
Arnolds *et al.* (1979a)	Active behaviour; transitions to more active behaviours (spontaneous or reward-induced)	RSA frequency shifts from 5 Hz to 6 Hz
Arnolds *et al.* (1979b)	Walking on belt at various speeds	Frequency and amplitude of RSA increased with speed
Arnolds *et al.* (1979c)	Spatial discrimination and its reversal	RSA amplitude and narrowness of frequency, vigour of lever press, and correctness of response all intercorrelated
B Theta activity in absence of locomotion or during immobility		
Storm van Leeuwen (1967)	Dogs in tense posture	RSA present
Lopes da Silva and Kamp (1969)	Stationary after delivery of reward; or on response ommission	5 Hz RSA
Urban *et al.* (1974)	Stationary while waiting for signal of reward	5–6 Hz RSA
C Failures of correlation between movement and RSA		
Ellison *et al.* (1968) Konorski *et al.* (1968)	Instrumental lever press compared with waiting for food, signalled classically	RSA similar in the two situations
Black and Young (1972)	Lever pressing	A minority of lever presses occurred without RSA
Preobrazenskaya (1974)	Appetitive or avoidance lever-pressing	RSA sometimes occurred before lever press; RSA most prominent for partially learned task

Table 8. (continued)

Arnolds *et al.* (1979b)	Comparison of steady-maintained walking with sudden increase to the same speed	RSA frequency higher for sudden increase in speed
Bartel's and Urayev (1982)	Repeated performance for the same motive; or switch of motives	RSA increased during performance for the same motive; fell when motive is switched
D Miscellaneous correlations		
Yoshii *et al.* (1967)	Orienting response	6–7 Hz RSA
Storm van Leeuwen (1967)	Eye movements	RSA occurs
Konorski *et al.* (1968)	Intertrial interval	4–4.5 Hz RSA
Yoshii *et al.* (1967)	Seizing food, defaecation,urination, barking, unpleasant odour, food becomes inaccessible, "give paw" command	Desynchronized hippocampal EEG
Konorski *et al.* (1968)	Eating in classically conditioned paradigm	4–4.5 Hz RSA
Lopes da Silva *et al.* (1969)	Eating	Desynchronized hippocampal EEG

The enigmatic nature of the hippocampal EEG in monkeys has led most investigators to use more sophisticated methods of analysis of the raw EEG trace (such as bandpass filtering, or spectral analysis). Most of the behavioural correlations thus obtained concern frequencies in the 3–4 Hz range, a very low frequency by comparison with RSA in subprimate animals. It is unclear whether activity detected at these frequencies by bandpass filtering is to be classed with the higher frequency RSA bursts, or is of different origin and significance.

Crowne and Radcliffe (1975) make several further observations on the basis of analysis of frequency spectra. During quiescence the peak frequency is below 3 Hz, but with significant contributions over a wide range of higher frequencies. During movement (reaching for a grape, turning the head with or without orienting, vigorous struggling elicited by threatening gestures by the experimenter), the dominant frequency increases to 3–4 Hz. In corroboration, Joseph and Engel (1981) found an increase in the 4–8 Hz band during struggling. According to Crowne and Radcliffe, neither orienting to a sudden sound without movement nor exposure to a novel visual stimulus, with no response demanded, elicited this upward spectral shift; but if the same visual stimuli are used as part of a delayed-match-to-sample task, the upward shift does occur from around 2 Hz to 3–4 Hz. This seems to imply that tasks requiring detailed visual processing can elicit an upward shift in EEG frequency. Moise and Costin (1974) obtained a similar result in a red/green discrimination. Crowne *et al.* (1972) found on GO/NO GO discrimination that EEG activity in the 3–5 Hz band was greater on "no go" trials than on "go" trials, regardless of whether these responses were correct or not. During the course of training, such activity progressively declined on the "go" responses and progressively rose during the "no go" responses. During left/right spatial discrimination activity in this band was elevated similarly to that during "no go" trials. The authors point out that to perform a "no go" response is a very difficult task for a monkey, requiring continuing attention, with constant hovering above the lever, without actually pressing it. The "go" response on the other hand is very easily acquired, and (presumably) easily automated, unless it requires discrimination between signals.

4.3.7 *Man*

The technical and ethical problems concerning recording from the human hippocampus have imposed severe restriction on the quantity and value of data available. Brazier (1968) recorded from the hippocampus in 30 patients prior to surgery (28 cases of temporal lobe epilepsy, 2 non-epileptics). The hippocampal EEG showed peak power in the 2–4 Hz range, but peak of coherence with EEG from other sites (e.g., parahippocampal gyrus) was in 4–8 Hz range. Lieb *et al.* (1974) recorded from the hippocampus, amygdala and parahippocampal gyrus. A small peak was seen at 8–10 Hz, variable between patients, with a broad peak at slower frequencies. Halgren *et al.* (1978) in a single case study made some behavioural observations. The hippocampus was synchronized at 5–6 Hz during quiet resting, when the patient tensed all his muscles, or made simple alternating movements. Increasing synchrony was also seen during testing the patient on a series of word meanings, or a tapping sequence. Desynchronization occurred during speech or rapid breathing, tying a bow, imitating movements of the experimenter, or while the patient was giving a verbal description of the ward. Arnolds *et al.* (1980) recorded the hippocampal EEG, using depth electrodes in a single case study of an epileptic woman. For the right hippocampus (whose potentials were of larger amplitude) a significant peak appeared in the 3–4 Hz range. The frequency and rhythmicity of this peak increased when the woman wrote on a blackboard, or in a word association task in the period of silence between question and answer.

4.3.8 *Summary*

In this section a brief cross-species comparison will be presented of the correlates between major categories of behaviour and hippocampal RSA. The section is based entirely on subprimate species, since no firm conclusions can be made in monkey or man. Moreover consideration of the more problematical observations constituting exceptions to the RSA correlations with overt behaviour will be deferred to the discussion in Sect. 4.3.9, and later in part 3 of this book.

Most obviously there is a strong correlation between large movements of the whole body (e.g., locomotion) and RSA. The species in which this has been observed include rat, mouse, gerbil, guinea pig, hamster, rabbit and dog. Distinct (though less dominant) correlations of the same sort have also been seen in cat. For locomotor movements, the theta rhythm occurring in such circumstances tends to be in the higher range of frequencies for each species (most clearly so in the rat). For smaller movements, from a fixed base, it has been seen in various small mammals that the theta rhythm drops to a somewhat lower frequency than that during locomotion. Whatever the origin of the correlation between RSA and locomotion, it does not concern the motor movements themselves or any sensory feedback from active muscles, since RSA can be seen in curarized animals, in response to stimuli that would normally elicit locomotion. Moreover, high frequency RSA can be trained to occur in curarized animals, and will be accompanied by increased body movement when the eliciting stimulus is given after the animal has recovered from the paralysis. The correlate of RSA thus seems to be with central precursors of locomotion, rather than with movement *per se*. For quick repetitive movements, such as sniffing, vibrissal twitching or lever pressing, RSA is sometimes phase-linked with these repetitive movements.

In many species, RSA of a lower frequency is seen sometimes in the immobile animal if it is alert in posture rather than relaxed. This immobility RSA is seen rarely in rat, and then under rather special circumstances. It has been reported in mouse, guinea pig, and rabbit, and there are suggestions that it also occurs in dog. In the immobile rabbit such RSA is not always seen even if the animal has an alert posture. However, it is seen especially in response to novel stimuli, and quite commonly in other states of immobility (but not in very relaxed immobility). When rabbits adopt a fearful posture (for instance on introduction to a new cage), a frequency lower than that of the RSA in locomotion is produced. In cat, RSA is often seen during immobility connected with important signals. There is evidence that the capacity of signals to produce RSA may sometimes depend on conditioning. In guinea pig, for instance, neutral stimuli given during immobility can acquire the capacity to elicit RSA if paired with a signal that regularly does produce RSA.

The correlation between specific orienting movements (in the absence of larger body movements) and hippocampal RSA has been investigated in several species. There is some evidence for such a correlation in rat and rabbit, and suggestions of one in dog. The correlation with orienting is strongest in the cat, where it applies both to unconditioned and conditioned orienting responses, the frequency increasing as degree of apparent attentiveness increases. Although this correlation does exist in a variety of species, orienting is neither a necessary nor a sufficient condition for the appearance of RSA.

In cats, RSA occurs during eye movements and during fixed staring. The frequency of RSA is higher in the former case than in the latter. There are suggestions of a similar correlation in dog. The significance of these observations is unclear. Is the correlation with the eye movements themselves, regardless of the other accompanying behaviour? The fact that high frequency RSA occurs during phasic bursts of eye movements in REM sleep suggests that this might be so. Alternatively, eye movements or fixed staring might be considered a part of the orienting response, or as a necessary part of whole body movements (which commonly are associated with eye movements).

There is agreement in most species that automatic behaviours such as eating, drinking, chewing, biting, grooming or face washing occur when the hippocampus is not generating RSA. However, there are some conflicting data, notably the evidence that low frequency RSA is seen in rabbit during such behaviours.

4.3.9 Discussion

In the evidence described above, the difficult areas are: (1) the various exceptional cases, where RSA does not apparently correlate with the usual specific behavioural items; and (2) the case of monkeys and man, where RSA is either of very different appearance from RSA in other species, or may even be non-existent. In the first subsection below (4.3.9.1) these difficult areas will be discussed in relation to possible technical problems of recording the hippocampal EEG. In the second subsection (4.3.9.2) discussion will focus more on the functional, ethological and ecological significance of the behavioural correlates described above.

4.3.9.1 Technical matters

1. Depending on the site of hippocampal recording, and whether monopolar or bipolar electrodes are used, the amplitude of RSA may vary considerably. Robinson (1980) reviews this topic, giving details of amplitude recorded in various species and ranging from 50 µV up to 3 mV. (Most amplitudes are below 1 mV in amplitude). There is a possibility that in some of the exceptional cases cited above where RSA is unaccountably absent, low amplitude RSA is below the amplitude threshold for detection. For instance Sainsbury (1970) has pointed out that Routtenberg's (1968) results were obtained for very low amplitude (200 µV) RSA. However, this argument does not seem to have any general applicability, either in rat or cat. Furthermore, most of the above studies are exceptional not so much because RSA is absent generally, but because it is absent in specific instances of body movement where it would otherwise be expected.

2. It is possible that some of the discrepancies in the evidence and apparent exceptions to otherwise clear behavioural correlations arise because of inconsistency between different laboratories of what frequencies constitute RSA.

3. Another possibility to explain some of the absences of hippocampal RSA in movement and other situations where it would be expected is that excess activation of the mechanism initiating theta activity actually changes RSA into some other form of EEG activity in the hippocampus. Gralewicz (1981) considers this possibility and cites several lines of evidence. Firstly, there is his own result in cats that in shuttling responses of short latency, small amplitude irregular activity tends to appear, in contrast to the uniform RSA in more typical responses. Secondly there is evidence that high intensity stimulation in the hypothalamus can desynchronize the hippocampus, although lower intensity stimulation leads to RSA. The paper cited is Paiva *et al.* (1976) in cat. Actually, desynchronization was found with only two stimulation sites. These were unusual in their anatomical placement compared to the others, but were not distinct in the intensity required to produce desynchronization, compared with that producing synchronization at other sites. Other work (e.g, Vertes 1981) shows clearly that sites in the brainstem and diencephalon from which hippocampal desynchonization can be produced are distinguishable from those from which synchrony is produced on *anatomical* grounds, rather than on the basis of stimulus intensity. Thirdly, Graliewicz cites Kramis *et al.* (1975) as finding that slow irregular activity can replace RSA in rabbits if they are running very fast. This again is a misreading: Kramis *et al.* illustrate that during fast running in rabbit RSA is reduced in amplitude, but continues nevertheless with a very clear 12 Hz rhythm. Thus there is actually no clear evidence that RSA turns to a desynchronized EEG when it is activated intensely, at least by naturally occurring means of activition.

4. Another possible reason why, even within a species, there are exceptions to the correlations between hippocampal RSA and behaviour is that in practice it is not possible to observe behaviour in sufficient detail, so that somewhat hidden responses which may correlate with RSA are not noticed. Arnolds *et al.* (1979c) make this point. However, the point should not be accepted without question.

Firstly, although it is true that much behavioural observation is relatively crude, it is somewhat *ad hoc* to postulate this as the explanation of failures in correlation of electrophysiological and behavioural data. Secondly, in the foregoing chapter, the aim was to review correlations between EEG and *overt* behaviour, in other words it was behaviourist in objective, if not in philosophical commitment. Philosophically, behaviourism has the advantage that any correlations can be freed, as far as possible, from pre-theoretical biasses, and is thus, in a sense, more objective. On the other hand, as soon as one considers correlations with complex internal processes such as learning, one is forced to adopt schemes of description which have implicit pre-theoretical biasses, which are not likely to have been validated, and may seem arbitrary or even subjective. To postulate unobserved behaviours (which are still, in principle, observable) as correlates of EEG data loses the objectivity of the behaviourist approach, without gaining the theoretical insights which might come from considering the shape of hypothetical internal processes. Thirdly, we know from the studies with curare and similar drugs that the hippocampal theta rhythm *has* a correlate with internal neural processes which is probably more basic than those with overt behaviour. So pure behaviourism must be inadequate. However, behaviourism is not the philosophical framework of the present work. It has been adopted in the above chapter merely as a way of presenting the easy correlative data, so as to demarcate the more difficult exceptions for further attention, and hypothesis formation, later in this work.

5. The elusive nature of RSA in the primate hippocampus may be explained in a number of ways, other than by proposing a radical species difference. Firstly, it may be noted that larger-brained animals such as cat and dog have lower frequency theta than smooth brained animals. Conceivably, this progression extends further with the primate, whose brains are even larger. The implication would then be that the equivalent of theta frequencies in primates is below the frequency usually associated with hippocampal RSA in subprimates. Against this possibility, however, there is clear evidence of theta driving in monkeys, at frequencies within the range found in other animals, and there are confirmed reports that RSA at these frequencies *can* occur spontaneously, albeit in short, infrequent bursts, whose correlation with behaviour is difficult to determine. A more likely reason underlying the difficulty in unravelling the theta rhythm in primates is that its behavioural correlates are more subtle, and more difficult to reproduce in the laboratory setting. This topic will be discussed further below.

4.3.9.2 Ecological Matters

Some of the above technical arguments to explain the exceptions and inconsistencies do not hold water. Others may. Regardless of this there does appear to be some *prima facie* evidence of species differences in the behavioural correlates of theta activity. How can we reduce the complex data to a more parsimonious form? It is necessary to consider the ecological significance of the behaviours correlating with theta in each species, to see if they represent epochs of similar significance to each species. In fact a good case can be made out that this is so. The common theme is that *those periods of an animal's activities when information important to that species requires to be gathered from the environment, are the times when RSA is most likely to be generated by the hippocampus.*

The *rat* and many other rodents are small animals, which therefore cannot obtain much information about distant regions by vision. By contrast, most of their information gathering must be done by actually going and looking. Observations of rats in the wild have shown them to have a home territory which they reconnoitre regularly, favouring places for visits which have least recently been visited (see Barnett 1981: p. 212). They also make periodic forays well beyond this home territory, and have been observed to locate and revisit sites as much as 1 km away from their nest if they provide a good food source (Ewer 1967: p. 66). All this implies that the time when such animals need to be gathering information is when they are travelling. This correlates with the fact that hippocampal RSA in rat is most prominent during locomotion.

The *rabbit* is also a territorial animal, which occasionally forays further afield than its home territory (Ewer 1967: p. 75). It is not strong in conflict with carnivorous predators, and relies for its safety on being able to hide in burrows, and to be alert to signs of danger. Thus the rabbit needs to be gathering information not only when it is travelling, but also in quieter moments, including eating and drinking, when danger signals may require it to run for cover. These ecological facts correlate with the evidence that hippocampal RSA appears not only during locomotion, but also during quiet immobility, in response to many unexpected or novel sensory stimuli.

The *cat* is territorial but also an astute predator. Its capacity to deliver a lethal neck bite to smaller mammals appears to be an innate response, released by appropriate stimuli, once the animal has reached a certain degree of excitement (Ewer 1967; Leyhausen 1965, 1979). On the first occasion on which this response is elicited, it is not aimed at fulfilling the hunger motive, it is poorly coordinated to the precise circumstances of the attack, and it is not accompanied by all the elaborate stalking characteristic of an adult animal. Adult cats make elaborate use of cover, cease moving when the prey is restless, and in some species in the wild are even reported to stalk from downwind if possible (see Chauvin and Muckensturm-Chauvin 1977, p. 195). All these matters have to be learned. It is therefore important that when kittens make a kill they be alert to all features of the prey and its environment, so that performance may be elaborated beyond the innately released response. The time when such information is to be gathered is during the periods of immobility during stalking and just prior to launching an attack. In addition, according to Glickman and Sroges (1966), carnivores (including both cats and dogs) have a highly developed sense of curiosity, and utilize the response patterns associated with predation in their investigatory activity. Thus, in keeping with the idea that hippocampal RSA appears at times of important information gathering in cat, the behavioural correlates of RSA are immobility, particularly when there are behavioural signs of an orienting response, as well as during locomotion.

Dogs are social animals, evolutionary relatives of wolves who hunt in packs, covering large distances in the process. However, their social instincts have been closely linked to human society for many generations. In accordance with their evolutionary history of hunting, an activity which requires constant information gathering, RSA is prominently produced during locomotion or other large body movements. There are a few indications that tie in with the need of dogs to recognize and learn social signals which may be subtle secondary reinforcers. For instance, RSA occurs in response to signals that predict reward, or to sounds indicating the presence of the humans with whom the dogs interact (Yoshii *et al.* 1967). However, the role of social reinforcing stimuli has not yet been specifically investigated in any detail.

Monkeys and to an even greater extent *man* are highly socialized, with well developed dominance hierarchies in most species. They are careful observers of each others' gestures, facial expressions and vocalizations, using these cues to permit individual recognition and to establish the social structure of their colonies (Chauvin and Muckensturm-Chauvin 1977: p. 22). In monkeys reared in isolation, the capacity to use such cues later in life is severely impaired, so there appears to be a great deal of learning/imprinting associated with these social signals (Miller RE 1974). Primates are also animals with a sense of curiosity more highly developed probably than any other group. Monkeys have been observed to work for hours at a time in order to produce a change in their sensory input, especially if this involves complex or socially relevant visual or auditory stimuli (Barnett 1981, p. 213). In comparative tests of a large number of zoo species, Glickman and Sroges (1966) found that primates (with carnivores) showed the greatest responsivity (of all major mammalian groups) to novel objects. This was especially so of the primate species which were omnivorous and partially land-based rather than exclusive tree-top dwellers. The common laboratory species of monkeys are all of the former habitat. Habitually, they investigate by holding the novel object in both hands and studying it visually.

It would therefore be predicted that the most potent way of triggering RSA in monkeys and man would be by the use of socially significant stimuli, or novel stimuli, particularly in the characteristic attitude described above. None of the published work explores such possibilities. Indeed, it is possible that the stress of an electrophysiological experiment in a conscious primate would preclude normal responses to such social or novel stimuli.

4.3.9.3 Correlations Between Hippocampal RSA and Overt Behaviour: Concluding Remarks

To decipher the significance of hippocampal RSA by studying its correlations with behaviour is, logically speaking, a difficult excercize. Ideally, one wants to establish a causal link between some behavioural manifestation or (more likely) a specific neural precursor of behaviour, and the EEG phenomenon. However, mere correlation never implies causation. There is always the possibility that any observed correlation may be no more than a subsidiary biproduct of a more fundamental correlation directly linked to causation. In principle, such correlative studies can exclude causal links but can never prove them. The most that can be hoped is that all plausible extraneous sources of correlation can be falsified leaving, one hopes, one (or more) hypothesis intact, though strictly speaking, unproven. The suggestions made above (Sect. 4.3.9.2) under the title "Ecological Matters", makes an attempt to formulate one such hypothesis, but research on the behavioural correlates of hippocampal RSA have a long way to go before this hypothesis can stand without challenge.

This unsatisfactory state of affairs reflects another factor which makes the subject more complex. Admittedly, with due attention to ecological variations between species, one may establish a useful generalization of the relation between hippocampal RSA and behaviour. But such an ecological generalization is intended to apply only to the *innate* correlations with behaviour emphasized in the above chapter. Moreover, it is not intended to resolve the exceptional data cited in this chapter. Beyond this generalization there is a possibility that RSA occurs not only in such innate behavioural correlations, but also in correlation with other neural processes which are strictly internal, always acquired, and perhaps with much closer cross-species parallels. By suggesting a relation between RSA and internal processes, it is implied that they have no immediate reflection in behaviour, unless that term includes inferential evidence about stages of learning, or internalized frameworks for information processing. Many of the data reviewed above, concerning apparent exceptions to the overt behavioural correlations, appear to fit such a possibility. In the chapters below, hypotheses will be generated as to some of these internal processes. In forming these hypotheses the exceptions cited above are important constraints. If the hypotheses cannot make sense of these exceptional data, they will have failed.

5 The Septum as a Pacemaker for the Hippocampal Theta Rhythm

5.1 Introduction

Before one can deal with the rhythmic interplay between hippocampus and isocortex, many questions need to be resolved about the physical mechanism of the theta rhythm. In the next two chapters the focus is on the mechanisms controlling the initiation and frequency of the theta rhythm. Chapter 5 concentrates on the theta rhythm *per se*, while Chapter 6 considers evidence that there is more than one variety of theta rhythm and control mechanism. Here, the following questions are addressed: What brain mechanisms initiate theta activity? What mechanism sets the frequency of the theta rhythm? It appears that the septal region is involved in both these processes; but is the septum sufficient to initiate rhythmic activity in the hippocampus, or does the hippocampus contribute any rhythmic capacity of its own to the theta rhythm? *How* does the septum control initiation and frequency of theta activity? Is it here that the frequency is set, or in some of its afferent signals arriving from the brainstem and elsewhere? In any case, how does the rest of the brain control the production and frequency of the theta rhythm?

5.2 The Effect of Septal Lesions

As mentioned in the historical introduction to the previous chapter, it has been known since the 1950's that lesions of the septal region will permanently abolish the theta rhythm recorded from the hippocampus. Later papers which corroborate this early result are those of Eidelberg, *et al.* (1959) in paralyzed rabbits, Brugge (1965) and Gray (1971) in free moving rats, Winson (1978) and Vinogradova and Braznik (1978) in conscious undrugged rabbits and Bland and Bland (1986) in freely moving rabbits. In three of these papers (Gray 1971; Winson 1978; Bland and Bland 1986) the lesions were small enough to localize the crucial area to the medial septal nucleus (see Fig. 18), rather than other parts of the septal region, though partial lesions of this nucleus might not completely abolish theta activity.

A significant exception to this consensus comes from Destrade and Ott (1982). Confirming the above results, theta activity occurring spontaneously during exploration was blocked by injection of local anaesthetic into the medial septum. On the other hand, very high-frequency theta activity elicited by electrical stimulation of the dorsomedial hypothalamus was not blocked by such injections. This result is not sufficient to challenge the view that the medial septum is the primary locus of control of theta activity, especially since Christie and Bland (1989) failed to replicate the result in urethanized rats. However, it will require further consideration as hypothesis-building takes place.

The most recent papers on this subject introduce another complication to the above apparent consensus. Stewart and Vanderwolf (1987a;b) made lesions in the septal region using ibotenic acid, with the intention of destroying cells alone but sparing axons of passage. In fact, the lesions also spared the cell bodies of the medial septo-hippocampal cells. Nevertheless, theta activity was altered significantly. The alteration consisted of a selective abolition of one pharmacological type of theta activity, while the other type was left moreorless intact. This pharmacological subdivision of theta has yet to be considered (Chap. 6). Therefore the details of these experiments, and their implication will be reserved to the later section.

5.3 Rhythmic Electrical Activity in the Septum, and Behavioural Correlates of this Rhythm

Petsche *et al.* (1962) followed up the early lesion evidence of septal involvement in the theta rhythm by recording from units in the septum of curarized rabbits. Theta activity was elicited by sensory stimulation or electrical stimulation of the reticular formation. A proportion of septal units showed regular rhythmic bursts of impulses, in phase with the hippocampal theta rhythm (see Fig. 25). Such rhythmic bursting cells were seen most abundantly in the nucleus of the diagonal band (part of the septal complex; see Fig. 18).

These findings have been confirmed many times since: during theta activity elicited by eserine or nicotine (Stumpf *et al.* 1962; Gaztelu and Buno 1982); in curarized rats (Macadar *et al.* 1970); and in chronic undrugged rats (Morales *et al.* 1971). Petsche *et al.* (1965) found that all cells recorded in the nucleus of the diagonal band could produce theta-locked bursts. They remark that hippocampal theta activity cannot occur unless there is rhythmic septal unit activity, and the frequency is, without exception, the same in both locations. Vinogradova and Zolotukhina (1972), studying alert rabbits, found that only one third of neurons in both medial and lateral septum showed theta bursts, either spontaneously or evoked by sensory stimuli. The evoked effects also included overall inhibition of firing rate (more commonly than excitation) but this did not abolish the rhythmic discharge. Apostol and Creutzfeldt (1974), using curarized rabbits, recorded simultaneously the septal and hippocampal EEG and the spike activity of individual septal units during spontaneous rhythmic activity or that evoked by sensory stimuli. The units were described as of several types. Some units could show steady (non-rhythmic) discharge which increased when theta frequency increased. Others had a discharge rate proportional to theta amplitude. The third group showed a consistent phase relation with the rhythm of the septal or hippocampal EEG, the phase relation being consistent within a cell but varying from cell to cell. Ranck (1976) obtained records from units in the medial septum of free moving rats, and found that the majority of cells increased firing during hippocampal theta activity. According to his illustration, such units were also rhythmically bursting in phase with hippocampal theta activity. Wilson *et al.* (1976), using curarized cats, recorded the hippocampal RSA and rhythmic septal bursts. Most of the rhythmically active septal neurons were in the nucleus of the diagonal band. In summary, it is clear that rhythmic activity in phase with the hippocampal theta rhythm is seen best in the medial septum and nucleus of the diagonal band. This conclusion is consistent with the data reviewed above on the effect of lesions in the septal region on the theta rhythm. The evidence can be drawn together by the hypothesis that *the*

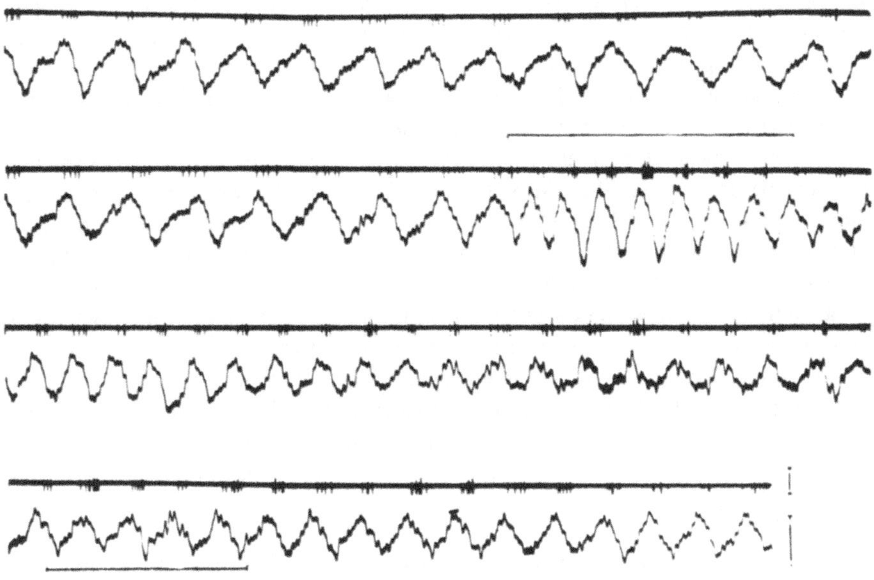

Fig. 25. Single unit recording from a medial septal neuron, showing rhythmic bursting of the neuron (*upper trace*) phase-locked to the hippocampal theta rhythm (*lower trace*). Theta activity is induced by esserine. Sensory stimulation (*horizontal bar*) accelerates the rhythm. Calibrations: time: 1 s; voltage 1 mV for unit; 500 µV for EEG. (Petsche *et al.* 1962, Fig. 2) (From *Electroencephalography and Clinical Neurophysiology*, vol. 14, p. 202–211, 1962)

septo-hippocampal pathway, which originates from just these nuclei (see Chap 4) relays rhythmic messages from the septal region to the hippocampus.

Although the septal units' rhythm and the hippocampal EEG rhythm were obviously closely correlated, and the hippocampal RSA could not occur without the septal rhythmic discharge, there are observations from several sources that occasionally septal rhythmic bursts can occur in the absence of hippocampal RSA. Petsche *et al.* (1962) found that the spontaneous septal rhythm could continue during brief periods when hippocampal theta activity had stopped. Indeed, when seizure discharge was generated by the hippocampus the septum might continue its rhythmic discharge both in the early part of the seizure activity, and after the seizure when the hippocampus itself was silent. During the height of the seizure, however, the septal activity itself was disrupted. Both Stumpf *et al.* (1962) and Macadar *et al.* (1970) have made the pharmacological observation that the hallucinogen LSD-25 can abolish hippocampal theta activity without abolishing the rhythmic septal bursts. Ranck (1976) found a variety of rhythmically active cells in the medial septum which produced theta bursts all the time, even when there was no hippocampal theta activity.

The phase relation between septal rhythmic unit bursts and hippocampal EEG has been reported in several papers. Phase relations of a given unit stay constant in the short term, but vary from unit to unit (Petsche *et al.* 1962; Gogolak *et al.* 1968;

Wilson *et al.* 1976), and may vary within a unit if the rhythm stops, and starts again after an interval, or with different methods of elicitation of theta activity (Wilson *et al.* 1976). Superimposition of phases of many septal units resulted in a curve similar to the hippocampal theta rhythm, implying that there must have been a preferred phase relation for the bursting septal cells (Gogolak *et al.* 1968).

Several papers give values for the phase difference between septal burst and hippocampal theta waves (Petsche and Stumpf 1960; Petsche *et al.* 1965; Apostol and Creutzfeldt 1974; Gaztelu and Buno 1982), but these vary considerably. Similarly, Brown (1968) showed that the septal electrographic rhythms bore a very variable relation to the theta rhythm recorded in the dorsal hippocampus. However, Petsche *et al.* (1962) point out that absolute values for phase lag are meaningless because the theta activity is a travelling wave, whose lag depends on where it is measured.

There are a few reports concerning the behavioural correlations of septal rhythmic bursts (Morales *et al.* 1971; Vinogradova and Zolotukhina 1972; Ranck 1976; Wilson *et al.* 1976; Zin *et al.* 1977). Although quite fragmentary compared with the large literature on similar correlations with hippocampal theta, the basic correlations appear to be similar.

5.4 Rhythmic Activity in the Hippocampus not Dependent on a Rhythmic Input from the Septal Region

Rowntree and Bland (1986), using urethanized rats, showed that the muscarinic agonist carbachol, injected into the hippocampus, could initiate rhythmic activity similar in frequency (3.5–8 Hz) and amplitude to spontaneously occurring theta activity. This effect was produced with injections into any of the part of the hippocampus which was rich in cholinergic markers, viz. strata oriens and radiatum of CA1 and CA3, and strata moleculare and granulosum of the dentate gyrus, as well as the infragranular portions of the hilus. Eserine could produce effects similar to carbachol, but the results with carbachol were more interesting because this drug, being a direct agonist, would bombard cholinergic receptors continuously in a non-rhythmic manner and yet still evoked a rhythmic response. This suggests that under some circumstances the hippocampus can produce rhythmic activity within the theta range of frequencies, but without mediation by rhythmic signals from the septum. This is also implied in the results of Gray (1971), who made the lesions while the rats were conscious and moving and found that hippocampal RSA did not cease until at least 15 min after the lesion had been made.

There are two results which disagree with these implications: the first is that of Vinogradova and Braznik (1978). In rabbits with septo-hippocampal connections severed, the rhythmic activity of hippocampal neurons was abolished. It could not be reinstated by administration of physostigmine, although this increased the overall activity of these neurons. Similarly, Olpe *et al.* (1987) found that various cholinomimetics (arecoline, oxotremorine, pysostigmine) could evoke theta activity in normal animals (as the early workers found), but in those in which large septal lesions had been made 3 weeks previously, this action was prevented.

Despite this conflicting evidence, the result of Rowntree and Bland was followed up, and a more dramatic demonstration of a similar effect was carried out by MacIver *et al.* (1986) in hippocampal slices. It was shown that application of the carbachol into the bathing fluid around a hippocampal slice preparation (without the septum) could also produce rhythmic activity (1–12 Hz) similar in some ways to RSA in the intact animal. These results were confirmed by Konopacki *et al.* (1987a), where frequencies of 2–12 Hz were seen, and the rhythm induced by carbachol could last a full hour of recording time (see also Fig. 26). These effects were produced over a rather limited concentration range (80 to 400 μM). A wider range of concentration could influence evoked responses. Thus, Konopacki *et al.* (1987b) found that in such slice preparations carbachol also depressed field EPSP's produced by perforant path stimulation, but not the population spikes. It was suggested that carbachol had a depolarizing effect on neurons in the slices. All these effects were antagonized by atropine, but not by d-tubocurarine.

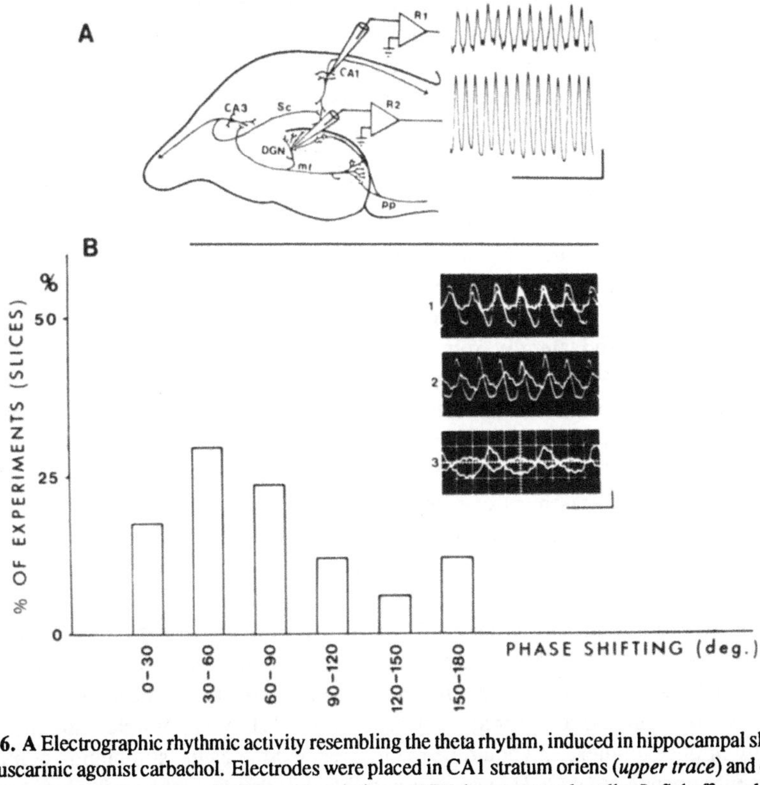

Fig. 26. A Electrographic rhythmic activity resembling the theta rhythm, induced in hippocampal slices by the muscarinic agonist carbachol. Electrodes were placed in CA1 stratum oriens (*upper trace*) and dentate stratum moleculare (*lower trace*). Other abbreviations: *DGN* dentate granule cells; *Sc* Schaffer colaterals; *pp* perforant path; *mf* mossy fibers; R1 and R2: two recording amplifiers. Calibrations: 1 s; 200 μV. **B** Illustration of the variety of phase relationships between the rhythm in CA1 and the dentate gyrus, shown as a histogram of phase differences between these two sources over a number of slice experiments. Typical records shown in *insert* (*1* Traces roughly in phase; *2* phase difference is about 70°; *3* phase difference is about 180°). Calibrations: 1 s; 200 μV. (Konopacki *et al.* 1987c, Figure 1) (From *Brain Research.* Vol. 417, pp. 399–402, 1987)

Thus the hippocampus does seem to have some inherent capacity to generate rhythmic activity, in addition to the rhythmic activity relayed to it from the septal region. Nevertheless, the vast majority of the lesion evidence implicates the septal region as the immediate controller of hippocampal RSA, the medial septum/diagonal band being the most likely centre of this control. The conditions under which the hippocampus can exert its own rhythmic activity are not understood at present.

5.5 Brainstem, Diencephalic or Septal Stimulation and Rhythmic Activity in the Septum and Hippocampus

5.5.1 Septal Stimulation

Brucke *et al.* (1959a) showed that low frequency stimulation near the midline in the septal regions could drive an EEG rhythm in the hippocampus in a manner very similar to the spontaneous theta rhythm, but locked to the stimulation frequency. However, rhythmic driving from the septum was possible over a range of frequencies considerably greater than the theta range, from 4 Hz to 30 Hz, or occasionally and for brief instants, up to 100 Hz. This result was confirmed by Kramis and Routtenberg (1977). Direct stimulation of the hippocampus at the same frequencies could also drive the hippocampal EEG. With intense stimuli or high frequencies (> 10 Hz), septal stimulation led to hippocampal desynchrony (blocking) of rhythmic activity, a finding also reported by Brucke *et al.* (1959a), Torii (1961), Ball and Gray (1971), Deupree *et al.* (1982) and Holt and Gray (1983), amongst others.

Other results show that rhythmic activity can be driven only over a more limited range of frequencies, typical of the true theta rhythm (Brucke *et al.* 1959a; Kramis and Routtenberg 1977). These results have usually been obtained with electrode positions misplaced dorsally and laterally from the medial septal/diagonal band complex. Kramis and Routtenberg (1977) suggested that electrodes in these positions were probably activating *afferents* to the medial septum, rather than neurons and efferent axons of the medial septum. This conjecture was addressed by Braznik *et al.* (1985). They applied low frequency (3–17 Hz) stimuli to medial septal afferents close to the nucleus itself in rabbits. Rhythmic entrainment of medial septal neurons occurred preferentially in the frequency range 4–9 Hz, and most strongly at 5–6 Hz (the commonest spontaneous theta frequency in rabbits). At higher or lower frequencies a higher or lower harmonic was adopted, or a combination of two frequencies, or no rhythm at all might occur.

The view that rhythmic entrainment limited to the theta frequency range only occurs when electrodes are misplaced from the medial septal/diagonal band complex is not supported by all the evidence: Kramis and Vanderwolf (1980) found that driving of theta potentials limited to the frequency range 5.5–12 Hz was possible with electrodes actually in this complex, and could be maintained as long as the stimulation continued. Sometimes driving at 14–20 Hz was also possible, but this could be maintained only for 0.5–3.0 s.

Some studies have found that the rhythmic entrainment induced by septal stimulation grows over a number of stimulus repetitions. Brucke *et al.* (1959a) sometimes observed that the septally driven rhythmic EEG only emerged after a few pulses of the stimulus train had already been given, though this phenomenon was not seen for equivalent stimuli to the hippocampus. Braznik *et al.* (1985) also observed gradual entrainment of medial septal neurons. From their data it appears that full entrainment might take about six cycles. This process might consist of entrainment "by the pause" (interburst interval reset by the stimulus) or "by the burst" (bursts become time-locked to the stimulus). Braznik and Vinogradova (1988) made similar observations, showing that entrainment "by the burst" or "by the pause" was generally shown only by those neurons that already had a strong rhythmic bursting pattern.

One study comments on the relation between spontaneously occurring theta rhythms and those evoked by septal stimulation. Kramis and Routtenberg (1977) found that when 6–12 Hz septal stimulation was given during spontaneous RSA it would supplant the spontaneous rhythm only if it was at a higher frequency. If behaviour associated with high frequency RSA (e.g., running) appeared during low frequency driving, the higher frequency spontaneous RSA would appear, rather than the driven frequency.

In conclusion, it appears that the hippocampus itself can be driven over a substantialy wider range of frequencies than the naturally occurring theta range. This is possible with stimuli applied to the hippocampus itself, according to Kramis and Routtenberg (1977), and sometimes with stimuli in the medial septal nucleus. In the latter case, it is likely that the stimuli directly excite efferent links of the medial septum projecting to the hippocampus. In other experiments, when stimuli are applied in or near the medial septal nucleus or the nucleus of the diagonal band, then the range of frequencies over which driving is possible becomes limited to the natural theta range. It is plausible to suggest that in these cases, stimulation excites the afferents to neurons in these nuclei, rather than the projection neurons themselves. This implies that *the rhythm of naturally occurring hippocampal theta activity derives from some property of the medial septal/diagonal band neurons, rather than of the hippocampus, or septo-hippocampal feedback loops.* Another effect sometimes seen in these neurons is gradually, over about six cycles of the stimulus, to build up a "resonance" to afferent stimuli in the natural theta range.

5.5.2 Brainstem and Diencephalic Stimulation

A number of papers report that stimulation of various sites in the brainstem or diencephalon can elicit rhythmic activity within the theta range of frequencies, either in medial septal neurons (Eidelberg *et al.* 1959; Petsche *et al.* 1965; Apostol and Creutzfeldt 1974) or in the hippocampal EEG (Yamaguchi *et al.* 1967; Klemm 1972; Apostol and Creutzfeldt 1974; Radil-Weiss *et al.* 1976). These effects are produced with high frequency stimulation, usually 100 Hz (see below). Although many papers merely report these phenomena, without giving any details of precise stimulation sites, several papers report anatomical details of stimulation sites which produce synchronization of septal or hippocampal rhythms, or desynchronization. The most detailed of these studies is that of Vertes (1981) in rats (see Fig. 27). He found that the most potent site in the pons and midbrain for eliciting RSA was the nucleus reticularis pontis oralis (nRPO). Additional RSA-producing sites were found projecting rostrally from here along several identifiable pathways through midbrain and diencephalon. Desynchronization was best obtained by stimulation of the medial raphe nucleus, with more rostral desynchronization sites continuing from here to the ventromedial part of medial forebrain bundle.

Vertes also cites anatomical studies showing that the medial raphe nucleus projects to the medial septum, and that the nRPO has projections compatible with the above electrophysiology, but not yet traced to the medial septum, or hippocampus. In a more recent paper using anatomical techniques (retrograde transport), Vertes (1988) shows that there are indeed direct pathways from the median raphe nucleus to the medial septum/diagonal band complex, these presumably being the ascending route which mediates the desynchronizing influence on the septo-hippocampal system. No direct

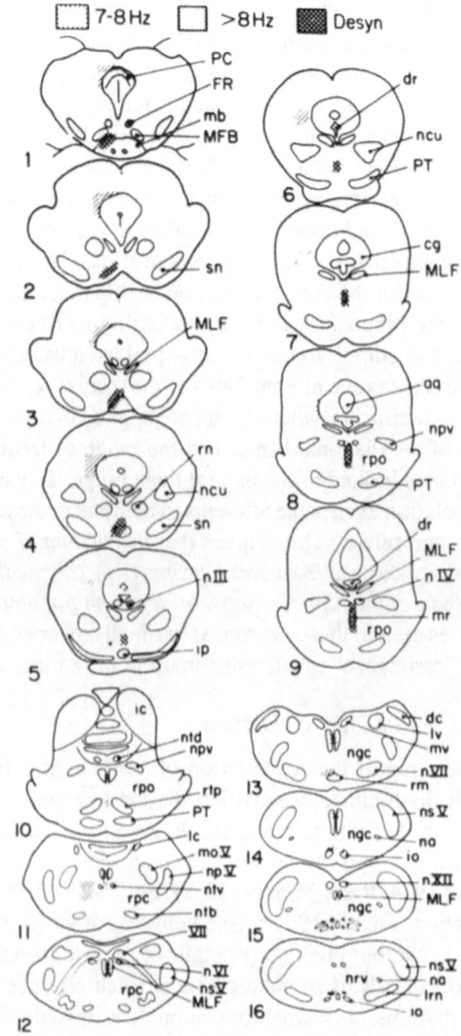

Fig. 27. Representative cross-sections through the brainstem of the rat, showing sites producing desynchronization (Desynch) and rhythmic activity in the hippocampal EEG (7–8 Hz and > 8 Hz). Abbreviations aq, aqueduct; cg, central gray; dc, dorsal cochlear nucleus; dr, dorsal raphe, FR, fasciculus retroflexus; ic, inferior coliculus; io, inferior olive; ip, interpeduncular nucleus; ic, locus coeruleus; lrn, lateral reticular nucleus; lv, lateral vestibular nucleus; mb, mammillary bodies, MFB, medial forebrain bundle; MLF, medial longitudinal fasciculus; moV, motor Trigeminal nucleus; mr, medial raphe; mv, medial vestibular nucleus; na, nucleus ambiguus; ncu, nucleus cuneiformis; ngc, nucleus gigantocellularis; npv, ventral parabrachial nucleus; nrv, nucleus reticularis ventralis; nsV, spinal Trigeminal nucleus; ntb, nucleus of the trapezoid body; ntd, dorsal tegmental nucleus (Gudden); ntp, pontine tegmental nucleus; ntv, ventral tegmental nucleus (Gudden); III, oculomotor nucleus; nIV, trochlear nucleus; nVI abducens nucleus; PC, posterior commissure; PT, pyramidal tract; rm, nucleus raphe magnus; rn, red nucleus; rpc, nucleus pontis caudalis; rpo, nucleus pontis oralis; sn, substantia nigra. (Vertes 1981: Fig. 3) (From *Journal of Neurophysiology* Vol 46 pp. 1140–1159 1981.)

pathways were found from nRPO to the medial septum/diagonal band. However, there were pathways from neurons in the supramammillary nucleus, which Vertes postulates might be an intermediate relay nucleus from nRPO.

These data are amplified by various other studies. Macadar *et al.* (1974) using cats under nitrous oxide found nRPO to be the lowest threshold region in the brainstem for eliciting theta activity. However, Robinson *et al.* (1977) found, in free moving rats, that nucleus reticularis pontis *caudalis* was the most potent synchronizing site in the pontine tegmentum. This is a nucleus which Macadar *et al.* (1974) identify as a desynchronizing locus, though in Vertes' results it appears as a synchronizing site (see Fig. 27). Other pontine sites from which elicitation of RSA has been described are the locus coeruleus, though at higher threshold than for nRPO (Macadar *et al.* 1974; Robinson *et al.* 1977) and the giant cell regions of the pontine tegmentum.

Several reports add to Vertes' data concerning synchronizing sites in *midbrain and diencephalon* (Torii 1961; Yokota and Fujimori 1964; Anchel and Lindsley 1972; Macadar *et al.* 1974; Wilson *et al.* 1976; Robinson *et al.* 1977; Kramis and Vanderwolf 1980; Gaztelu and Buno 1982). There is some consistency in anatomy of synchronizing sites in brainstem and diencephalon, but also some serious discrepancies, not all of which can be explained as species differences in layout of fibres. Yokota and Fujimori, 1964), in curarized cats, identify the medial hypothalamus and medial preoptic area as synchronizing sites. Anchel and Lindsley (1972) confirm this, adding that cold block anterior to the medial hyothalamic site, or dorsal fornix lesions could prevent elicitation of RSA. However, Destrade and Ott (1982) find 8–12 Hz hippocampal RSA, evoked by stimulation of the dorsomedial hypothalamus, was not blocked by local anaesthetic injected into the septum, a result which receives comment later. Wilson *et al.* (1976) found that medial hypothalamic stimulation elicited not only rhythmic bursts in diagonal band neurons, but also definite *responses* in these neurons. These were of long latency and rather indirect (as judged by wide latency variation).

With respect to the sites from which desynchronization can be elicited, Macadar *et al.* (1974) confirm the result of Vertes (1981) that the median raphe nucleus stimulation can produce desynchronization. Of relevance here is the observation of Maru *et al.* (1979) that median raphe lesions in free moving rats lead to low frequency RSA, lasting at least 20 days. The brainstem control of hippocampal EEG thus appears to be in a dynamic equilibrium between antagonistic synchronizing and desynchronizing components. In the diencephalon, the lateral hypothalamus has been identified as a desynchronization site by Torii (1961), Anchel and Lindsley (1972) and Wilson *et al.* (1976). Additional sites mentioned in medullary and pontine tegmentum have been reported by Macadar *et al.* (1974), Yokota and Fujimori (1964). Desynchronizing sites have also been seen in hypothalamus and amygdala (Yokota and Fujimori 1964; Wilson *et al.* 1976). Wilson *et al.* describe the effects evoked on diagonal band neurons by desynchronizing lateral hypothalamic stimuli: either an increase or a decrease in overall firing rate was observed, but with, in addition, unit responses of latency 1.7–4.5 ms, and little latency variation, implying quite direct linkage (as opposed to the effects these authors found for synchronizing stimuli on these neurons). In contrast to this, Braznik and Vinogradova (1988) find the neurons in medial septum/diagonal band show single spike responses of much longer latency (12–16 ms) on stimulating the most rostral end of MFB. These responses were seen preferentially in neurons which were not already rhythmically active.

The actual rhythmic phenomena produced by such stimulation, and its dependence on the parameters of stimulation are described in a few papers. In Torii's (1961) paper RSA at 5–7 Hz was evoked by stimuli in rabbits, while a species-typical frequency (4.8 Hz) was also found in cats by Wilson *et al.* (1976). On a few occasions, low intensity lateral hypothalamic stimuli led to low amplitude RSA. Macadar *et al.* (1974) find that the evoked RSA frequency varied according to the site of stimulation, 3.5 Hz in the case of giant cell regions of the pontine tegmentum and ventrolateral PAG, and 5 Hz for stimulation of nRPO or locus coeruleus. These are again frequencies typical of the species used (cats). The frequency at threshold stimulation intensity could be increased by 1–1.5 Hz by intensity increase, and caused theta activity to outlast the stimulus by 2–30 s. When RSA was studied as a function of stimulus frequency, no RSA was found at frequencies below 40 Hz, 80–100 Hz was optimal, and above 100 Hz the amplitude of RSA tended to diminish. Radil-Weiss *et al.* (1976) also found that brainstem stimulation produced RSA at a frequency which increased with increasing voltage, and which habituated unless voltage was increased gradually. Likewise Braznik *et al.* (1984) found that reticular formation (and to lesser extent MFB) stimulation increased the frequency, regularity and density of septal bursts in intact rabbits, the burst frequency increasing logarithmically with stimulus intensity, up to the natural maximum of 9–10 Hz. Short electrical stimuli could reset the burst activity, in a regular pattern, reproducible on next stimulus application. Gaztelu and Buno (1982) also found that with stimulation of the dorsal fornix, supramammilary nucleus, midbrain reticular formation or PAG the rhythm of the hippocampal EEG and septal rhythmic bursts was reset. The first four or five waves after reset were larger in amplitude, and accompanied by higher frequency and longer neuronal bursts in medial septum/diagonal band.

In the few experiments where free moving animals have been subjected to brainstem stimulation, there is a little data on behavioural correlations of stimulation. Vanderwolf (1975) found that different sites in the midbrain produced RSA at different frequencies. When low frequency RSA was produced, it usually occurred while the animal was standing motionless. When high frequency RSA was evoked, running or jumping was also elicited. Similarly, Robinson *et al.* (1977) found that stimulation of synchronizing sites also produced head movements or locomotion, at about the same threshold. Anchel and Lindsley (1972) found in cats that stimulation of synchronizing sites also evoked orienting and search (as did blockade of the lateral hypothalamic desynchronizing sites), while stimulation of the desynchronizing sites in the lateral hypothalamus or cold block of the medial hypothalamus caused arrest of behaviour and "attentive fixation of gaze".

On the basis of these few observations, the possibility is raised that there is a closer linkage between hippocampal EEG and behaviour when RSA is elicited by brainstem stimulation, than when it is elicited by direct frequency-specific driving of the septo-hippocampal system [see Sect. 4.3.1.3: (Kramis and Vanderwolf 1980) above]. The behavioural correlates of brainstem-driven theta activity appear to have the same species typicality as for spontaneous theta activity. The correlation between theta activity and species-typical varieties of behaviour may thus arise because both are controlled from common sites in the brainstem, rather than because there is any fundamental link between theta activity and these types of behaviour.

5.6 Conclusion and Discussion

From the above evidence, two conclusions come to mind initially: (1) The septum, specifically the medial septal nucleus and/or the nucleus of the diagonal band generate the rhythmic pulses which set the hippocampal theta rhythm into resonance. (2) Cells in medial septum/diagonal band convert trains of impulses from the brainstem, which themselves have no specific rhythm, into rhythmic bursts at the theta frequency. In the paragraphs below the evidence concerning each these two points will be summarized.

5.6.1 The Septum as the Pacemaker for the Theta Rhythm

Concerning the first conclusion, three arguments may be used. The first, though least specific, line of evidence is the fact that septal lesions, most particularly those of the medial septum, if made *in vivo* permanently abolish spontaneous hippocampal theta rhythm. The second line of evidence is the various incidental observations that septal rhythmic bursts can occasionally occur spontaneously in the absence of hippocampal theta activity, but the latter cannot occur without frequency-matched septal rhythmic bursts. The third line of evidence is that stimulation of the hippocampus itself can generate rhythmic activity over a wide frequency range, but if the septo-hippocampal system is stimulated on the afferent side of the medial-septal efferents, thus recruiting the properties of medial septal neurons themselves, the frequency range for rhythmic driving is limited to the natural theta range.

Against these arguments it has been suggested by McLennan and Miller (1974, 1976) that rhythmic activity in the hippocampus may depend on reverberating circuits between this and the septum. These authors draw attention to the loop of connections from medial septum to hippocampus to lateral septum and then back to the medial septum. They also present evidence that the lateral septum contains long-lasting inhibitory processes, which might act as a frequency-gating mechanism to control the regular rhythms seen in medial septum and hippocampus. However, their suggestion appears not to be true: According to lesions studies, disconnection of septum and hippocampus abolishes rhythmic activity in the hippocampus but leaves intact that in the septum, with relatively minor changes in the proportion of medial septal neurons showing such rhythmic bursts (Vinogradova and Braznik 1978; Vinogradova *et al.* 1980).

The evidence in Section 5.4. above shows that in non-physiological situations (carbachol injected into the hippocampus, carbachol acting on hippocampal slices) a rhythmic input to the hippocampus is not necessary for a hippocampal rhythm to be produced. If this applied to more normal circumstances, it would imply that hippocampal theta rhythms could be initiated and controlled via a non-pulsatile input from the septum. Ford *et al.* (1989) believe that this possibility cannot yet be excluded. However, until the evidence available confirms clearly that these results *do* apply to more normal situations *in vivo*, it should be regarded as primarily a phenomenon of slice preparations.

5.6.2 The Nature of the Control from the Brainstem

Concerning the second conclusion, it is clear from the evidence discussed in Section 5.5.2 that trains of impulses having no low-frequency rhythm, applied to the brainstem or diencephalon, can generate septal or hippocampal rhythms in the theta frequency range. Moreover, there is relevant evidence that neurons in pontine and medullary regions closely linked with control of septo-hippocampal theta rhythms (viz. the gigantocellular fields of pons and medulla) do not spontaneously discharge in rhythmic bursts (Vertes 1979). Further evidence that it is the septum, rather than its afferent pathways, which generates the rhythms comes from Vinogradova *et al.* (1980): deafferentation of ascending pathways to the septum in rabbits lowered burst frequency to 3.3 Hz but did not abolish the rhythmicity. In addition, Bragin and Vinogradova (1983) studied isolated explants from hippocampus or septum, obtained from foetal rats and grown for 4–6 months in the anterior chamber of the eye. They found that neurons in the septal grafts tended to be rhythmically active (at a frequency of 0.5–5 Hz) and incapable of following high frequencies. On the other hand, neurons from hippocampal grafts were not rhythmically active and would follow frequencies of up to 30–50 Hz.

When theta activity is elicited by electrical stimuli applied to the brainstem, it is common to see behavioural correlates similar to those seen during spontaneous theta activity. This is not true for theta activity elicited by electrical stimulation of the septal region. It therefore appears that the correlations described in Chapter 4 between theta activity and the innate repertoire of behaviour are probably coordinated at the brainstem level, rather than in the forebrain. It is reasonable to regard the brainstem as containing the programmes for innate behaviour and associated neural processes, while the forebrain is more concerned with behaviours acquired during learning. The fact that these behavioural correlates of theta activity do appear to be organized at the brainstem level therefore supports the argument put forward at the end of Chapter 4: the varied behavioural correlates of theta activity in different species represent innate and species-specific ways of "switching on" the theta rhythm. The manner of *triggering* a pattern of neural activity does not tell us what information processing task is accomplished by that pattern. Therefore we should not be seeking the significance of the theta rhythm primarily in the details of these triggers, but rather in the evidence we can obtain about the details of the neuronal relations in the forebrain which are under control of the theta rhythm.

6 Subdivision of Theta Types According to Sources of Extra-Hippocampal Control

6.1 Introduction

In Chapter 4, particularly when dealing with the smaller-brained species, there was some evidence that different frequencies of hippocampal RSA correlated with different types of behaviour. The faster frequencies for a given species were associated with whole body movements, such as locomotion, especially if it was vigorous, while the slower frequencies were associated with smaller movements, confined to a section of the body, or various other conditions linked with immobility. Hippocampal RSA has also been subdivided in other ways, for instance according to its susceptibility to various drugs (transmitter agonists or antagonists, and anaesthetics) and various lesions. These lines of evidence generally concur: subdivision of theta rhythms according to one criterion correlates with that according the others. The evidence as a whole has been taken as indicating the existence of two broad types of theta, differing in the manner of extrinsic control of the hippocampus. However, this conclusion has come under challenge recently, and needs therefore to be scrutinized quite carefully. These topics will be dealt with in Chapter 6. The evidence considered will concern both the anaesthetised and the free moving animal, but, as in Chapter 4, excludes that of the awake animal during learning task, a topic reserved until later (Chapter 10). It should also be pointed out that there is another important way in which hippocampal theta activity can be subdivided, that is according to its precise locus of generation within the hippocampus. However, the subdivisions revealed by the evidence on this topic cut right across that dealt with here. Hence it will be discussed separately, in Chapter 7.

6.2 Control from the Brainstem

6.2.1 Evidence for Dual Control of the Theta Rhythm from the Brainstem

Kramis *et al.* (1975) presented evidence appearing to show, in rats and rabbits, that hippocampal theta activity consisted of two types from the pharmacological standpoint. If *atropine* was given to a rat or a rabbit, the *high* frequency theta activity seen during locomotion and other whole body movements survived, but the lower frequency theta activity seen during immobility or grooming was abolished. On the other hand, if *anaesthetics such as urethane or ether* were administered, they produced long trains of *low* frequency (< 5 Hz) theta activity. This prolonged theta activity subsided eventually if the animal was left undisturbed for several minutes, but could then easily be reinstated by sensory stimuli. If atropine was given to animals already under the effect of either of these two anaesthetics, theta activity under *any* circumstances was abolished. Thus the two components, atropine-sensitive and anaesthetic-sensitive theta activity, together appeared to comprise the full theta phenomenon.

The dissociation also appeared to apply to theta activities elicited by brainstem or diencephalic stimulation. Stimulation of the midbrain tegmentum elicited abnormal maintained postures ("freezing") and widening of the palpebral fissure, accompanied by low frequency (about 6 Hz) theta. This was abolished by atropine, but survived under ether. On the other hand, stimulation of the lateral or posterior hypothalamus led to quasi-normal exploratory movements, sniffing, vibrissal twitching and high-frequency (7–10 Hz) theta, the EEG and behavioural components never occurring apart in this study. Anaesthesia reduced the theta activity in these circumstances (see also: Gray and Ball 1970). If both atropine and anaesthetics were given, neither type of theta activity could be driven from brainstem/diencephalon.

Table 9. Types of Theta Activity in the Rat

	Type 1	Type 2
Behavioural correlate	Vigorous whole body movements (1,2)	Movements of small segments of body (1,2)
Frequency range	6–12 Hz (2)	4–7 Hz (2)
Blocking drugs	Anaesthesia (especially urethane) (1)	Muscarinic blockers (1)
Blocking lesions	Septum (3,4); fornix(4) disconnection of cortex from hippocampus (5,6,7)	Septum (3,4); fornix (4)

Sources:
(1) Kramis *et al.* (1975)
(2) Vanderwolf (1969)
(3) Green and Arduini(1954); Winson(1978); Bland and Bland (1986)
(4) Vinogradova and Braznik (1978)
(5) Vanderwolf *et al.* (1985)
(6) Vanderwolf and Leung (1983)
(7) Montoya and Sainsbury, (1985)

In view of these data, Vanderwolf *et al.* (1978) reached the conclusion *that there are two distinct types of input pathway from the brainstem capable of generating RSA in the hippocampus.* One of these is blocked by atropine and stimulated by eserine. During immobility this is the only source of control of theta. The other functions without cholinergic elements and is active if and only if certain classes of movement are occurring. The characteristics of these two components of theta activity, (according to Vanderwolf) are summarised in Table 9.

Two further issues arose from these conclusions, which Vanderwolf and colleagues have clarified with more recent work: firstly it may be asked what transmitter(s) are used by the pathways which control atropine-resistant theta activity. This has been a difficult question to answer. Since high-frequency, atropine-resistant theta activity is depressed by phenothiazines, and activated by d-amphetamine (Vanderwolf 1975) it was suggested that this component of theta depends on catecholamines. However, later, Vanderwolf *et al.* (1978) showed that catecholamine-blocking drugs attenuate high-frequency RSA only in so far as they depressed locomotor activity. High-frequency RSA could still occur on the rare occasions when vigorous movement took place. In addition, neonatal 6-hydroxydopamine severely depleted noradrenaline but had little effect on RSA

(Robinson *et al.* 1977). Recently Vanderwolf and Baker (1986) suggested that atropine-resistant theta activity might have a serotonergic component, since large doses (1.5 g) of para-chloro-phenylalanine (PCPA), a serotonin-synthesis inhibitor, could reduce or abolish such activity. In addition rats under the influence of this drug could produce low-frequency RSA during immobility, a phenomenon not usually seen in undrugged rats. There are severe doubts, however, about whether these effects are due to serotonin depletion. In other animal experiments, the dose given to block 5HT synthesis is about 200–400 mg/kg, tenfold lower than in the above experiments (see Gibson *et al.* 1970; Milliaressis *et al.* 1975). There is thus the possibility of many other pharmacological influences being responsible for the effect described at the higher dose. Vanderwolf *et al.* (1978) also showed that reserpine does not abolish atropine-resistant theta activity, and since this drug depletes all monoaminergic transmitters, it seems that none of them mediates this form of theta rhythm. The only other two clues to the pharmacological nature of atropine-resistant theta is that it is abolished by morphine (Vanderwolf *et al.* 1978), and also by phencyclidine (a drug with incompletely understood actions, combining anaesthetic/analgesic and stimulant properties: Vanderwolf and Leung, 1983).

A second issue concerns the precise cholinergic pathway involved in the atropine-sensitive component of theta activity. Vanderwolf *et al.* (1978) do not commit themselves on this but speculate that it may be the the ascending cholinergic elements originating in the reticular formation (as defined by Shute and Lewis, 1967). Recently, Stewart and Vanderwolf (1987a, b) provided more definite evidence about the nature of the cholinergic elements in the atropine-sensitive theta activity. They made ibotenic acid lesions of the septal region, intended to destroy cells, but leave intact any fibres of passage. In fact, the septo-hippocampal cell bodies, located in the medial septum, were also spared by the lesion. Nevertheless, theta activity was significantly changed. Theta activity seen during urethane anaesthesia (i.e., the atropine-sensitive component) was greatly reduced. That seen in the undrugged animal was only slightly reduced, and showed prominent increases during walking or struggling, as in the unlesioned animal. Pharmacological studies showed that the latter theta activity was atropine-resistant. Thus the lesion appeared to selectively reduce the atropine-sensitive component of theta activity. In the second of the two papers (Stewart and Vanderwolf 1987b) this result was also found to apply to the sleeping rat: the atropine-sensitive theta activity which occurred during the tonic phase of REM sleep was abolished, but not that which occurred in the phasic part of REM sleep (an atropine-resistant component). For both sets of results, the impairment in atropine-sensitive theta activity correlated with cell damage in the *dorsal lateral* part of the septum, rather than in the medial septum. The implication was that the atropine-sensitive component of theta activity was a function not of the septo-hippocampal projection neurons, but of a group of intrinsic septal neurons, presumably cholinergic. In this context, the authors cite Macadar *et al.* (1970) who found that atropinic drugs abolished not only hippocampal rhythms but also rhythmic firing of septal neurons. This latter effect must also have involved a cholinergic effect intrinsic to the septum.

6.2.2 Evidence for Single Source of Control of Theta Activity

Some of the evidence suggests that the idea of two independent sources of theta control may be too simple. This alternative conclusion comes partly from the studies with the anaesthetics described above. For instance, the suppression of high frequency theta activity produced by urethane could occur even with subanaesthetic doses which left walking and other whole body movements possible (Kramis *et al.* 1975). Thus the link between whole body movements and high-frequency theta activity is apparently not invariable. The anaesthetic pentoparbitone also presents some problems. Kramis *et al.* (1975) found in rat that subanaesthetic doses (10 mg/kg) had roughly similar effects to ether and urethane. However, slightly larger doses (producing light anaesthesia) completely abolished theta activity. Later, Bland *et al.* (1981) showed that there were some species differences in the actions of pentobarbitone between rat and rabbit. High frequency, movement-related theta was abolished in both species. In the rat, low frequency, immobility-related theta, as well as the high frequency theta was abolished, though only the high frequency theta was abolished in the rabbit. This applied whether the low frequency theta was elicited by sensory stimulation or eserine injection, though in rabbits, its frequency was reduced even further (to about 2.5 Hz) after pentobarbitone. Thus the pharmacological characteristics of anaesthetic-sensitive theta, as defined with urethane or ether in rats, are not well shown with pentobarbitone in the rat, though they are in the rabbit.

A second difficulty with the "two-pacemaker" idea comes from study of behavioural correlates of theta activity during brainstem stimulation. Some work suggests that the distinction between atropine-sensitive and atropine-resistant theta activity correlates better with behavioural features than with frequency of theta. Vanderwolf *et al.* (1978) report that during *immobility*, theta driving by septal stimulation was possible over the full frequency range for spontaneous theta (6–12 Hz), and was abolished by atropine over this whole range. During movement, theta driving from the septum over the same range was also possible but was atropine-resistant. This evidence is less easily accommodated within the "dual control" idea than is the data of Vanderwolf *et al.* (1978). The authors take it as tentative evidence that the pathways for control of both atropine-sensitive and atropine-resistant theta pass through the septum but find it difficult to understand why the frequency ranges of the two components are wider than for spontaneous theta activity. One possibility relies on the suggestion mentioned in Chapter 5, that the hippocampus, if stimulated by its direct afferents from the septum, can be driven over a range of frequencies much wider than the theta range (Kramis and Routtenberg 1977). It is unclear from Vanderwolf *et al.* (1978) whether this is the explanation of their results. If the septal pacemaker *was* bipassed, one would expect that driving would have been possible well above 12 Hz. This was not mentioned in their paper. If, however, the septal neurons were being activated via their own afferents, control of theta would have imposed tighter limits to frequency of each type of theta. The similar ranges for atropine-sensitive and atropine-resistant components would then suggest that there was just a single pacemaker.

More forceful evidence for the "single pacemaker" hypothesis comes from Braznik *et al.* (1985). These authors showed that rhythmic pulses applied to the immediate afferents of medial septal neurons can produce rhythmic entrainment of the neurons' firing, and this is possible throughout the frequency range of spontaneous theta, but not beyond it. They therefore find unnecessary the idea of two sources of control of theta, each with different frequency characteristics. Further evidence for this view was presented by Braznik and Vinogradova (1987). In alert rabbits it was shown that intravenous atropine or scopolamine eliminated (or decreased the stability of) the hippocampal theta rhythm and the rhythmic bursting of cells in the medial septum and diagonal band. A small proportion of septal neurons retained their rhythmic bursting, even under maximal doses of atropinics, but this was insufficient to maintain hippocampal theta activity. Significantly, *the frequency of the rhythm of these remaining cells was unaltered by atropinic drugs.* Conversely, physostigmine could increase the stability of septal unit bursting and the hippocampal theta rhythm, and could sometimes reinstate these rhythms when they had been suppressed by phenobarbitone. Again *the frequency of the rhythm was unaltered by the drug.* In animals under the influence of atropinic drugs, sensory stimulation or electrical stimulation of the reticular formation could increase the proportion of septal cells which were bursting, and could sometimes reinstate the hippocampal theta rhythm.

Braznik and Vinogradova formulated a concept of the control of the theta rhythm which is rather different from that of Vanderwolf's group: *There is just one group of cells in the medial septum/diagonal band which controls the hippocampal rhythm. A certain threshold proportion of these cells (possibly about 30%) must generate rhythmic bursts before the hippocampal theta rhythm is elicited. A single source of control from the brainstem can simultaneously control their burst frequency, and to a lesser extent the proportion of rhythmically bursting neurons. A cholinergic mechanism in or near the bursting neurons controls mutual recruitment of these cells, but not their frequency.* Atropinic drugs prevent the mutual recruitment within this group of cells, and so lower the proportion of rhythmically bursting cells. This may prevent the occurrence of the hippocampal rhythm, but does not alter the frequency of the pacemaker. Increased activity in the ascending drive to this group of cells can reverse this process, so that enough septal cells are bursting to allow the hippocampal rhythm to emerge again. However, an incidental effect of this increased ascending drive is to increase the frequency of the bursting and the theta rhythm. Thus, the reason why theta activity which survives atropinic drugs tends to have a higher frequency than that which survives anaesthetics is because the ascending drive in the former case is greater than in the latter. [Both Stumpf *et al.* (1962) and Macadar *et al.* (1970) have showed that anaesthetics can slow down and in higher doses abolish the rhythmic bursts occurring spontaneously in septal units.] Thus there is no need to postulate independent sources of control for different frequency ranges.

The idea that cholinergic mechanisms are involved in mutual recruitment of medial septal cells is supported by another recent paper (Braznik and Vinogradova 1988). It was found that increase of the efficacy of cholinergic mechanisms, produced by drugs such as physostigmine, would increase the spontaneous rhythmicity of medial septal units, and decrease the efficacy of elicitation of single spike responses by stimulation of their afferents. Cholinergic blockers, on the other hand, reduced or abolished the spontaneous rhythmic activity, but restored or even enhanced the single unit responses from afferent pathways. Presumably, when cholinergic mechanisms are potentiated, the mutual recruitment is stronger, and can occlude responses produced from afferent sources.

6.2.3 Dual or Single Pacemaker? Discussion

Since the theta rhythm and rhythmic septal bursting can, under certain circumstances, be seen throughout the range of theta frequencies in the atropinized animal, the original conclusion of Vanderwolf *et al.* (1978) that the low-frequency component of the theta rhythm is controlled by the cholinergic reticulo-septal pathway and the high-frequency one by non-cholinergic pathways must be substantially modified. From the evidence presented, the only certain role for cholinergic neural elements is in mutual recruitment of bursting cells in the medial septum/diagonal band. The results described above of Stewart and Vanderwolf (1987a, b) in which septal lesions selectively abolish the atropine-sensitive component of theta activity could be taken as evidence for just such a role. The neurons in dorsolateral septal region which appeared to be critical in this effect could well be part of the circuitry for mutual recruitment of bursting cells in medial septum/diagonal band. It will be recalled that it was in this region that Kramis and Routtenberg (1977) found that electrical driving stimuli entrained hippocampal rhythms within the theta frequency, whereas nearer to the septo-hipocampal fibre pathways, a wider range of frequencies could be driven, as when the septum is bipassed altogether.

At this point an obvious question should be raised: Does the cholinergic component of the septo-hippocampal pathway (reviewed in Chap. 3) carry rhythmic theta-modulated signals from septum to the hippocampus? There is no clear evidence on this, and to obtain direct evidence would require the technically difficult combination of ChAT staining and physiological characterization of cells in the medial septum/diagonal band complex. However, it is very likely that the rhythmically bursting cells are both cholinergic and non-cholinergic long-axon neurons projecting to the hippocampus. The relative role of cholinergic and non-cholinergic fibres is probably interchangeable, according to the pharmacological status of the animal. According to this idea, high-frequency theta activity seen in atropinized animals would be controlled by non-cholinergic fibres, while in the undrugged preparation, both cholinergic and non-cholinergic neurons would act synergistically. In the urethane- or ether-anaesthetized preparation, the two components of the septo-hippocampal pathway would also act together, and could usually generate only the lower frequencies. This is not because of an inherently low-frequency rhythm for the cholinergic component, but because, after anaesthetics, the ascending drive will not sustain a high frequency one.

A last possibility is that the cholinergic and the non-cholinergic septo-hippocampal neurons are capable of rhythmic bursting but with different optimal frequencies. This would recreate something of the "two-pacemaker" hypothesis. However, the cholinergic and non-cholinergic populations of cells are closely intermingled. It is difficult to envisage one population giving one frequency range at one time, and the other producing a higher frequency range at another, without any interaction or mixing of the frequencies. Therefore this possibility will not be taken seriously until it is supported by definite evidence.

To summarize the argument so far, there is no conclusive evidence for two pacemakers for the theta rhythm, with different frequency ranges, different pathways and different transmitters. Both cholinergic and non-cholinergic pathways probably act synergistically under normal circumstances throughout the frequency range. Under the influence of drugs, incidental correlations between theta frequency and avalaible transmitters are obtainable, but these correlations do not reveal fundamental subdivisions of theta activity.

This conclusion is not the end of the matter, however. There is a small but important body of evidence of the effect of lesions in cortex other than the hippocampus, which raises quite other issues about the pathways controlling the theta rhythm. This subject is dealt with below.

6.3 Control from Within the Hemisphere

6.3.1 The Effect of Extra-Hippocampal Lesions Within the Forebrain (see Table 9)

The earliest studies of the effect on theta rhythms of forebrain lesions other than those to the septo-hippocampal axis itself were made by Azzaroni and Parmeggiani (1967). They were investigating the role of the Papez circuit on theta rhythms. In locally anaesthetized cats, theta rhythms, produced by sciatic nerve stimulation would outlast the period of stimulation by anything from 10 to 60 s. Lesions to the anterior thalamic nuclei, if made bilaterally, reduced this period of "reverberation" to about 4–10 s, and reduced the frequency of the rhythms. Similar effects were made by lesions to another part of the Papez circuit—the cingulate gyrus—but in this case an ipsilateral lesion reduced the reverberatory activity only ipsilaterally. Control lesions outside the Papez circuit did not have such an effect. Trains of high frequency stimulation delivered at 4/s to either the anterior thalamus or the cingulate gyrus elicited theta activity even when the septum or fornix had been ablated. When such rhythms were elicited by stimulation of the anterior thalamus they could be suppressed by lesions of the cingulate gyrus.

Chronister *et al.* (1974) made lesions in the entorhinal cortex, and Kolb and Whishaw (1977) made lesions of the lateral hypothalamus or orbitofrontal cortex. Their effects were transient and need not be discussed in detail. Vanderwolf and Leung (1983) made lesions in the entorhinal cortex of rats. Subsequently, much of the hippocampal EEG phenomenology was normal, but in two respects was different from controls. Firstly, RSA and walking were not invariably linked in the lesioned animals, in the way they are in intact rats. Secondly, and more dramatically, atropinic drugs completely abolished RSA in the lesioned animals. This is in sharp contrast to the situation in normal rats where the atropine-resistant component of RSA survives. Both these observations were corroborated by Montoya and Sainsbury (1985) using guinea pigs.

To analyze this question further Vanderwolf *et al.* (1985) studied the hippocampal EEG in rats in which lesions had been made which disconnected the remaining bulk of the cortex from the entorhinal cortex, but leaving the latter region intact. Animals with such lesions had some of the EEG abnormalities found in rats with entorhinal cortex lesions: Hippocampal RSA existed, which was unaffected by urethane, but abolished by atropine. In other respects, however, these animals were closer to normality than those with entorhinal cortex lesions: the full range of theta frequencies was seen, in the normal relationship to behaviour. In particular the close relation between vigorous whole body movements and high-frequency theta activity was preserved better than in entorhinal cortex-lesioned rats. Clearly in these preparations atropine-sensitive mechanisms can govern theta activity over its full frequency range. This corroborates the conclusion reached in Section 6.2.3 (above) for the unlesioned animal.

In further experiments, lesions were made which disconnected the entorhinal cortex from the rest of the cortex by all pathways except the cingulate cortex/cingulum bundle on each side. In this case theta activity after atropine administration remained intact. If the cingulum/cingulate cortex alone were lesioned on both sides, some atropine-resistant theta activity remained, interspersed with irregular activity. However, this was less than in the intact animal (reduced by 53% at 8 Hz peak on power spectrum, compared with the intact atropinized rat). With unilateral entorhinal cortex disconnection it was shown that the cingulate/cingulum region on one side could maintain atropine-resistant theta on both sides of the hippocampus. Lesions of the limbic thalamus (anterior or medial nuclear groups), amygdala or cerebellum, were without effect.

There are two other relevant publications on this issue. Destrade and Ott (1980, 1982) studied theta activity in rats, either during spontaneous exploration or elicited by high-frequency stimulation of the dorsomedial hypothalamus. During exploration a theta rhythm of frequency 6–8 Hz was produced, while hypothalamic stimulation elicited a somewhat higher frequency (8–12 Hz). Local anaesthetic injected into the cingulate cortex had no effect on the lower frequency associated with exploration, but reduced the peak frequency during hypothalamic stimulation from 8.8 Hz to 7.5 Hz (Destrade and Ott 1982). Similarly injection of a glutamate antagonist (glutamate diethylester) into the hippocampus did not influence the lower frequency theta activity recorded during exploration, but during hypothalamic stimulation replaced the higher by the lower frequency (Destrade and Ott 1980). The most likely pathway upon which this drug was acting is the perforant pathway, part of the sequence of fibre tracts by which the isocortex can influence the hippocampus. The two varieties of theta rhythm defined in these papers do not represent the usual type 1/type 2 dichotomy. Nevertheless, there is a correspondance with the results of the lesion studies described above: theta activity generated in the hippocampus, especially in the upper part of its natural frequency range requires normal functioning of the cingulate cortex and intact transmission on the afferent pathway from the entorhinal cortex.

6.3.2 Discussion of Section 6.3.1

Two issues deserve discussion: firstly, is the lesion evidence described above compatible with a single locus of control from the brainstem/septal region, the idea favoured in Sections 6.2.2 and 6.2.3? Or is one forced back to the dual control concept of Vanderwolf and colleagues? Secondly, given that the neocortex also has some influence on hippocampal theta activity, what can we conclude about the nature of this influence?

6.3.2.1 Single or Dual Control from Brainstem/Septal Region and the Cortical Lesion Data

The evidence on the effects of cortical lesions described above has been produced largely by Vanderwolf and his colleagues. It is, however, susceptible to interpretations other than those given by these workers. The same dichotomy between dual control and single control arises here as it did in discussing the brainstem control of theta activity.

According to Vanderwolf, evidence from entorhinal lesions, and lesions disconnecting the hippocampus from the neocortex indicates that the lesions have abolished atropine-resistant theta activity, leaving only the atropine-sensitive component. In the case of the entorhinal cortex lesion, the lesion has also disrupted the correlation between walking and theta activity. Following the reasoning of Braznik and Vinogradova (1987), the observed results could be explained again on the basis of a critical proportion of septal cells needing to display rhythmic bursts before theta activity appears. In the intact animal 30% may be required, but perhaps more than this is required in entorhinal cortex lesioned or hippocampally disconnected animal. This may be more than can be provided by the non-cholinergic septo-hippocampal component alone. Atropine therefore suppresses all theta activity, though according to the argument, some septal cells would continue to discharge in rhythmic bursts. The failure of urethane to abolish theta activity in the lesioned animals can be based on the conjecture made above: this anaesthetic in rats may lower the frequency, but it does not lower the proportion of septal cells which show rhythmic bursts below the critical level for the lesioned animals. If this is so, it would follow that the cholinergic component of the septo-hippocampal pathway plays a somewhat larger role, quantitatively speaking, in the initiation of theta activity than does the non-cholinergic one, and this experiment shows up this relationship.

The choice between these two interpretations depends to a considerable extent on the conclusion of the discussion above about brainstem control. This favoured a single source of control, which could cover the complete range of theta frequencies. However, it should also be pointed out that Vanderwolf *et al.* (1985) themselves produce further evidence for this view, since in their preparations with hippocampus-disconnecting lesions, atropine-sensitive mechanisms could govern theta activity over its full frequency range. This corroborates the conclusion reached for the unlesioned animal (Sect. 6.2.3 above).

6.3.2.2 The Nature of the Influence from the Neocortex

The evidence on lesions of the entorhinal cortex, and the discussion presented above suggest the following conclusion: the entorhinal cortex seems to be required for theta activity when mutual recruitment of rhythmically bursting cells in the septal region is limited by atropine. It thus seems as though signals generated in (or relayed through) the entorhinal cortex can substitute for the partial loss of bursting septal cells. After entorhinal lesions these signals can no longer influence the hippocampus, and atropine completely prevents the appearance of a theta rhythm. This conclusion can be explained on the basis of a number of different hypotheses:

1. The entorhinal cortex may be an intervening site in part of the pathway between the septal pacemaker neurons and the hippocampal theta generators. Anatomically speaking, this raises two possibilities: (a) Septo-hippocampal fibres may induce theta activity in the entorhinal cortex, and this rhythmical activity is then relayed to the hippocampus, and collaborates with the rhythmic signals relayed directly to the hippocampus to allow the macroscopic hippocampal theta rhythm to emerge. This hypothesis is consistent with the anatomical data reviewed in Chapter 3, that there are axonal projections from the medial septum/diagonal band to the entorhinal cortex, which itself has strong projections to the dentate gyrus and other parts of the hippocampal complex. It is also consistent with data to be reviewed (Chaps. 7 and 10) that the entorhinal cortex can generate its own theta activity. (b) Alternatively, the pathway from the septum to the entorhinal cortex may be a less direct one via the cortex, that is in front of and above the corpus callosum.

2. A second hypothesis is that the entorhinal cortex may be part of a local functional loop with the hippocampus allowing recruitment of the hippocampal theta rhythm by some form of resonance around the loop. The loop thus permits a positive feedback process, which ensures the emergence of a macroscopic theta rhythm, even if septo-hippocampal rhythmic signals are relatively weak, when it would

die out without such feedback. According to this scheme, atropine-sensitive theta activity can occur without this loop, but atropine-resistant theta activity relies on its intactness.

3. There is also a third possibility, that rhythmic signals from isocortex to hippocampus help to maintain some aspects of the hippocampal rhythm. This seems at first sight to be absurd, since there is no rhythm generator in the isocortex which is detectable in the obvious way that the hippocampal theta rhythm is detectable. However, it is central to the hypothesis developed in Part 3 of this book, that there be rhythmic signals returning to the hippocampus from the isocortex. The specific idea is advocated that there be a series of functional loops, between hippocampus and cortex. In these loops there would be the possibility of *resonance*, just as (in the above paragraph) the entorhinal cortex and the hippocampus are envisaged as being able to resonate. When resonance is established this produces a positive-feedback effect and allows weak rhythmic input to the hippocampus (e.g., after atropine) nevertheless to give rise to a macroscopic rhythm, provided the loop is intact. This possibility is compatible with the anatomical evidence, reviewed in Chapter 3, of abundant reciprocal connections between hippocampal formation and the neocortex. The details of this idea will be worked out later. (In particular it is necessary to explain why there is so little evidence of macroscopic theta activity in most of the isocortex. See Sect. 9.2.4.3.)

In evaluating these various possibilities, one immediate point is that the lesions of the entorhinal cortex produce the same apparent reduction in the atropine-resistant component as do lesions disconnecting the hippocampus from the cortex. Both types of lesion thus produce the same apparent reduction of theta drive to the hippocampus. It could thus be argued that a role for the direct septo-hippocampal influence on the entorhinal cortex [hypothesis 1(a) above] cannot be critical for the appearance of the atropine-resistant component of theta activity. Likewise it could be argued that local resonant circuits involving the entorhinal cortex and hippocampus, but not the neocortex generally, are unlikely to be crucial (hypothesis 2).

The idea of a supra-callosal septo-hippocampal pathway as a mediator of atropine-resistant theta activity has been advocated by Vanderwolf *et al.* (1985). They speculate that serotonergic fibres passing supracallosally in the cingulum bundle might play a part in mediating atropine-resistant RSA. (A role for serotonin in high frequency atropine-resistant RSA had been suggested earlier, on the basis of the experiments of Maru *et al.* 1979). However, the idea of a supracallosal route for the serotonergic influence on theta activity seems implausible: section of fornix or fimbria, leaving such a putative pathway intact, usually abolishes theta activity completely (see Sects. 4.2 and 5.6.1), a more severe deficit than that produced by atropine.

Another possibility is that the Papez circuit may somehow be involved in relaying atropine-resistant theta activity from the septum. Evidence on this is somewhat contradictory. Azzaroni and Parmeggiani (1967) produce clear evidence of its involvement in reverberatory theta activity, but this seemed to be possible even when the septal region had been destroyed. On the other hand, Vanderwolf *et al.* (1985) found that lesions of limbic thalamus leave atropine-resistant theta activity unaffected.

6.4 Conclusions

Despite substantial evidence for two components to the ascending control of theta activity, there is no convincing evidence that these are independent, and specific to different ranges of theta frequency. It is more plausible to suggest that a single population of medial septal/diagonal band neurons, under control from the brainstem, generates rhythmic bursts and can relay rhythmic signals at frequencies throughout the theta range to the hippocampus, along both cholinergic and non-cholinergic axons. Fornix lesions therefore usually abolish completely all varieties of hippocampal theta activity recordable *in vivo*, while atropine has less severe effects. A cholinergic mechanism also controls mutual recruitment of rhythmically bursting neurons in these septal regions. Atropinic drugs act mainly to prevent mutual recruitment, while non-rhythmic signals ascending from the brainstem influence frequency of bursting, and also the proportion of bursting neurons (though less powerfully in the latter respect than the cholinergic recruitment mechanisms). All the evidence used to advocate a dual control system can be explained by these conclusions, while some of the evidence used to advocate a single control mechanism cannot be explained on the basis of dual control. Single control thus seems to be the more viable concept.

The isocortex also influences the hippocampal theta rhythm, but can promote such a rhythm only in so far as the septal neurons are already rhythmically active. The entorhinal cortex appears to have a role which may sometimes be of greater quantitative importance than that of the rest of the isocortex. The influence of the entorhinal cortex can be explained on the basis of a direct relay of rhythmic signals from septum to hippocampus via the entorhinal cortex. Resonance between hippocampus and entorhinal cortex may also play a part. It is, however, difficult to explain the role of most of the isocortex on the basis of a pathway from septum, above the corpus callosum, to the hippocampus. This leaves us with the alternative hypothesis *that there is a component of the overall theta rhythm which is maintained by recursive loops reciprocally connecting the hippocampus and cortex via a direct posterior route behind the corpus callosum.* Thus the anatomo-pharmacological dissection of theta activity leads us towards a *concept of theta activity in which reciprocal interaction between hippocampus and cortex is necessary to maintain at least some components of the overall theta activity.*

Although atropine-resistant theta activity may involve pathways in addition to the septo-hippocampal link, the medial septum/diagonal band region can still be regarded as the central focus for this, as it is for atropine-sensitive theta. Likewise, the septo-hippocampal fibres in the fornix, with both cholinergic and non-cholinergic contributions, can still be regarded as the primary pathway for initiation of theta activity. However evidence that these statements might not always be true comes from Destrade and Ott (1982), who showed that local anaesthetic injected into the septal region failed to abolish theta activity elicited by high-frequency hypothalamic stimulation. Azzaroni and Parmeggiani (1967) also found that the circuit of Papez could produce rhythmic activity at the theta frequency even when the septal region was destroyed. These exceptional results can however be incorporated into the theory of cortico-hippocampal interaction developed in Part 3 of this book.

7 Evidence for Multiple Sources of Theta Generation in the Hippocampus and Related Structures

7.1 Introduction

In earlier chapters, the topics dealt with have included the anatomical basis for hippocampo-cortical interaction, the general behavioural correlates of the rhythmic neural activity in the hippocampus which is envisaged to allow that interaction, and, in the last two chapters, the mechanisms of extrinsic control of that rhythm. The present chapter looks more deeply into the physical basis of the hippocampal theta rhythm itself. It is concerned with description of the profiles of electrical potentials which form the theta rhythms seen in different fields of the hippocampus and related structures, and in the different preparations that have been used. The primary data to be considered are, of course, the theta potentials generated in the hippocampal complex itself. However, towards the end of this chapter the data available on theta rhythms outside this structure are also described. In addition, details of experiments in which lesions have been made in the hippocampal complex and their influence on theta rhythms are described, because these data are important in interpreting the EEG profiles. However, the interpretation itself, which is complicated, is left to Chapter 8.

7.2 The Discovery of Two Theta Generators in the Hippocampus

The first indication that the source of hippocampal theta activity was heterogeneous came from Petsche and Stumpf (1960), using paralyzed rabbits. They observed that there was a 180 degree phase shift in the theta rhythm between that recorded in the cortex and that recorded in the hippocampus and diencephalon. Early corroborations of this findings came from Green *et al.* (1960) using rabbits and Porter *et al.* (1964) using cats. The observation has been repeated many times subsequently, some of the more recent investigators presenting full laminar profiles of the shape and phase relations of theta activity throughout the different fields of the hippocampus. These data are reviewed in the next section.

7.3 Laminar Profiles of Theta Activity in the Hippocampus

7.3.1 Phenomenology of Laminar Profiles

7.3.1.1 Introduction

Laminar profiles for theta activity vary in different behavioural states, and to some extent in different species. The first indication of this came from Winson (1974), who found that in the free moving rat phase reversal took place *gradually* over the length of electrode tracks passing downwards through the stratum radiatum of CA1. This gradual phase reversal had occurred completely by the level the hippocampal fissure, but at no point on the track did the amplitude of the potentials become zero. On the other hand, in the curarized, eserinized rat the phase reversal occurred suddenly in the stratum radiatum of CA1 at a level slightly below the pyramidal cell layer. At this location there was a definite null point. Subsequently, laminar profile analyses for theta potentials have been conducted in the following preparations: curarized rat (Winson 1976a; Bland and Whishaw 1976); rats anaesthetized with urethane (Bland and Whishaw 1976; Feenstra and Holsheimer 1979; Leung 1984a; Buzsaki *et al.* 1986), ether (Leung 1984a) or pentobarbitone (Leung 1984a); urethane anaesthetized rabbits (Bland *et al.* 1975); urethane anaesthetized cats (Bland *et al.* 1979); atropinized rats (Leung 1984a; Buzsaki *et al.* 1986); eserinized rats (Leung 1984a); reserpinized rats (Leung 1984a); rats walking while drugged with phencyclidine (Leung 1984a); free moving awake rats (Bland and Whishaw 1976; Leung *et al.* 1982; Buzsaki *et al.* 1983; Buzsaki *et al.* 1986); awake but immobile rats (Leung *et al.* 1982); rats during grooming (Leung *et al.* 1982); rats during REM sleep (Monmaur *et al.* 1979; Monmaur 1982; Leung *et al.* 1982; Leung, 1984b), or slow-wave sleep (Leung *et al.* 1982); rabbits during voluntary movement (Winson 1976b); immobile rabbits during sensory stimuli (Winson 1976b); rabbits during REM sleep (Winson 1976b). In addition, some data on hippocampal sources of theta generation after partial septal lesions have been provided by Monmaur *et al.* (1979), Monmaur (1982) and Monmaur and Thompson (1983) in rats during wakefulness or REM sleep.

7.3.1.2 Curarized and Urethanized Preparations

In *curarized rats* there is a peak of theta activity in stratum oriens of CA1, and a decline of amplitude through the pyramidal cell layer, to a null point in the proximal part of the stratum radiatum. Deep to this point theta amplitude builds up again, but with reversed phase, reaching a second amplitude peak at the level of the hippocampal fissure (Winson, 1974; 1976a)[6].

Under *urethane anaesthesia* the rat hippocampus has a very similar laminar pattern of theta activity, according to Bland and Whishaw (1976), Feenstra and Holsheimer (1979), Leung (1984a) and Buzsaki *et al.* (1986). This is true both for spontaneous theta activity in such preparations and for theta activity elicited by reticular formation stimulation (Bland and Whishaw 1976; Green and Rawlins 1979). The progression of amplitude along a track through the hippocampus is shown in sample traces in Fig. 28, while the change of both amplitude and phase along such a track are shown graphically in Fig. 29. *Etherized* rats, (Leung 1984a) and *urethanized rabbits* (Bland *et al.* 1975) *and cats* (Bland *et al.* 1979) also show this laminar profile. *Undrugged rabbits* under all circumstances investigated (voluntary movement, REM sleep, immobile during sensory stimulation) also show this profile (Winson, 1976b), a fact upon which comment will be made in Chapter 8. The amplitude of the deep source of theta activity is greater (up to 2 mV) than that of the superficial source (up to 1 mV), and amplitude declines as the electrode tip passes from the buried blade of the dentate gyrus to the exposed blade. Feenstra and Holsheimer (1979) sometimes found a second reversal point in stratum

[6]This profile of theta activity endured longer than the paralysis itself, suggesting that the curare was having definite central influence (perhaps cholinergic) on theta activity.

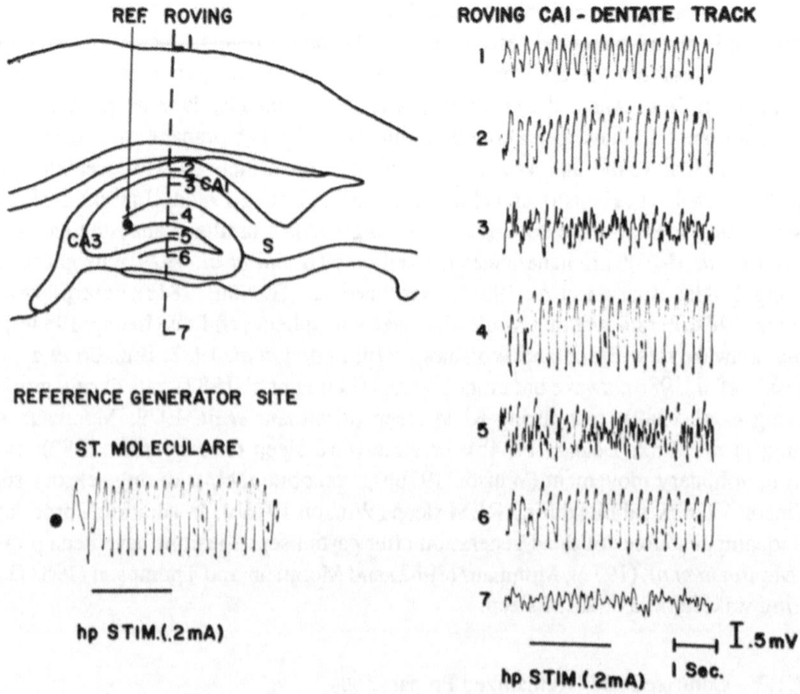

Fig. 28. Depth profile of amplitudes in an electrode track passing from the isocortex, into the dorsal hippocampus, through field CA1 and on through the dentate gyrus. A reference electrode was located in stratum moleculare of the buried blade of the dentate gyrus (*black dot*) while the traces *1–7* were recorded at various depths with a roving electrode. *1* Surface of isocortex; *2* stratum oriens; *3* null zone in stratum radiatum; *4* stratum moleculare, buried blade; *5* fast activity in the hilus; *6* stratum moleculare, exposed blade; *7* thalamus. The different records shown to the right are not matched to a reference record with respect to phase, and are intended to shown amplitude changes only. Calibration: 0.5 mV; 1 s. *Solid black bars* indicate the duration of electrical stimulation of the posterior hypothalamus, in this case used to initiate theta rhythms. (Bland and Whishaw 1976, Fig. 2) (From *Brain Research* Vol 118 pp. 259–280 1976)

moleculare of the dentate gyrus. However, the amplitude of the theta waves for which this was observed was quite small.

The most detailed study of this type of laminar phase profile reported to date is that of Buzsaki *et al.* (1986), in urethanized rats. As just mentioned, this confirms previous findings for regions around and dorsal to the hippocampal fissure. The most interesting additional details concern the dentate gyrus, and help to clarify the occasional findings of a second reversal point in this region. Averaging techniques were adopted, but since the frequency of the theta waves to be averaged fluctuated somewhat the results should not be regarded as fully quantitative. In the outer molecular layer of the dentate gyrus, phase was 180 degree reversed compared with the CA1 pyramidal cell layer (in agreement with earlier findings). Between the inner third of the molecular layer and the granule cell layer, abrupt phase changes took place, both in the buried and the exposed blade of the dentate gyrus. In Buzsaki *et al.*'s Fig. 4, this phase change amounted to 270 and 320 degrees

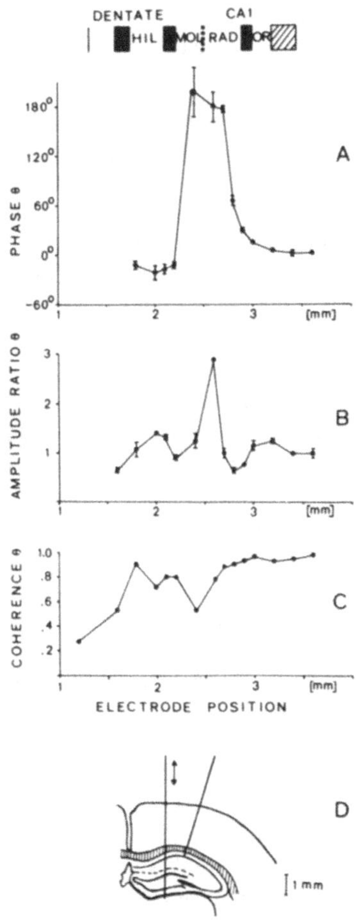

Fig. 29. Changes of phase (**A**), amplitude (**B**) and coherence (**C**) for theta potentials recorded along an electrode track through the isocortex, hippocampus and dentate gyrus. *Inset* above **A** shows diagramatically the laminae at each depth level in the roving electrode's track (see also *diagram* in **D**). Phase difference is obtained from comparisons of traces from fixed electrode in stratum oriens (**D** *oblique line*) and roving electrode (**D** *double arrow*). Amplitude is expressed as the ratio of amplitudes recorded in the fixed and roving electrodes. (Feenstra and Holsheimer 1979) (From *Electroencephalography and Clinical Neurophysiology* Vol 47 pp. 532–538 1979)

in the buried and exposed blades respectively. However, in the corresponding text the authors write: "the magnitude and steepness of the phase change across the granule cell layer was similar to that of normal rat". In the same figure the phase change in this region for normal (free moving) rat is shown as about 360 degrees. If this is so, then in effect the granule cell layer potentials are roughly in phase with those at the fissure, i.e., about 180 degrees phase advanced with respect to the CA1 pyramidal cell layer. However, bearing in mind the data from Buzsaki *et al.*'s Fig. 4, it is safer to assume that the potentials in the granule cell layer range between 0 and 90 degrees behind potentials at the hippocampal fissure and outer molecular layer, that is between 90 and 180 degrees ahead of the potentials in the CA1 pyramidal cell layer.

According to Rall (1962), a synaptic input to a neuron produces extracellular potentials of opposite sign to the synaptic potentials recorded intracellularly in the same region of soma or dendrite. However, the synaptic input also generates a neuronal dipole along the length of the dendrite. Therefore, at a distance from the synaptic input, the extracellular potentials generated by the dendrite may be of opposite sign to those recorded in the locality of the input. This means that, in principle, a laminar profile showing sudden phase reversal and a null point could be explained in terms of a neuronal source for theta potentials located within a narrow lamina in the hippocampal architecture. Such a simple hypothesis could not explain the more complex laminar profiles seen in free moving animals, but must be considered for the above examples.

There are, however, some reasons for thinking that even in curarized or urethanized animals, a more complex origin of theta activity must be postulated: 1. In cats, phase reversal is not abrupt, but takes place over about 200 microns below the null point (Bland *et al.* 1979). 2. The two sources of theta potentials sometimes vary independently: in some tracks in Green and Rawlins' experiments only the deep source was found. These workers, and others (Bland and Whishaw 1976; Bland *et al.* 1975) find generally that the superficial source of theta is more variable than is the deep one. This may, in part, reflect the greater vulnerability of the CA1 region to mechanical disturbance, especially in anaesthetized preparation where overlying cortex is removed. There are hints, however, that the CA1 source also shows greater variability in the free moving animal (see below). In addition to these findings, Green and Rawlins report on one urethanized rat (out of 17) where reticular formation stimulation was applied to initiate theta activity. In this animal the superficial source started theta activity within 1–2 s of start of the electrical stimulus and stopped immediately the stimulus stopped. On the other hand, in the deep generator 10–20 s of continuous stimulation was required before theta activity would start, and when the stimulus was stopped theta activity continued for some time afterwards. Kirk and McNaughton (1989) have also described separate initiation of RSA in the superficial and deep generators from different sites in the brainstem of urethanized rats.

Such evidence implies the existence of more than one dipole as generators of theta potentials. Nevertheless, the basic relation between superficial and deep theta activity still appears to be one of complete 180 degrees phase shift rather than some lesser or greater value. Thus Leung (1984a) found (his Fig. 2b) that as urethane anaesthesia is deepened from 0.5 g/kg to 1.0 g/kg, the frequency of RSA falls from 6.7 Hz to 5.5 Hz, but the relative phase shift remains constant at 180 degrees. Likewise in his Table II, Leung presents data on theta activity in the immobile urethanized rat, with frequency varying from 5.1–6.7 Hz, although phase between superficial and deep potentials varied only between 166 and 187 degrees (reproduced here as Table 10).

In view of the above findings, it seems likely that, even in the anaesthetized preparations, there is both "a superficial theta generator" and "a deep theta generator". Further evidence for this is presented below (7.3.1.3, and discussion in Chapt. 8). The precise nature of the two generators will be considered later.

Several studies have reported on the variation in amplitude and relative phase of theta activity recorded at different points along the length and width of the hippocampus, in anaesthetized preparations (Bland *et al.* 1975; Bland and Whishaw 1976; Winson

Table 10. Action of urethane (1 g/kg i.p.) on theta phase and frequency (c/s). Compare with Table 11. (reproduced from Leung, 1984a)

Rat's code	Control walk	Immobile under urethane	"Walk" under urethane
E45	70°(8)	170°(6.5)*	70°(7)
E46	146°(7.8)	181°(5.1)*	158°(6.1)*
E68L	150°(7.8)	166°(5.4)	165°(6.2)*
E68R	133°(7.8)	179°(5.4)*	157°(6.2)*
C988	143°(8.2)	184°(6.7)*	146°(7.4)
C290	135°(7.8)	187°(5.5)*	177°(5.5)*
C293R	155°(7.2)	178°(5.1)*	166°(5.8)*

* 95% confidence limits do not overlap with values during normal (control) walling.

1976a; Green and Rawlins 1979). It appears that large areas of the dorsal hippocampus show coherence of their theta activity. There are also significant differences in phase of theta activity recorded at various parts of length and width of the hippocampus, described in greatest detail by Green and Rawlins (1979).

In field CA3 theta activity is rarely recorded in the urethanized preparation. Bland *et al.* (1975) found no theta activity there. Bland and Whishaw (1976) concluded, on the basis of amplitude and phase relations between theta activity recorded in CA3 and elsewhere in the hippocampus, that in almost all of field CA3, theta activity was volume-conducted from other regions. The exception was a small area near the alvear surface of CA3. Green and Rawlins (1979) clarified this matter by showing that theta activity could be recorded near the alvear part of CA3 only sometimes, when the electrode was being advanced, and when the alvear surface was covered with CSF. This implies that theta potentials here, when present, are also volume conducted from distant regions, via the low resistance pathway of the CSF.

In the CA4/hilar region of the dentate gyrus of urethanized rats theta activity cannot be recorded with normal techniques (Bland and Whishaw 1976), though in both urethanized rabbits (Bland *et al.* 1975) and cats (Bland *et al.* 1979) theta activity is detectable. Green and Rawlins (1979) attribute this absence to the fact that within the dentate gyrus of rats, starting at about 0.1–0.2 mm below the hippocampal fissure, high frequency activity tends to obliterate theta activity. This high frequency activity was seen also by Vanderwolf *et al.* (1978) and Bland and Whishaw (1976). This interpretation is supported by some of the data obtained with intrahippocampal lesions (see below). However, Buzsaki *et al.* (1986), using extensive averaging to abstract theta activity from higher frequencies, found there was only very low amplitude theta in the hilar region. Its apparent absence in unaveraged records is thus not merely due to the obscuring effect of high frequency activity.

Lastly, Feenstra and Holsheimer (1979) found in urethanized rats, that coherence between deep and superficial EEG sites was high not only at the theta frequency, but also at twice the theta frequency. This is presumptive evidence of a "harmonic" to the main frequency, an observation implying that resonance may be a factor in initiating or maintaining theta activity (see below). Harmonics were also detectable in other studies, e.g., in the spectral analyses of theta activity elicited by cholinomimetics, illustrated by Olpe *et al.* (1987).

7.3.1.3 In Vitro Preparations

In the hippocampal slice preparation, theta-like activity can be induced by carbachol (or similar drugs) applied in the bath (see Chapt. 5). In these cases, theta activity occurs in both the dentate gyrus and CA1, usually at the same frequency (Konopacki *et al.* 1987c). However, Konopacki *et al.* (1987d) illustrate one slice (their Fig. 2C) in which both regions generate theta activity, but their respective frequencies are not the same. A more consistent result is that the phase relation is disrupted in the slice preparation compared to the *in vivo* theta rhythm: According to Konopacki *et al.* (1987c) the phase relation between the two sites seen in a variety of slices covers the whole range from 0 to 180 degrees (see Fig. 26).

Konopacki *et al.* (1987d) have also studied carbachol-induced theta-like activity in slices of the hippocampus which have been transected horizontally, so that different portions contain either field CA1, or the dentate gyrus, in isolation from one another. Both regions in isolation could generate theta-like rhythms in response to 50 μM carbachol, and in both cases the rhythm was abolished by atropine. As found *in vivo* the rhythm recorded from the dentate gyrus had a higher amplitude than that from CA1; and it also had a higher range of frequencies. Clearly, both regions of the hippocampus have independent capacities to generate electrographic rhythms within the theta range of frequencies. In contrast, when carbachol is injected into small regions of the hippocampus *in vivo*, theta activity is initiated ipsi- and contralaterally, and in both CA1 and the dentate gyrus, but frequency in all regions remains the same and the 180 degrees phase reversal between the two sites on the same side occurs apparently normally, as for spontaneous theta activity (Rowntree and Bland 1986). The authors suggest that this evidence implies that an extra-hippocampal mechanism (e.g., one in the septal region) normally sets the frequency and relative phase between CA1 and dentate gyrus. This is one of a number of possibilities which will be discussed in Chapter 8.

Most recently, Bland's group has conducted experiments on smaller subdivisions of hippocampal slices (Konopacki *et al.* 1988). In slice portions containing subfields CA3c and CA3b, theta potentials were recorded in both subfields, the amplitude being much higher in field CA3c. On the other hand, in slice portions containing subfields CA3b and CA3a, no theta potentials were recorded. In intact slices, the amplitude was greatest in field CA3c (apparently in the pyramidal cell layer and stratum lucidum). The authors therefore argue that subfield CA3c is capable of acting as a third generator of theta potentials. This requires further confirmation.

7.3.1.4 Free Moving Rats and REM Sleep

In the free moving rat, Winson's (1974) observation of gradual phase shift, without a definite null point has been supported by several recent studies. Leung *et al.* (1982) conducted spectral analyses (in rats) of both superficial and deep theta sources (recorded respectively from stratum oriens and hippocampal fissure) during waking epochs classified according to emitted behaviour. In most of these epochs it was found that spectral peaks within the theta frequency range showed phase differences other than complete 180 degree reversal. Usually the phase difference was 135–150 degrees, with the deep source leading. In a later paper (Leung 1984b) theta rhythms at various frequencies were recorded in different stages of waking and REM sleep. It was observed that *phase difference between deep and superficial theta activity shifted from about 180 degrees for the highest frequencies to about 120 degrees for the lowest.* The tabulated evidence for REM sleep is reproduced as Table 11 (see also Leung 1984b, Figs. 4 and 6.) This evidence suggests that, in the free-moving rat, and in REM sleep, the relation between superficial and deep theta rhythms may be one of *constant delay*, rather than of constant phase angle. *Leung (1984c) estimates this delay as roughly 30–50 ms.*

Table 11. Radiatum theta phase as a function of frequency during REM (phase estimates for df > 18 only). (Leung 1984b). Compare table 10

Frequency (c/sec)	Rat's code						
	35	41	53L	53R	56	58	1153
5.47	128.2	128.7	157.0	167.7		104.4	
6.25	131.6	143.4	164.4	170.6	135.2	106.0	120.5
7.03	145.9	146.7	168.2	171.9	151.9	124.4	140.9
7.81	153.0	152.5	171.8	178.5	157.7	124.5	151.3
8.59		151.5	169.2	175.9	155.4	133.2	
9.38		139.5					
Range of coherence	0.9 – −0.97	0.89– −0.97	0.97– −0.99	0.96– −0.98	0.91– −0.98	0.74– −0.94	0.86– −0.92
Magnitude of change[a]	24/8	23.8	14.8	10.8	22.5	28.8	30.8

[a] Maximum-minimum phase.

This relationship turns out to be of critical importance to the theory developed in Chapter 9. In the case of Leung's evidence, the regular relation between theta frequency and the phase difference between deep and superficial generators was seen both in waking and during REM sleep. In the latter condition, however, theta power was considerably greater, and the phase differences, at comparable frequencies, were consistently higher (by 5–30 degrees, that is closer to complete reversal of phase) than in the waking state. Is there any other evidence for this relationship? Two other papers have suggested that there may be some special significance in the phase relation in different parts of the hippocampal formation in free moving animals: Brown (1968) made this claim for relations between the rhythms recorded in the dorsal hippocampus and the parahippocampal gyrus of free moving cats. Buzsaki *et al.* (1979b) made a similar claim for relations between deep and superficial generators in the hippocampus of free moving rats. Neither of them specifically defined the relationship to which Leung alluded. The single trace published by Buzsaki *et al.* (1979b) does not help in establishing this relationship. However, a recent article by Buzsaki (1985) gives support to the idea of a constant delay with evidence about the phase of preferred firing of hippocampal units in different fields. This is described in the next chapter.

Apart from these intriguing details about phase (which will be followed up later) detail of a different sort is provided by Buzsaki *et al.* (1986). They conducted a full current-source density analysis on theta activity of rats running in a wheel for food reward. The electrode was advanced in 82.5 μ steps, and averaged records of 64 theta waveforms were obtained at each position. Throughout the CA1/dentate gyrus region, potentials had a greater amplitude in this preparation than in urethanized preparations.

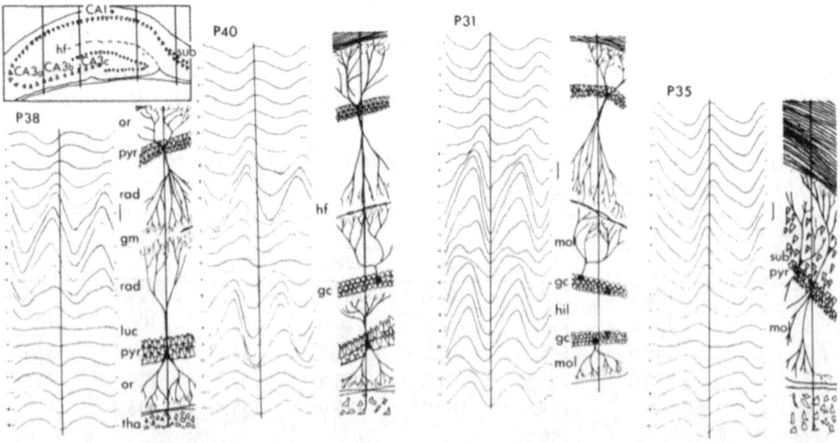

Fig. 30. Representative tracks through the hippocampus and dentate gyrus showing averaged theta waveforms locked to the phase of the theta rhythm recorded from a stationary reference electrode common to all points on the same track. Four different medio-lateral penetrations, at similar septo-temporal levels, were made in four different rats, running at similar speeds. Each *curve* represents the average of 64 different samples, each of 256 ms duration. The *inset* shows the four electrode tracks, and the position of the stationary electrode (*black dot*). *Vertical* lines represent the peak of the positive wave in the stationary electrode (trace not shown). Also shown are histological reconstructions of each track. Abbreviations : gc, granule cell layer; gm, stratum lacunosum-moleculare; hf, hippocampal fissure; hil, hilus; luc, stratum lucidum; mol, dentate molecular layer; or, stratum oriens; pyr, stratum pyramidale; rad, stratum radiatum; sub, subiculum; tha, thalamus. (Buzsaki *et al.* 1986, Fig. 1, enlarged) (From *Brain Research* Vol 365 pp. 125–137 1986)

In stratum radiatum, the theta wave was far from sinusoidal, having at some depths a slow negative-going phase and a more rapid positive-going phase, while near the amplitude minimum the slow phase was positive-going. The gradual phase shift was again confirmed. Between the pyramidal cell layer and the hippocampal fissure there was a region of minimum amplitude, but no null point, and as the electrode passed deeper, amplitude built up to a maximum at the fissure. At the same time phase gradually advanced to give a phase lead ranging from 130–180 degrees for the theta potentials recorded near the fissure compared with that in the pyramidal cell layer. This is depicted in Fig. 30 especially track P31.

On crossing the hippocampal fissure into the dentate gyrus, maximum amplitude was seen in the outer third of the molecular layer. In both the buried and exposed blades of the dentate gyrus a minimum was seen between the inner and middle thirds of the molecular layer. In the granule cell layer and hilus, theta activity was present (unlike the situation in a urethanized rat, where amplitude in the hilus is very low). Its detection depended on the averaging of records at each point in the track. In unaveraged records high frequency activity tended to obscure it, as described above. In phase analyses it appeared that this hilar theta activity was approximately 360 degrees phase-shifted compared with that in the molecular layer. However, this may be a misinterpretation, since the amplitude in the phase transition zone was low, and results were admitted as being only semi-quantitative. This study will be referred to again in discussing the sources and sinks for hippocampal theta activity.

Buzsaki *et al.*'s (1986) paper is also the only one to provide detailed evidence about theta activity in field CA3 of the free moving rat. In contrast to the urethanized preparation, it appears that CA3 *does* generate theta fields in the free-moving rat. This conclusion is based on the fact that the amplitude and phase gradients may be quite sharp in CA3, and that cells in CA3 may show phase-locking to local theta. The profile in detail is revealed by Buzsaki *et al.*'s Fig. 1, showing track P38, passing vertically down through CA1 or CA2, across the region of apposition of the distal dendrites of CA2 and CA3, and on through the depth of CA3 (reproduced as Fig. 30). In stratum oriens of CA2, there is low amplitude theta activity. On passing deep to the pyramidal cells and through stratum radiatum, there is a gradual phase change with no null point, as in CA1. In the stratum oriens and the pyramidal cell layer of CA3 there are also low amplitude theta potentials, and superficial to this there is a null point in stratum lucidum, and a reversal of sign. The region of the amplitude maximum is between the two regions just mentioned. This profile appears to be evidence of a double dipole similar to that found in tracks passing from CA1 into the dentate gyrus. However, Buzsaki *et al.* do not present any analysis of the relative phase between superficial and deep waveforms in this track.

Two of the above papers comment on behavioural correlates of the laminar phase profile for theta potentials. Leung *et al.* (1982) find no major difference in phase relationship between the two sources amongst the following categories of behavioural epochs: walking/rearing; grooming; awake immobility; REM sleep. In this study it should be noted that the similarity in phase relations amongst the different types of behaviour does not imply similarity in power or frequency, as discussed in Chapter 4. In fact four aspects of the EEG analysis—peak frequency, resonance (reciprocal of bandwidth), dorsoventral coherence at theta frequency, and number of harmonics—all increased in a similar way over a series of behaviour types. This series was from slow wave sleep to awake-immobile, to grooming, to walking to REM sleep. Buzsaki *et al.* (1985) obtained laminar theta profiles during both running and lever pressing. The gradual phase change across the depth of stratum radiatum of CA1 occurred in both profiles, but there was a more rapid phase change across the lamina in the case of lever pressing.

The detection of harmonics in both CA1 and the dentate gyrus in the study of Leung *et al.* (1982) (up to the third was seen) is a significant fact, not evident unless spectral analyses are conducted. A similar observation, though using a different technique, was made in Buzsaki *et al.*'s (1986) paper: In the region of minimum amplitude in the dentate gyrus the averaged waveforms often had a subsidiary peak bisecting the normal theta period, as if it were the first harmonic of the main frequency. Reference will be made to such data when developing a model of cortico-hippocampal interaction.

7.3.1.5 Laminar Profiles in Other States

Leung (1984a) has presented evidence on a number of drug states in which the phase shift between superficial and deep sources of theta activity is intermediate between that of the urethanized rat (about 180 degrees) and the free moving rat (120–150 degrees). In order of decreasing phase shift, these drug states are: immobility after eserine; walking after phencyclidine; walking after eserine; walking after reserpine; walking after pentobarbitone. A phase shift *less* than that in the undrugged free moving animal was found in rats after atropine or scopolamine. Buzsaki *et al.* (1986) also present laminar profiles in the atropinized rat. The profiles were similar to those of the free moving undrugged animal, except that power at all parts of the vertical track was reduced, particularly deep to the hippocampal fissure. In the stratum radiatum of CA1, the gradual phase change occurred somewhat more rapidly over space than in the free moving animal. The decreased phase shift in the atropinized rat compared with the undrugged one, reported by Leung, was not seen in this study.

7.3.1.6 Conclusions

From the above review, it is clear that there are multiple sources of theta-modulated potentials in the hippocampus. A detailed analysis of these data will be discussed in Chapter 8. Without pre-empting that discussion, there is already evidence that the different sources can sometimes vary independently, though deep and superficial sources do seem usually to be closely linked by either constant phase reversals or delays. The most prominent sources of theta activity are in field CA1 and the dentate molecular layer, the latter having the larger amplitude. In field CA3 there is no independent generation of theta activity in the urethanized rat, though there is in the free moving rat. In the hilus of the dentate gyrus theta activity is present in the free moving rat, but probably not in the urethanized rat. The source of theta activity in the dentate gyrus seems to be more robust and invariable, both in anaesthetized and free moving animals, and there is a little evidence to suggest that theta activity there is less easily initiated, and once initiated, is less easily silenced in the dentate gyrus. In both atropinized and urethanized rats, amplitude of theta potentials is reduced compared to the free-moving animal, the deep source being especially affected under atropine. For both superficial and deep sources of theta rhythms, activity at the frequencies of the lower harmonics has been reported by several groups, suggesting that some form of resonance is involved in either the initiation or maintenance of theta activity.

In the anaesthetized rat the deep source of theta activity is phase reversed compared with the superficial source. In the free moving or atropinized rat this is approximately so, but a more accurate description is that theta activity in the dentate gyrus has a constant time-lead over that in CA1, so that phase angle differences vary according to theta frequency. Taken together with the fact that field CA3 (and apparently the dentate hilus also, at least in rats) generate theta activity only in the free moving rat, one reaches the following tentative conclusion: *that the transverse pathway from dentate gyrus to field CA1 becomes more efficacious in the free moving rat than in the anaesthetized one, so that the dentate gyrus comes to control theta rhythms in CA1*. Thus, at the time when hippocampus and isocortex show some interdependence in their rhythmic activity (as indicated by the lesion experiments described in Chapter 6), pathways through the hippocampus are "opened" as well. This idea is discussed more fully in the next chapter, and is of central importance to the concept of resonance, hinted at here by the evidence of "harmonics" of the basic rhythm, and advanced more strongly in Chapter 10.

It also appears that the results in free moving rabbit, and the curarized rat are both more like the urethanized rat than the free moving rat. This is an untidy conclusion to reach. However this conclusion is based on laminar analysis in only one paper (Winson 1976b). In this paper no details are given of the precise phase relation between theta waves recorded at different depths, there being merely a general statement that there was a null point and a sharp reversal of phase. It would therefore seem important that analyses be performed for the free moving rabbit similar to those carried out by Leung and by Buzsaki's group in free-moving rats.

7.3.2 Intrahippocampal Lesions

There are several reports of the effects on theta activity of small intrahippocampal lesions. The lesions have been made in a variety of ways: electrolytically (Porter *et al.* 1964; Chronister *et al.* 1974); using focal irradiation in immature rats, which adversely effects the still-developing dentate gyrus (Whishaw *et al.* 1978); using kainic acid lesions (Whishaw and Sutherland, 1982); or colchicine (Monmaur and Thompson, 1985; 1986). When the dentate gyrus is damaged, abnormalities in theta profiles are remarkably small, a finding which suggests that theta potentials can be accounted for mainly on the basis of a generator in CA1. However, this strong conclusion may be unwarranted in view of other evidence already cited (such as the studies with slice preparations). Theoretical analyses of laminar phase profiles (to be considered in Chap. 8) suggest that the dentate gyrus plays an active part in generation of theta potentials. Lesions restricted to CA3/CA4 do not cause obvious disruption of the profile of theta potentials recorded from the dentate gyrus or CA1 in anaesthetized rats, but subtle changes of phase relationships in the free-moving rat have not been ruled out. High frequency activity in the dentate hilus seems to survive lesions of the dentate gyrus. Lesions confined to the superficial or deep generators do not appear to differentially block the atropine-sensitive or the atropine-resistant components of theta activity.

7.4 Theta Potentials Outside the Hippocampal Complex

7.4.1 Theta Activity in the Entorhinal Cortex

A key issue for studies of theta rhythms outside the hippocampus is whether these rhythms are generated locally, or are volume conducted from CA1 or from the dentate gyrus. A number of experimental observations provide evidence for theta activity generated in the entorhinal cortex. Holmes and Adey (1960), recording from the entorhinal cortex alone, describe theta activity in cats performing a delay task, and give some detail of its precise behavioural correlates. Bennett (1970), also using behaving cats, recorded theta activity in both hippocampus and entorhinal cortex. Theta activity generally occurred in similar situations in the two sites. In both these papers there were insufficient data on laminar profiles and phase relations to rule out volume conduction to the entorhinal cortex from the hippocampus proper. However, two early papers from Adey's laboratory provide some indication of separate generation of theta activity in the entorhinal cortex. Adey *et al.* (1960) recorded from cats, during approach learning. Theta activity was recorded in both entorhinal cortex and hippocampus. The phase relation between the two was observed to change during the course of training. This data will be considered in more detail in Chapter 10. Adey *et al.* (1961) and Adey and

Walter (1963) recorded from cats during T-maze discrimination learning: there were phase differences between theta activity in hippocampus and entorhinal cortex, which varied systematically according to whether the animal performed correctly or incorrectly on that trial. These interesting findings will also be discussed further in Chapter 10. In the present context, these results indicate that the entorhinal cortex probably has an independent capacity to generate theta activity.

As mentioned above, Brown (1968) recorded theta activity simultaneously in the dorsal hippocampus and entorhinal cortex of free moving cats. There was a suggestion of some special significance in the phase relation between the two rhythms. In her Fig. 5, she presents the filtered EEG records from the dorsal hippocampus and parahippocampal gyrus, recorded simultaneously in a free moving animal together with the instantaneous frequency, and phase difference between the upper two traces. These data are reproduced below in Fig. 31/1, with further graphical representation of the same data in Fig. 31/2 A and B. In Fig. 31/2A it appears that the phase advance of the parahippocampal gyrus with respect to the dorsal hippocampus is greater at higher frequencies than at lower ones. In Fig. 31/2B the relationship is more complex but the distribution of phase and frequency is again not random. Unfortunately there are not enough data of good quality in the paper by Brown to submit to a statistical analysis. In addition, the anatomical sites from which the data is derived is not specified precisely by Brown, though it is different from that of Leung. These data are discussed again in Chapter 9, where they assume further significance.

Three more extensive studies of theta activity in the entorhinal cortex have been reported recently. Mitchell and Ranck (1980) concentrated on the medial entorhinal cortex of free-moving rats. Alonso and Garcia-Austt (1987a, b) examined most of the entorhinal cortex of paralysed rats. Both pay attention to laminar phase profiles. In Mitchell and Ranck's study, in laminae III and IV, theta activity had a similar shape and phase to that in CA1, with an amplitude maximum in lamina III. Superficial to this there was a null zone, varying in precise location amongst animals between the outer third of lamina III and inner half of lamina I. The deep part of lamina II was the most accurately localized reversal point. In laminae I and II an approximately phase-reversed theta rhythm was recorded. The precise phase relations were as follows: the theta activity recorded in the deeper layers of entorhinal cortex was about in antiphase with dentate gyrus theta potentials, usually leading the inverted dentate theta by 13–43 degrees (5–17 ms). In other words, this deep entorhinal theta rhythm was 137–167 degrees ahead of the dentate gyrus (or equivalently 193–223 degrees behind the dentate gyrus. In one out of four cases, however, it showed a small phase lag (3.5 ms). The superficial source of entorhinal cortex theta potentials was approximately in phase with dentate theta. The phase relation between deep and superficial entorhinal theta amounted to a phase lead by the superficial source of 120–167 degrees.

Alonso and Garcia-Austt (1987a, b) used paralyzed rats in which theta activity was induced with physostigmine. In this preparation laminae V and VI, in all parts of the entorhinal cortex, also showed a theta rhythm approximately in phase with that in stratum oriens of CA1. An amplitude minimum was seen between the outer third of lamina III and the inner half of lamina I. In laminae I and II theta activity was approximately phase-reversed compared with that in deeper laminae. In some regions of the entorhinal cortex, the superficial theta activity was less than 180 degrees ahead of the deep theta activity,

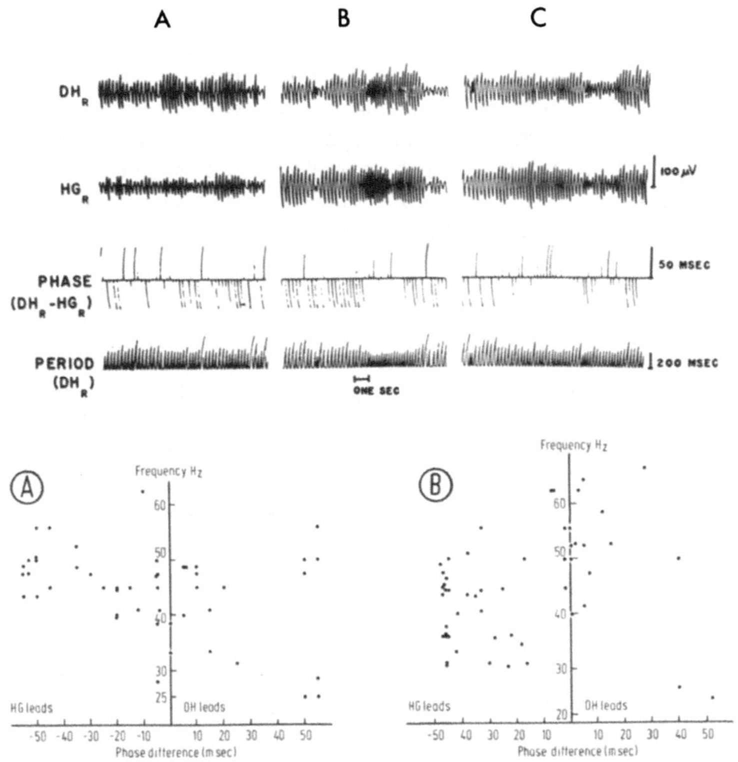

Fig. 31. 1: Data reproduced from Brown (1968 Fig. 5), showing EEG tracing (after passing through 3–7 Hz filter) from the dorsal hippocampus (DH′) and (para) hippocampal gyrus (HG′) of a freemoving cat, as well as the period (ms) for the dorsal hippocampus and relative phase difference (ms) between the two rhythms (DH′-HG′). Upward deflection of phase difference trace indicate a phase lead by the dorsal hippocampus. **A** During orienting **B, C** during REM sleep. (From *Electroencephalography and Clinical Neurophysiology* Vol 24 pp. 53–62 1968) 2 **A** and **B** Plot of the relation between instantaneous frequency (reciprocal of period) as the ordinate and phase difference as the abscissa, derived from the data presented in **1A** and **1B**

and in these cases, gradual phase shift, such as is found in CA1 of free moving rats, was illustrated. In the second of these two papers, details are given of the phase relationship of unit discharges in the entorhinal cortex to the local theta activity. Two types of cells with a rhythmic autocorrelogram were seen. One had low frequency discharge (less than 1 impulse per cycle), and a multimodal interval histogram. The other had a higher mean discharge frequency and a unimodal interval histogram. These types of cells fired maximally during the negative phase of CA1 (stratum oriens) theta, and were mainly located in lamina II of the entorhinal cortex. A further type of cell had a non-rhythmic autocorrelogram, but still fired in consistent phase relation to local theta, though its discharge frequency was low. Such cells fired maximally during the positive phase of CA1 (stratum oriens) theta.

The interpretation of these findings will be considered in Chapter 9. At present, however, it is safe to conclude that theta potentials recorded in the entorhinal cortex are generated there, at least partly, (rather than volume conducted from the hippocampus), since there is locking of unit firing to the theta potentials recorded in the entorhinal cortex. In addition, in the free moving animal the relationship between theta potentials in the entorhinal cortex and those in the hippocampal complex are not accurate phase reversal. It is possible that there may be a regular delay between the two sites, which stays constant as frequency changes, similar to that found between deep and superficial generators in the hippocampal complex itself. However the data available at present do not definitely show this.

7.4.2 Retrosplenial Cortex

In an early study Petsche and Stumpf (1960), using curarized rabbits, found that theta activity in the cortex overlying the hippocampus showed mainly a complete phase reversal with respect to that within the hippocampus, and the cortical theta waves correlated on a beat-by-beat basis with the hippocampal theta rhythm. These data could be explained in terms of volume conduction. In addition, however, they found that in a region of medial and basal cortex before transition to the hippocampus (viz: in area retrosplenialis) phase shift was gradual, rather than an abrupt reversal. This suggests that there may have been independent sources of theta potentials in this region.

7.4.3 Theta Activity in the Cingulate Cortex

A number of papers have investigated theta activity in the cingulate cortex. Feenstra and Holsheimer (1979), using urethanized rats, recorded theta waves which sometimes showed a phase reversal in lamina V and VI of the posterior cingulate cortex. Holsheimer (1982), in the same preparation, recorded from units in the cingulate cortex whose firing was phase locked to local or to hippocampal theta rhythm. The precise phase lead or lag of the cingulate units or EEG, with respect to the hippocampal theta rhythm is not given in detail. One unit is illustrated however with 147 degrees difference from hippocampal CA1 theta, a phase difference other than simple reversal of a potential originating in the hippocampus. Holsheimer's results clearly indicate theta generation within the cingulate cortex itself. He attempts to explain this finding in terms of a direct projection from septum to cingulate cortex, by which theta-modulated signals are transmitted. However, the anatomical support for this, a paper by Swanson and Cowan (1979), is inadequate, because it refers to septal projections limited to the very anterior part of the cingulate cortex, whereas Holsheimer's recording was obtained from the posterior part. Colom *et al.* (1988) have also studied theta activity in the cingulate cortex of urethane anaesthetized rats. No reversal of sign was seen across the laminae of the cingulate cortex. Of units recorded in the cingulate cortex 80% were of non-rhythmic type, and only four cells (9%) showed a rhythm at the frequency of the hippocampal rhythm. These authors therefore conclude that in the urethane anaesthetized rat, no theta potentials are generated in the cingulate cortex, a conclusion in conflict with that of Holsheimer.

Two recent papers have studied theta activity in the cingulate cortex of free moving rats (Leung and Borst 1987; Borst *et al.* 1987). The first of these papers reports on the behavioural correlates of cingulate theta rhythms. These rhythmic potentials accompany walking, rearing, postural shifts, head movements and rapid eye movement sleep, as do the hippocampal rhythms. During grooming, eating and drinking, however, theta activity is replaced by irregular slow activity. This is different from the hippocampal rhythm, at least for grooming, when low frequency theta activity is seen (see Chap. 4). In this study there was no evidence of phase reversal across the laminae of the cingulate cortex, and the authors believed that cingulate theta activity was at least partly accounted for by volume conduction from the hippocampus. However, local mechanisms were also thought to contribute, since there were small unit spikes in the area of recording whose firing was phase-locked to local theta activity. Phase relation to local theta activity varied amongst units, though little detail on this is presented. The second of these two papers presents further clear evidence that mechanisms local to the cingulate gyrus are involved. Lesions of the medial septum/vertical limb of diagonal band usually abolished or decreased theta activity in both the hippocampus and the cingulate cortex. However, in some cases (five rats), the cingulate theta activity remained, or was even increased in amplitude, with many negative spikes appearing during the peak negativity, while hippocampal theta activity was abolished. Other details provided by this paper concern the effect of muscarinic drugs. Atropine increased the irregular slow activity in the cingulate cortex, while pilocarpine induced a regular 4–7 Hz theta rhythm there. These results are similar to the effects these drugs have on the hippocampus. The authors consider the pathways by which pacemaker signals may reach the cingulate gyrus, including both septo-cingulate and hippocampo-cingulate paths in their discussion. Concerning the former, it was found that the loss of cholinesterase staining in the cingulate cortex after lesions of the septal region did not always predict a reduction of theta activity in the cingulate cortex. However, cholinesterase staining may be the wrong marker for these septo-cingulate pathways, since the posterior cingulate gyrus contains very little cholineacetyl transferase, at least in lamina III (Lysakowski *et al.* 1986). In any event, there is no evidence which forces the conclusion that septo-cingulate fibres carry the pacemaker signals to the cingulate gyrus. The alternative route, from the hippocampus, is not discussed in detail.

In summary, it appears that theta activity can be generated in the cingulate cortex, both in anaesthetized and in free-moving rats, but this is not invariably seen. Theta activity in the cingulate cortex occurs during running and similar behaviours, as it does in the hippocampus, while during grooming it occurs only in the hippocampus.

7.4.4 Theta Activity in the Prefrontal Cortex

As mentioned above, theta activity recorded from the hippocampus has been difficult to find in human subjects, and this may be mainly because of the technical/ethical difficulties of human central electrophysiology. However, there is some recent evidence that the *midline prefrontal region* of the cortex can generate theta activity, in certain cognitive states. This was reported by Mizuki *et al.* (1980), EEG rhythms of 5–5.5 Hz frequency appearing with some regularity during the performance of simple repetitive mental arithmetic tasks. In later short reports, some pharmacological effects on this frontal midline theta activity were also described (Mizuki *et al.* 1981; Mizuki *et al.* 1982). It has also been recently reported that such frontal theta activity is even more common during REM sleep (Hayashi *et al.* 1986). In addition Lang *et al.* (1987), using spectral analysis of frontal EEGs, showed that theta frequencies (3–7 Hz) were increased during motor or verbal learning tasks. All these data are compatible with the thesis that theta activity in frontal regions is associated with a theta activity in the hippocampus. It might be expected that the prefrontal cortex should show theta activity at the same time as the hippocampus, in view of the strong conections between the two structures (see Chap. 3). If such a relation could be established, one would have a method of determining the behavioural correlates of at least one aspect of hippocampal theta activity in man, which would be of general importance. As yet, however, these studies have not provided any definite information that frontal theta activity is related to that in the hippocampus, except in so far as it tends to occur at the times when hippocampal theta activity might be expected.

7.4.5 Theta Activity in the Neocortex Generally

A variety of reports indicate that, apart from cortical areas such as the cingulate and prefrontal cortex, which have direct links with the hippocampus, the neocortex more generally may carry potentials within the theta frequency range (Anokhin 1961; Parmeggiani 1967; Yamaguchi *et al.* 1967; Landfield *et al.* 1972). In some of the illustrated examples, concomitant hippocampal activity was also recorded, but the waveform and amplitude was not sufficiently different between cortical and hippocampal sites to rule out volume conduction as the source of the cortical theta activity. Gerbrandt *et al.* (1978) showed, in curarized rats, that volume conduction could be an adequate explanation of theta activity recorded in the cortex. Theta activity was recorded in many areas of the cortex using epidural or depth electrodes, but no phase reversals or locking of unit activity to the theta trace could be detected. More decisively, cessation of all neural activity in the cortex, by elicitation of cortical spreading depression, did not abolish the theta potentials. The consensus from this evidence was therefore that theta potentials in the isocortex are volume conducted from the hippocampus.

Nevertheless, the matter should not be considered settled. For instance, in the paper of Anokhin (1961) two of the illustrations (their Figs. 16 and 17) show cortical theta activity occurring when the hippocampus has either no theta activity or theta activity at a much lower amplitude than that recorded in the cortex. In addition, several studies have detected cortical theta activity which is specific to particular behavioural states or particular stages of acquisition of learned tasks. Pickenhain and Klingberg (1967) in rats showing active or passive motivated behaviour, found that theta activity could appear in the visual cortex (but not the motor cortex) at the same time as the hippocampus was showing theta activity. Typically hippocampal theta activity occurred only after the first few presentations of the CS, and visual cortical theta activity did not then start until one or two cycles after it started in the hippocampus, and at the same frequency (see their Figs. 3 and 5). Moreover (their Fig. 5) the amplitude of the theta activity in the visual cortex was not congruent with that occurring at the same time in the hippocampus. McFarland *et al.* (1975) showed, also in rats, that hippocampal theta activity in the 6–8 Hz band progressively increased in amplitude as speed of running on a runway increased. This activity could also be detected by electrodes positioned in the occipital cortex, but this occurrence was limited to the highest runway speeds. It should be pointed out that in rat the occipital (including visual) cortex is an area directly connected to the cingulate cortex (see Chap. 3). Elazar and Adey (1967b), using cats, acquiring a light/dark discrimination task, found, during an advanced stage of training that there was coherence at theta frequency between the hippocampus and visual cortex (their Fig. 4). Efremova and Trush (1973), working with chronically prepared rabbits, carried out spectral analyses of EEG from visual, auditory and motor cortices, during elaboration of a conditioned defensive reflex. At about the time of first appearance of the conditioned response, and particularly at the time of response to the CS, spectra of all three cortical areas became less variable and contained a narrow band peak in the theta range, with increased coherence between areas. Aleksanov *et al.* (1986) found in dogs undergoing alimentary conditioning, that CS presentation caused an increase in amplitude of theta activity recorded in the sensorimotor cortex, along with a general increase in frequency (by 1.1–2.2 Hz) of theta recorded from the hippocampus. These results of isocortical theta activity occurring only during some of the epochs of hippocampal theta activity suggest that the isocortex may be sometimes capable of generating theta activity.

8 Towards a Neuronal Model to Account for the Laminar Waveforms of Theta Activity

8.1 Introduction

The data presented so far, on the pacemaker and on the control and laminar waveforms of theta rhythms in the various structures from which they can be recorded, are quite complex. The task of theory building is in part to reduce complex data to simpler relationships which account for as much as possible of the data. One way of reducing the data presented is to consider the possible relationships of neurons which best account for the evidence considered above, in other words to construct a neuronal model of the hippocampus during generation of the theta rhythm. This is a major objective of the present chapter (which will not be achieved competely). In part this involves a deeper discussion of the laminar profiles of theta potentials described above. However, a variety of other evidence is drawn upon, which is also relevant to building a neuronal model. A related aspect of this model building, also dealt with below, is the relationship between theta potentials recordable in the hippocampus, and those recordable in isocortical regions, based on evidence reviewed above. A second way of bringing parsimony to the data has been hinted at several times already. In both Chapters 6 (Sect. 6.3.2.2) and 7 (Sect. 7.3.1.6) we obtained suggestions of a promising relationship: Theta rhythms, in some circumstances at least, may involve resonant circuits between the hippocampal formation where the rhythms originate, and neocortical loci which may be entrained to the hippocampal rhythm. However, the evidence considered so far does not *compel* this interpretation of the data. Therefore, to evaluate this hypothesis, it must be tested more critically against all available data, including that discussed in this chapter. Towards the end of the chapter, more substantial support is obtained for the concept of resonant circuitry being involved in theta activity. In fact, some of the necessary conditions for completion of such resonant circuits will be defined. In the first instance the evidence concerns the completion of circuits *within the hippocampal formation*. However, more hypothetically, the same sort of reasoning is applicable to other parts of the potential resonant circuits, in the links between hippocampus and isocortex. This opens the way for more definite theory formulation in Part 3 of this book.

Construction (or partial construction) of a neuronal model of the hippocampus during theta activity is constrained by many lines of argument. The crucial ones considered below are: 1. Methods of neuronal modelling used to analyze laminar profiles of theta potentials (discussed in Sect. 8.2.1). 2. Evidence of phase relations between hippocampal RSA and firing of principal neurons and interneurons in various regions of the hippocampus. 3. Intracellular recording studies. 4. The biophysics of the action of relevant transmitters. Other lines of evidence (such as the phase variation of evoked potenitals, or the phase relations between septal electrical activity and unit firing in the various types of hippocampal neuron) will not be covered here, because evidence available so far is subject to difficulties of interpretation which limit its value.

These methods must then be brought to bear on at least two experimental situations: that of the urethane anaesthetized rat, and that of the free moving rat. Since the urethane anaesthetized rat has a simpler laminar profile, it will be considered first (Sect. 8.3.1), then the free-moving rat (8.3.2), and finally other experimental preparations (8.3.3), on which less information is available. In concluding (Sect. 8.4.) brief reference will also be made to the relationship between theta potentials in the hippocampus and those in the isocortex. The incorporation of these data into a fully fledged theory of hippocampo-cortical relations is, however, a task for Chapter 9.

8.2 Technical Matters

8.2.1 Theoretical Analyses of Laminar Phase Profiles

Only certain combinations of dendritic location, sign and phase relation of active inputs can explain the observed laminar phase profiles. The relations between these aspects of the input to the potential fields generated are by no means simple. However, theoretical techniques have been devised which make these issues tractable. Assuming that EEG potentials are reflections of the postsynaptic potentials generated intracellularly (Creutzfeldt and Houchin, 1974), one need consider only the passive properties of neuronal membranes rather than the action potentials. Two methods are available. In one, the complex branching pattern of dendrites is approximated to a "cable equivalent", whose mathematical description is well established (Rall 1962; Jack *et al.* 1975). Another method is to treat a neuron as consisting of a number of isopotential regions, each resistively coupled to its neighbours (Perkel *et al.* 1981). This is mathematically simpler than the cable theory approach. It is the second of these two methods that has been used in simulations of the hippocampal theta potentials, i.e., compartmental models using realistic parameters for individual hippocampal neurons.

An obvious aspect of these theoretical approaches is that when a synaptic input becomes active, the locally recorded extracellular potential is of opposite sign to the intracellular one, since current is flowing from one to the other. A less obvious and also important conclusion is that the extracellular potentials recorded at a distance along the dendritic membrane from the active synapses are of the opposite sign to those recorded locally. This conclusion arises from the fact that when current flows in at an active excitatory synapse, it charges the membrane capacitance in the membrane within a short distance of that synapse, leading to intracellular depolarization and extracellular negativity. To charge this capacitance it requires that on the extracellular side of this membrane current flows away. Some of this flows to more remote regions of the dendrite, which produces an extracellular positivity. Similar but opposite potential changes occur for active inhibitory synapses.

As far as extracellular fields are concerned, the above methods apply only to neural networks with a high degree of organization, with dendrites tending to run in parallel. This is because in fields of dendrites whose orientation is at random, the field potentials generated by different dendrites would cancel each other out. The mathematical methods for deriving field potentials from intracellular potentials in such ordered arrays of dendrites have been described in several papers (Nicholson and Freeman 1975; Mitzdorf and Singer, 1978; Rappelsburger *et al.* 1981). Combining these methods with compartmental modelling, it has been possible to simulate the fields arising from various configurations of active inputs to the hippocampus. From these can be derived descriptions of the laminar profiles of theta activity for various types of rhythmic synaptic drive.

8.2.2 Identification of Interneurons and Projection Neurons

Before discussing the relation between rhythmic impulse bursts in hippocampal neurons and the concurrent electrographic theta rhythm, it is necessary to describe earlier single unit studies in the hippocampus, which have produced various criteria for identification of hippocampal projection cells and interneurons.

Ranck (1973), using unrestrained rats, found that hippocampal and dentate gyrus units fell into two distinct groups, on the basis of their firing pattern. One group of cells fired with single spikes, whose temporal width was 0.15–0.25 ms, and whose frequency was always above 5/s with maximal rates up to 30–120/s. These cells increased their firing rate when the hippocampus was generating theta activity, whether during wakefulness or paradoxical sleep, but fired at a lower rate during slow-wave sleep, or "automatic" behaviours. Ranck gave them the name "theta units"[7]. The other type of neuron (more than 75% of the total units recorded) always produced at least some "complex spike" groups. These are characteristic groups of high-frequency spikes, often with consistent inter-spike intervals, spike amplitude tending to decline through the sequence of spikes, with the group as a whole tending to occur in an all-or-none fashion (see Fig. 32). Individual spikes had a temporal width of 0.3–0.5 ms, and the firing rate was always less than 12/s and usually below 2/s. These cells had complex and individual behavioural correlations which will be described in Chapter 9 (Sect. 9.3.2.1 and 9.3.2.2). Since the theta cells had a higher discharge frequency, they were more easily detected, and so, Ranck suspected, were overrepresented in the total population of recorded cells. In Ranck's classification, the production of spike activity which was rhythmic at the theta frequency was not a distinguishing criterion: the firing of both types could have a phase relation consistent to concurrent theta activity, though this was more obvious for the theta cells, because of their higher firing rate. For both groups, Ranck comments that phase relation could vary from cell to cell, though details of this were not given. In a second paper (Feder and Ranck 1973), it was shown that the behavioural correlates of the theta cells were similar to those of the theta rhythm itself.

Two years later (Fox and Ranck 1975), it was reported that the distribution of the two types throughout the laminae of the hippocampus was different. Theta cells were found widely distributed, both within and outside the layers of principle projection neurons. On the other hand, complex spike cells were confined to these cell layers (i.e., stratum pyramidale of the CA fields, and stratum granulosum of the dentate gyrus), or the regions of their most proximal dendrites. On the basis of these results it was

[7]The realization that there is more than one sort of theta activity has necessitated some modifications to Ranck's original classification, especially in rabbits (Sinclair *et al.* 1982). The two types of theta potentials are generated by the same populations of neurons in the hippocampus. However, in type 2 behaviour in rabbits, cells fulfilling most of Ranck's criteria for theta cells failed to increase discharge rate during theta activity.

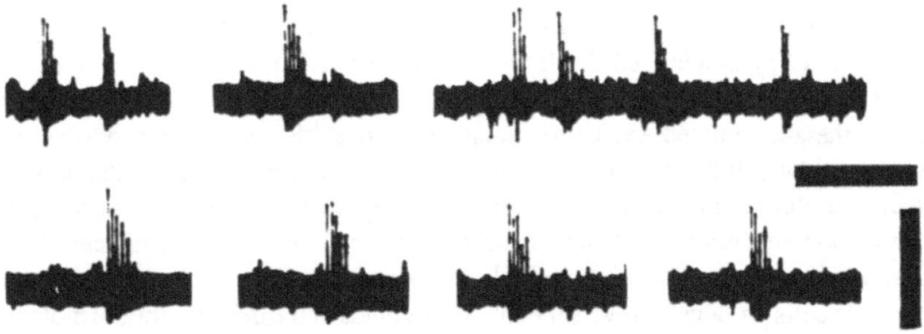

Fig. 32. Neuronal complex spike responses. Calibration: 360 μV; 0.1 s. (Ranck 1973, Fig. 1) (From *Experimental Neurology* Vol 41 pp. 461–531 1973)

suggested that complex spike cells are the projection neurons (pyramidal and granule cells) while theta cells are mainly the interneurons (local circuit neurons). This view was supported (Fox and Ranck 1981) when it was shown that complex spike neurons located in CA3 could usually be activated antidromically from distant stimulation sites, whereas for theta cells only 1/25 neurons could be so activated. Morphological support for the identification of theta cells with interneurons has been provided by Schwartzkroin and Mathers (1978), Schwartzkroin and Kunkel (1985), Lacaille *et al.* (1987), Misgeld and Frotscher (1986) and Lacaille and Schwartzkroin (1988a, b). These studies have involved impaling and labelling neurons in hippocampal slices, a preparation in which the full range of Ranck's criteria cannot be applied. Nevertheless, the identification of theta cells with non-pyramidal cells still seems largely correct.

In other studies, criteria other than those of Ranck have also proved useful in identifying neuronal types. Pattern of spikes, their latency, relation to wave responses, and laminar location of responses to afferent stimuli or antidromic activation have been used in several studies. However, the details are not relevant here.

8.3 Towards a Neuronal Model for Each Preparation

8.3.1 Urethane Anaesthetised Rat

8.3.1.1 A Theoretical Analysis of Laminar Phase Profiles in Urethanized Rat

In order to clarify the origin of the laminar phase profile in urethane anaesthetized rats Holsheimer *et al.* (1982) constructed a set of compartmental models of the hippocampus, with representation of the CA1 pyramidal cells, the dentate granule cells, and the hilar cells of the dentate gyrus. The only part of the hilar cells modelled in this study was the part of their dendritic trees which extended into the dentate molecular layer, because it was only here that their dendrites lie in parallel and can generate significant extracellular fields. Various combinations of theta-modulated excitatory synaptic influences were used as inputs to the model. The cellular membrane potentials were a key dependent variable, from which it was also possible to calculate the corresponding field potentials. No attempt was made to explore the fields possible with inhibitory inputs. The principle aim was to discover the type of input from which field potentials similar to those in the curarized or urethanized rat could be generated. The more complex laminar profile seen in free moving rats was given less attention.

It was found that a *single* population of theta-modulated excitatory inputs anywhere along the CA1 pyramidal cell dendrites, or anywhere along the granule cell dendrites, was incapable of generating the observed null point and phase reversal. Admittedly, a single theta-modulated excitatory input directed at the proximal dendrites of the CA1 pyramidal cell layer produced a reversal of the sign of the field potentials in the proximal stratum radiatum (corresponding to experimental fact); but it also produced an amplitude maximum well above the hippocampal fissure (in disagreement with experimental results). Likewise, a single theta-modulated excitatory input directed at the granule cells produced a reversal point for the fields in the dentate stratum moleculare (observed occasionally), an amplitude maximum at the fissure, but no reversal of sign in stratum radiatum. However, features corresponding to experimental findings in urethane-anaesthetized rats were simulable if synchronous excitatory inputs were delivered to both neuron populations simultaneously. This could be done with two alternative laminar arrangements of active inputs: activation at *either* at the proximal dendrites of both pyramidal cells and granule cells, *or* the distal dendrites of the two cells types (i.e., adjoining the hippocampal fissure on each side) could equally well acount for the observed laminar profile. One should, of course, add to these two the further possibility that theta-modulated IPSP's play a part in generation of theta potentials. These IPSP's could come either from long-axon connections, or as a feed-forward of inhibition from interneurons which themselves receive a rhythmic excitation.

Holsheimer *et al.* (1982) make a few additional points. Both the above two alternatives led to the prediction that there should be a second reversal of sign in the dentate molecular layer. As detailed already, this has sometimes been observed, but the more usual finding (neglecting the obscuring effects of high-frequency hilar discharge) is that there is a gradual decline in theta amplitude as an electrode passes vertically through the dentate molecular layer to the granule cells, without phase reversal. The authors explain this by postulating a theta modulated input to the hilar cells, whose dendrites extend to the level of the proximal dendrites of the granule cells. However, the evidence reviewed in Sections 7.3.1.2 and 7.3.1.4 suggested that in urethanized rats, the hilus generates only very low-amplitude theta activity. Fortunately, there are other possible explanations of the usual finding of an absence of a phase reversal in the dentate gyrus (see below: Sects. 8.3.1.5 and 8.3.2.1). A further point should also be made clear. Holsheimer *et al.* base their simulation on the premise of *simultaneous* theta activity in the two input pathways. They suggest, but without going into detail, that the gradual phase shift observed in free moving rats might be explained by *non-synchronous* theta activity in the two pathways (see Sect. 8.3.2 below).

There are thus three possible configurations of rhythmic synaptic inputs which could account for the laminar phase profile in urethanized rat: a double proximal excitatory input, a double distal input or the third possibility, inhibitory input, to the principal cells of the hippocampal complex. It might be thought that the anatomical evidence reviewed in Chapter 3 would contain details allowing one to narrow down these possibilities. However, this is not so. In CA1 the septo-hippocampal input is the most likely pathway to carry the rhythmic signals which drive theta activity. Evidence on the laminae of termination of the cholinergic component of this pathway is compatible with either of the first two possibilities. Laminae of termination of non-cholinergic (putatively inhibitory GABAergic) fibres are quite unkown. There is a third possibility, namely that a theta-modulated septal input is relayed directly to interneurons in CA fields, which interneurons then can relay theta-modulated inhibition to the projection neurons in CA1. However, the synaptic terminals of different classes of interneurons are distributed in both stratum pyramidale (Schwartzkroin and Kunkel, 1985; Lacaille *et al.* 1987) and stratum lacunosum/moleculare (Misgeld and Frotscher 1986; Lacaille and Schwartzkroin 1988a, b). There are almost as many anatomical possibilities for the dentate gyrus. In particular, in addition to possible cholinergic and non-cholinergic septo-hippocampal projections, with different laminar terminations, there are, according to Lacaille and Schwartzkroin (1988a), axonal projections from interneurons in stratum lacunosum/moleculare of CA1 which cross the hippocampal fissure and terminate in the dentate molecular layer. *These patterns of termination do not unequivocally point to either of the alternatives* in the model of Holsheimer *et al.* (1982) using excitatory inputs, or the third one involving feed-forward inhibitory theta-modulated inputs. It is possible that in the dentate gyrus one may be dealing with different inputs altogether from those considered by Holsheimer *et al.*

8.3.1.2 Evidence Concerning Phase Relations Between Unit Firing and Local Theta Activity

Although one cannot define precisely the possible synaptic populations giving rise to theta waves on the basis of evidence presented so far, there is an important difference between the two situations in Holsheimer *et al.*'s simulation which may aid the analysis: in the case of the double proximal input the activation of action potentials by the excitatory inputs should occur at the *negative trough* of the extracellular theta potentials recorded at the level of the pyramidal cell somata, while in the case of the double distal input it should be at the *positive peak.* What is the verdict of the experimental literature on this point? Most of the evidence suggests that, in urethane anaesthetized rats, the pyramidal cells of CA1 and the granule cells of the dentate gyrus fire during the *negative* phase of their respective local theta rhythms. Such a finding was made by Bland *et al.* (1980), Fox *et al.* (1983), Buzsaki and Eidelberg (1983) and Fox *et al.* (1986). Bland *et al.* (1980) comment that the dentate granule cells were more tightly locked to the negative phase of the local theta potentials there, than were the CA1 pyramidal cells to their local theta activity.

A discrepant finding is that of Holsheimer *et al.* (1983), who found that 70% of pyramidal cells in CA1 discharged during the positive phase of local theta potentials. However, they also found, in agreement with the majority view, that all dentate gyrus granule cells discharged during the negative phase of local theta potentials. A likely explanation for the discrepancy lies in the criteria used for neuron identification. In Bland et al.'s (1980) paper, granule cells were identified by orthodromic activation from the perforant path, and CA1 and CA3 pyramidal cells were identified by antidromic activation from alveus and Schaffer collaterals respectively. Buzsaki and Eidelberg's (1983) projection cells in the two regions were identified by low spontaneous activity, wide spike and absence of repetitive discharge to commissural stimuli. Fox *et al.* (1983) identified projection cells in CA1 on the basis of wide spikes, and presence of complex spikes, and in the dentate gyrus if perforant path stimuli gave only a single discharge, synchronous with the population spike. Holsheimer *et al.* (1983) reported on cells recorded in the pyramidal cell layer or granule cell layer, merely demanding that they have a phase relation to the local theta wave, without obtaining details of duration of spikes or their burst-pattern, or of their antidromic or orthodromic activation. These are far less definite criteria than in the other three studies.

In field CA3, Bland *et al.* (1980) find that, for the few pyramidal cells studied, firing was only loosely coupled to the local theta potentials. This might be expected, because in the urethanized rat such potentials appear to be volume-conducted from other fields, rather than generated locally. On this topic another paper should be referred to briefly. Kuperstein *et al.* (1986) recorded from cells in CA2 and CA3 of rats, using a 24-channel microelectrode to obtain records from several cells simultaneously. Phase relations of unit firing to CA1 theta activity were assessed. They were found to vary considerably from cell to cell, although 70% of the cells recorded from fired "within 45 degrees of 247 degrees after peak CA1 positivity". The big drawback with this paper, despite its technical advantages, is that the animals were anaesthetized with pentobarbitone and given atropine in addition. As premedication the animals also received the tranquilizer acepromazine. With such a complex pharmacological status, it is difficult to compare this paper's results with any of those discussed hitherto.

Interneurons in the CA fields appear, according to most studies, to have a phase relation to local theta activity the reverse of that for the projection neurons. Fox *et al.* (1983), Buzsaki and Eidelberg (1983) and Fox *et al.* (1986) find that CA1 "theta cells" fired near the *positive* peak of the local theta waves. Bland *et al.* (1983), on the other hand, found that cells in CA fields tentatively identified as basket cells were only loosely locked to local theta waves. Likewise, Colom *et al.* (1987) found that the majority of 13 "theta cells" in CA1 fired maximally during the negative phase of the local theta waves. However, there was considerable variation between cells, including three cells which fired during the positive phase of local theta. This study may not be in serious conflict with the majority view because the criteria for selection of these cells was looser than in the other studies: the group of cells just described merely had to increase firing during RSA. With regard to interneurons in field CA1, it should be noted that none of the studies attempts to identify the class of interneurons whose cell bodies are located in stratum lacunosum/moleculare, as described in slice preparations by Lacaille and Schwartzkroin (1988 a,b).

In the dentate gyrus, presumed interneurons have been identified in only a single study: Buzsaki and Eidelberg (1983) found equal numbers of neurons in this class which fired during the positive and the negative phase of local theta waves.

In summary, evidence on the phase of unit firing in urethanized rats is compatible with the idea that CA1 pyramidal cells and dentate granule cells are activated by rhythmic EPSP's directed to the proximal parts of the dendrites. There may, however, be other possible locations of rhythmically active inputs, if IPSP's are taken into account. In fact the idea of feed-forward inhibition of rhythmic signals receives some support, at least in CA1, in view of the inverse phase relations of interneuronal firing compared with pyramidal cell firing. [There is also a variety of other evidence, unrelated to the phase of theta activity, which shows that the principal (pyramidal) cells of the CA fields decrease their firing under the circumstances when theta activity is prominent, and, inversely, that interneurons ("theta" cells) increase their firing under these circumstances. This is reviewed by Buzsaki *et al.* (1983). It is not dealt with in detail here because it has been obtained in a wide variety of preparations, whose equivalence to the urethanized rat is uncertain].

8.3.1.3 Relation of Intracellular Potentials to Theta EEG Activity

Intracellular studies contribute further evidence relating to the above issue, though with some difficulties of interpretation. Fox *et al.* (1983) recorded intracellularly from several pyramidal cells in CA1 and CA3 in urethanized rats, and found, during spontaneous theta activity, that the maximum hyperpolarization occurred just before the positive peak of dentate gyrus theta, i.e., just before the negative trough of CA1 (local) theta activity. Leung and Yim (1986) made similar findings in this preparation. In quantitative terms, negativity of the rhythmic membrane potential lagged behind the negativity of the extracellular theta rythm by about 260 degree (or in alternative terms was 100 degrees in phase advance over the extracellular EEG). A third paper (Nunez *et al.* 1987), using urethanized, curarized rats, showed for rhythmic potentials that the negative phase of the extracellular theta activity was always locked to the positive phase or the rising depolarization phase of the intracellular rhythm. These three results for pyramidal cells in CA fields are in rough agreement, as illustrated below (Fig. 33).

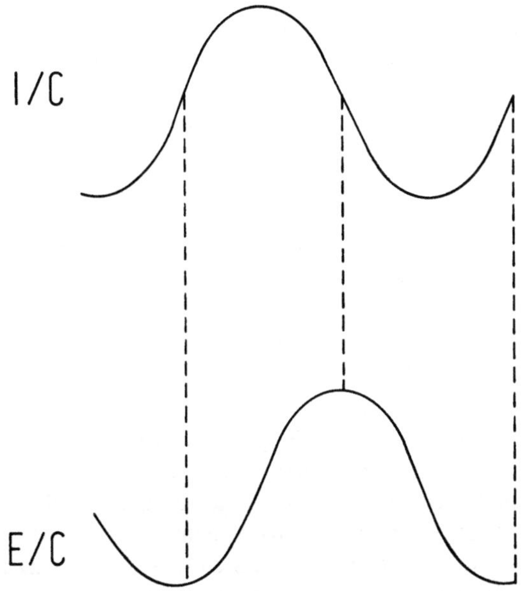

Fig. 33. Schematic drawing of the relative phases of sinusoidal potential waves, as recorded intracellularly (*above*) and extracellularly (*below*). Intracellular potentials are 90 degrees phase advanced over extracellular ones (or equivalently, intracellular depolarization lags extracellular negativity by 90 degrees)

The theoretical phase relation between intracellular and extracellular theta oscillations merits discussion. In theory, if currents flowing into or out of a neuronal element are the sole source of the local extracellular fields, the time of maximum extracellular negativity should coincide with the time of maximum inward flow of positively charged ions. In other words, extracellular negativity should be 90 degrees phase advanced compared with intracellular depolarization. This was, in fact, the observation of all three of the above-mentioned studies, allowing for some experimental error. However, this prediction from theory may not be rigidly true. There may be a contribution of interneuronal EPSP's to the field potential. [This is plausible in view of Shaban's (1969) finding that with sciatic nerve stimulation an early negative phase of the EP coincided with firing of cells outside the pyramidal cell layer (see below)]. Nevertheless the consensus of these three intracellular studies is compatible with the idea of Holsheimer *et al.* (1982) that *pyramidal cells generate the largest part of theta potentials recorded in CA1*, and if EPSP's are responsible for those potentials, they arise in or close to the somata of those neurons. The alternative is that IPSP's arising in the distal dendrites produce the theta rhythm. From what we know of available connections and transmitters, either or both of these are possible.

All of these three intracellular studies also address the issue of sign (IPSP or EPSP as generators) by providing evidence to determine whether the rhythmic potential fluctuations are controlled by EPSP's or IPSP's. Fox *et al.* (1983) and Leung and Yim (1986) report on phase shifts after intracellular injection of chloride ions. Leung and Yim found a 180 degrees phase reversal. However, Fox *et al.* (1983) found a smaller phase shift than this, amounting to about 90 degrees advance with respect to the extracellular record compared with the situation before chloride injection. Since IPSP's depend on chloride currents, Leung and Yim's study gives some support to the hypothesis that, in urethanized rats at least, the theta rhythm is generated by rhythmic modulation of IPSP's within projection cells. Fox *et al.*, in view of their finding a smaller phase shift after chloride injection, are led to suggest that other unspecified mechanisms in addition to rhythmic modulation of IPSP's may be involved. Nunez *et al.* (1987) measured membrane conductance at different phases of the intracellular potential oscillations. This was lower during the hyperpolarizing part of the cycle than the depolarizing part of the cycle. Moreover, the amplitude of the oscillations was increased by injection of hyperpolarizing currents, and decreased by injection of depolarizing currents. Both these findings suggest that rhythmical *EPSP's* were the predominant driving force for the potential oscillations observed.

Thus, although there is agreement about the cell type generating theta activity and the phase relation between theta activity and intracellular potential oscillations in urethanized rats, there is controversy about whether it involves EPSP's or IPSP's. The conflicting data could be resolved, however, if one assumed that there were both distal IPSP's and proximal EPSP's driving the theta rhythm in CA1. If the rhythm of the inputs at the two sites was coherent, the effects on local intracellular potentials would be opposed, so they would act synergistically on extracellular potentials recorded at the cell body level. Chloride injection, being a relatively slow process, would affect chloride concentration throughout the soma and its dendrites. If a considerable build-up of chloride occurred, it would be sufficient not only to reverse the IPSP's at the distal site. In addition the contribution of the reversed IPSP's to soma membrane potential could become enough to reverse the phase relation between somatic membrane potential and local theta rhythm. On the other hand, the assessment of the sign of the PSP's generating theta which uses membrane resistance at different phases of the cycle would be dominated by local (somatic) membrane resistance. Such a test should be dominated by the proximal EPSP's. In summary, the intracellular evidence for field CA1 obtained by these two different methods can be explained quite well by a combination of some of the possibilities mentioned in earlier sections: there seem to be both proximal rhythmic EPSP's and distal rhythmic IPSP's active upon CA1 pyramidal cells during theta activity. Others sorts of evidence, already considered, also point to each of these current sources, rather than a single one, being active.

The only intracellular study of dentate granule cells was reported very briefly by Andersen (1978, p. 312; unpublished work of Andersen and Schwartzkroin). It is, however, rather enigmatic. In rats under light anaesthesia intracellular records were obtained from a few granule cells during theta activity produced by brainstem or hypothalamic stimulation. Intracellular potential oscillations were seen "perfectly in phase" with local extracellular theta activity. The precise meaning of this is unclear since no records are presented: it is unclear whether "in phase" means synchrony of hyperpolarization with the negative or the positive phase of the extracellular theta wave. The hyperpolarizing part of the oscillations represented IPSP's, since it was reversed by chloride ion injections. The depolarizing part was said to be "odd", and on the basis of size and duration was judged not to be a simple EPSP. Andersen was of the opinion that the rhythm was generated by regularly occurring inhibition. However he also believed that the regularity and frequency of dentate theta activity was a function of the (non-rhythmic) depolarizing pressure on the granule cells.

8.3.1.4 Biophysical Effects of Transmitters Which Might Mediate the Theta Rhythm in the Hippocampus

The relation between interneurons, principal neurons, and signals in the septo-hippocampal input can also be tested by referring to the actual biophysical events occurring in the hippocampus in response to stimulation with the natural transmitters of that input pathway. Although GABA is a transmitter in the septo-hippocampal pathway, its role in generating theta potentials is difficult to ascertain since GABA is abundant in intrinsic hippocampal neurons as well as in this input pathway. However, the cholinergic component of the septo-hippocampal pathway has been investigated in several studies. The data does not force any definite conclusion, so it will be reviewed here only briefly.

The most thorough work is by Krnjevic and colleagues (Krnjevic *et al.* 1981; Ben-Ari *et al.* 1981; Krnjevic and Ropert 1982; Krnjevic *et al.* 1988). In the first of these papers it was shown using extracellular recording that the inhibitory effects produced in CA1 pyramidal cells of urethanized rats following low intensity stimulation of fimbrial/commissural afferents was attenuated if acetyl choline was administered iontophoretically. As a result, the extracellular positivity became less, and excitatory effects were more easily elicited. In the second paper the attenuation of inhibition was found to consist of two separate components. *One* was a mild depolarization accompanied by a rise in membrane resistance, due particularly to decreased potassium conductance. [Similar effects of acetylcholine have been described by Madison *et al.* (1987), and Bland *et al.* (1988) in hippocampal slices]. This effect was a slow one, requiring many seconds to become apparent. It is therefore unlikely to be involved in phasic modulation of inhibitory drive at a frequency as high as the theta rhythm. The *other* effect was an actual reduction in the size of IPSP's produced by fimbrial or entorhinal stimulation. This was quite substantial (average reduction of 62% in the conductance increase at the peak of the IPSP). It was not a postsynaptic effect, since the inhibitory efffects of iontophoretically applied GABA were unaffected by acetyl choline. Ben-Ari *et al.* (1981) consider two presynaptic mechanisms for reduction of inhibition: inhibition of inhibitory interneurons at their somata; or inhibition of transmitter release at the inhibitory terminals. The authors favour the latter possibility, since the effect was shown best for iontophoretic application near the layer of pyramidal somata, where inhibitory terminals are concentrated and because no interneurons could be found that are inhibited by acetylcholine. Neither of these two arguments is very rigorous. The role of the interneurons whose somata lie in stratum lacunosum/moleculare is not considered, these being of relevance since they have dendritic processes extending to stratum pyramidale. Since there *is* evidence that the firing of presumed interneurons and pyramidal cells can have opposite phase relations to the local theta rhythm, the first of the two possibilities must still be a strong contender.

In the third of this series of papers (Krnjevic and Ropert 1982), the biophysical effects of natural cholinergic influences (i.e., septal stimulation) on commissurally evoked population spikes in CA1 was studied. The net effect of acetyl choline on pyramidal cells, (viz enhancement of excitability), was confirmed by this method, confined to the pyramidal cell layer of CA1/2 but not found in the dentate gyrus or CA3/4. The facilitation of the population spike by septal stimulation was maximal at septal stimulation frequencies of 50–100 Hz, similar to the frequency within bursts of naturally bursting septal cells. Moreover, the effect was maximal at 20–50 ms after the septal train, and decayed over about 300 ms. These temporal characteristics are particularly significant, since they would permit modulation at frequencies in the theta range by a cholinergic mechanism. Presumably they reflect a disinhibitory influence rather than a direct effect on potassium conductance.

A later paper by Krnjevic *et al.* (1988) also studied the intracellular effects of medial septal stimulation on pyramidal cells in CA1–3 of urethanized rats. Under control conditions, a regular series of spontaneous IPSP's occurred in these cells. Medial septal stimulation (10 pulses at 100 Hz) produced a marked and prolonged (200–500 ms) depression of these IPSP's, especially when they were reversed using chloride injection. From illustrated records, this depression appeared to be a lowering of the frequency rather than of the size of spontaneous IPSPs, suggesting an inhibitory effect on firing of inhibitory interneurons, rather than on transmitter release from their terminals. These results confirm that septal stimulation (like iontophoretic application of acetylcholine) reduces inhibitory effects, and provides tentative evidence that the effect is exerted on interneuronal cell bodies rather than terminals.

8.3.1.5 Urethanized Rat: Conclusions and Discussion of the Phase Reversal Between the Two Generators

The coincidence of intracellular depolarization and unit firing in pyramidal cells and granule cells with the negative phase of the local theta potentials leaves little doubt that it is *these principal cell types which generate the bulk of the macroscopic theta potentials*.

From earlier chapters we have good reason to believe that there is more than one set of rhythmically active synaptic potentials, and more than one dipole involved in generating the observed theta potentials in the hippocampus. Holsheimer *et al.* (1982) discuss two possible arrangements of active inputs (double distal or double proximal excitatory inputs) as possible determinants of the laminar profile of theta potentials.

These two do not exhaust the possibilities since there are also plausible configurations involving inhibitory inputs.

Of the possibilities considered by Holsheimer *et al.*, the case of the double proximal input was a somewhat better fit to experimental data than the double distal input, since in the latter case the amplitude maximum in stratum oriens was of quite low amplitude, only about 1/4 of the amplitude calculated for the hippocampal fissure. In either case it was postulated that there was an additional component of the fields in the dentate gyrus, arising from distal dendrites of hilar cells excited coincident with the excitation upon granule cells. This postulate was necessary to produce the observed potentials in the dentate gyrus, lacking, as they usually do, a reversal of sign. From this it might appear that, although there need to be at least two sources of theta activity, there is no need to postulate that they be 180 degrees phase reversed. *However*, if this model were accurate, then dentate granule cells should generate impulses in phase with the CA1 pyramidal cells. In fact, in urethanised preparations, their firing is almost 180 degrees out of phase. Therefore one must conclude that *not only are there two sources of theta activity (in CA1 and the dentate gyrus respectively), but in addition, in the urethane preparation they are systematically 180 degrees out of phase, at the cell body levels.*

The 180 degrees phase reversal between superficial and deep theta rhythms found *in vivo* appears to hold even when frequency of the theta rhythm changes in an immobile urethanised animal (Leung 1984a, fig. 2b; see also Table 10 above). This means that *the phase relation between the two sources is definitely one of reversal, rather than of constant delay* (which latter would lead to variations of relative phase as frequency changed).

The type of synaptic potential (EPSP, IPSP or a combination of both) is not fully resolved. However, in field CA1, a consistent account of the intracellular studies, and other data, can be based on the hypothesis of *rhythmic EPSPs directed at the level of the pyramidal cell somata, and rhythmic IPSPs directed at the level of the distal apical dendrites.* Signals driving each type of synaptic potential would need to be phase-coherent, so that the resulting intracellular potentials would be phase-reversed. Either of these current sources, acting alone, would tend to show the laminar phase reversal, so the two types would act synergistically to generate the observed theta potentials.

This hypothesis is attractive and not in conflict with information we have about local circuit connections. According to this scheme, the proximal excitation could correspond to the proximally distributed fine axons which Nyakas *et al.* (1987) believe to be cholinergic. The distally located inhibitory interneurons described by Misgeld and Frotscher (1986) and by Lacaille and Schwartzkroin (1988a,b) could give rise to the distal IPSP's. Lacaille and Schwartzkroin (1988b) recorded simultaneously from these interneurons and from adjacent pyramidal cells and could correlate the events in the two cell types. IPSP's were recorded in both the somata and the dendrites of pyramidal cells. Since these IPSP's could have been conducted electrotonically within the pyramidal cells, it is unclear whether they were generated at the somata or in the dendrites. In favour of the latter possibility, however, was the fact that the dendritic membrane potentials were closer to the reversal potentials for IPSP's than those in the somata, so that the small IPSP's recorded in the pyramidal cell dendrites could have represented quite substantial inhibitory effects generated locally. Another interesting implication of this study is the fact that these CA1 interneurons have axonal and dendritic ramifications which reach to the pyramidal cell layers, where they would be exposed to different afferent influences from the lamina of their somata. In addition they have axonal ramifications which cross the hippocampal fissure into the dentate molecular layer, and could thus play a role in synchronizing the theta rhythms in the dentate gyrus with those in CA1 (see below).

As yet this model cannot be supported by evidence of the biophysical effects of acetylcholine on CA1, which is, to say the least, a little confused. Much of the evidence appears to show that acetylcholine has a net excitatory effect. Even this is debated however [see Buzsaki *et al.* (1983) and Olpe *et al.* (1987)]. Any excitatory effect of acetylcholine which is short-lived enough to allow rhythmic modulation is probably mainly a disinhibition. This is difficult to reconcile with the proposals about inputs to both the proximal and distal dendritic region in CA1, made on the basis of intracellular work: in the proximal region intracellular evidence suggests that there are rhythmic EPSP's, rather than a rhythmic disinhibition. In the distal region a rhythmic disinhibition is suggested by some of the evidence, but as yet it has not been shown to be under cholinergic control.

In the dentate gyrus the conclusion that there is a theta generator phase-reversed with respect to that in CA1 leads one to reconsider the origin of the laminar profiles of theta activity in the dentate gyrus, especially the results of Buzsaki *et al.* (1986). It will be recalled that this study gave evidence that theta potentials in the outer molecular layer were coherent with those in the deeper layers of CA1 above the hippocampal fissure. Potentials in the granule cell layer were often also coherent with those recorded at the level of the fissure, though they could lag behind by 0–90 degrees. Between these two regions of the dentate gyrus (i.e., in the region of the middle portion of the granule cell dendrites), there were abrupt phase changes, amounting to another phase-reversal on either side of the middle region. It is tempting to postulate a relation between EPSP's and IPSP's in the dentate gyrus the converse of that suggested above for CA1 on the basis of intracellular studies. *There could be a rhythmic EPSP source directed at the middle portion of the granule cell dendrites, and a rhythmic IPSP source directed at the somatic level.* Once again the signals driving these two sources would be phase coherent, so their effects on local intracellular potentials would be opposed, and their net effect on theta potentials would be synergistic.

This hypothesis could tie in with the data of Nyakas *et al.* (1987) on the distribution of fine (putatively cholinergic) septo-hippocampal fibres to the mid-dendritic region, and of the coarse fibres (possibly GABAergic) to the hilar and granule cell regions. However, once again it is difficult to integrate this idea with details of the number of sequential synapses, and their respective signs intervening between the septo-hippocampal fibres and the principal neurons in the dentate gyrus which generate theta activity. The hypothesis would, however, account for the fact that, at both the hippocampal fissure and (often) at the granule cell layer, potentials are the reverse of those in the CA1 pyramidal cell layer, while in the middle region of the granule cell dendrites, theta phase is close to that in the superficial CA1 layers. It would also explain the fact that the firing of the two principal cell types was phase reversed.

The conclusion that there are two phase-reversed sources of theta activity in the hippocampus raises an important issue: by what mechanism can this 180 degree phase reversal between the two theta sources be coordinated? There appear to be five possible answers to this question:

1. Building on the idea of feed-forward of inhibitory influences controlled by the septum discussed above, it might be speculated that there is one less (or one more) inhibitory synapse in the septal pathway to CA1 pyramidal cell somata compared with that to the dentate granule cells and the deep part of stratum radiatum. Exactly opposite phase relations between dentate gyrus and CA1 would then be inevitable. However, it is impossible to build a case for this relationship: evidence both on phase relation of firing of interneurons versus principal cells and that from intracellular studies is inadequate in the dentate gyrus. From biophysical work it appears that the net effect of the septal input on both CA1 pyramidal cells and dentate granule cells is excitatory (by complex and heterogeneous mechanisms, including disinhibition) (see above, and Bilkey and Goddard, 1985; 1987). In short, there is no adequate evidence for phase reversal being coordinated by differing numbers of sign reversals in septal paths to dentate and CA1 and there is some evidence against this proposal.

2. A second possible mechanism of phase reversal could rely on signals being relayed to the hippocampus along two parallel pathways, in which theta activity was mutually phase-coherent, but whose influence in CA1 pyramidal cell layer was of the opposite sign to that in the dentate gyrus and deep stratum radiatum. This idea receives support from Fantie and Goddard (1982) and Bilkey and Goddard (1985): although muscarinic antagonists abolish the effect of septal stimulation in CA1 (Krnjevic and Ropert, 1982), they do not affect the septal modulation of the perforant path population spike in the dentate gyrus. On the other hand, GABA antagonists do abolish this modulation. However, this possibility seems less likely in view of the fact that the net sign of the effect produced in both regions is the same, namely excitation.

3. A third alternative is that there may be separate pathways from the septum to the deep generator (dentate gyrus) and the superficial one (CA1), whose net sign of influence is the same in each region, but which carry theta-modulated signals which are mutually in antiphase. In other words, the phase reversal is already determined in the septum/diagonal band complex, rather than in the hippocampus. The strongest piece of evidence for this comes from slice preparations. As described above, either the CA1 portion or the dentate portion could generate theta-like rhythms on exposure to muscarinic agonists, and (as *in vivo*) the amplitude of the potentials generated by the dentate portion was larger than those of the CA1 portion. However, in slices (unlike in the urethanized *in vivo* preparation) there is no consistent phase relationship between the two generators (Konopacki *et al.* 1987c). Thus, when the septum is no longer capable of influencing the hippocampal tissue, consistent phase reversal between dentate gyrus and CA1 is no longer observed (Konopacki *et al.* 1987c). Corroboratory evidence comes from the histochemical findings (see Chap. 3) that in the medial septum/diagonal band cholinergic and GABAergic neurons coexist, both types projecting to the hippocampus. If the influence within the septum of the cholinergic cell bodies on the GABAergic ones was excitatory, and that of the GABAergic cell bodies on the cholinergic ones was inhibitory, the required phase reversal in the different pathways would occur. That the phase relations of single unit firing in the

medial septum/diagonal band complex is spread over the whole of the theta cycle, rather than confined to one half of it, is also compatible with there being separate populations of neurons there roughly in antiphase. Another crucial item of evidence which would be required to substantiate this mechanism of phase reversal would be to show histochemicaly that there was a preferential distribution of the GABAergic fibres to one of the sites of theta generation, and of cholinergic fibres to the other. While septal afferent axons using either transmitter appear to innervate the dentate gyrus, and cholinergic fibres innervate CA1, there is as yet no definite evidence whether GABAergic fibres do or do not project to CA1. Since muscarinic agonists can initiate theta-like rhythms in the isolated dentate region of a slice preparation, there is unlikely to be simply a GABAergic control in the dentate gyrus and a cholinergic control in CA1, but a more complex subdivision of control might obtain.

4. A fourth alternative is a development of the idea suggested above that both in CA1 and in the dentate gyrus there may be coordinated activation of neurons by IPSP/EPSP input combinations. It was suggested that in CA1 the IPSP input might be directed at distal dendrites, while in the dentate gyrus it was directed near the granule cell somata. This inversion of IPSP/EPSP spatial relationships could account for the phase inversion between these two fields. The necessary assumption about the neural activity in the septo-hippocampal pathway would be the simplest imaginable: that input rhythms controlling the IPSP's in each of *their* loci should be in phase with those controlling the EPSP's at each of their loci. This idea would also fit the datum that phase coordination between CA1 and dentate gyrus is lost in slice preparations, because the phase reversal in this schema would depend on approximate phase locking of the signals controlling all four synaptic populations. The chief difficulty is that the number and signs of sequential synapses controlling each of the four components is quite unclear, and it is at present difficult to reconcile this conjecture with what evidence there is about synaptic relations.

5. Lacaille and Schwartzkroin (1988a,b) discovered that there are inhibitory interneurons with somata in stratum lacunosum/moleculare of CA1 and with axonal and dendritic ramifications which reach to the pyramidal cell layer of CA1 and also extend across the hippocampal fissure into the dentate molecular layer. These interneurons appear to give inhibitory synapses to dendritic spines on granule cell and CA1 pyramidal cell dendrites, and to receive synaptic inputs from CA3 (bilaterally) and the entorhinal cortex (ipsilaterally)(Kunkel *et al.* 1988). This is a very intriguing finding which promises to have an important bearing on the mechanism determining the consistent phase reversal found between superficial and deep generators of theta activity. However, as yet the significance of these findings cannot be guessed, since the conditions in which the afferents in each lamina control these interneurons is unknown. Do excitatory terminals at the CA1 pyramidal cell layer excite the pyramidal cells in that layer, and simultaneously, via the ascending dendrites of these interneurons activate inhibitory processes directed at the distal apical dendrites of the pyramidal cells? Or do excitatory terminals in the dentate molecular layer activate both the granule cell dendrites and

the dendrites of these CA1 interneurons, the latter then inhibiting the pyramidal cells at the level of their somata? In either case, if the input was rhythmically active, it would set up rhythmic potentials which were phase reversed between the deep and superficial theta generators. Future work must clarify these issues.

In summary, the sign of septal influences on CA1 pyramidal cells seems to be the same as that on dentate granule cells, a net excitatory effect. There are a number of possible ways in which these effects may be mediated, both cholinergic and non-cholinergic. There is insufficient evidence to definitely decide the mechanism of the phase reversal between dentate and CA1 theta generators. The most likely possible explanation however is a combination of two of the mechanisms discussed above: (1) A phase coherent septo-hippocampal input activating pairs of synaptic populations, with proximal excitation and distal inhibition in field CA1, and proximal inhibition and distal excitation in the dentate gyrus. (2) Coordination of phase between the different hippocampal regions by the recently described interneurons in stratum lacunosum/moleculare of CA1. In addition, it should be remembered that although signals from the septum determine frequency, and may synchronize the reversed phase relationship, the hippocampus itself has the capacity to generate theta-like rhythms.

The solution to this complex problem is most likely to come from a combination of transmitter cytochemistry and electrophysiology. This approach is already being followed in some laboratories. In CA1, for example, Lacaille *et al.* (1987) and Lacaille and Schwartzkroin (1988b) have presented a preliminary circuit diagram of cell types, their mutual excitatory and inhibitory interactions, and the influence of some afferent pathways on these neurons. As more details become available, and as a similar approach is applied to the dentate gyrus, it may be possible to answer more definitely, in terms of circuit diagrams, how phase reversal between the two theta generators is achieved. Whatever the mechanism of the phase reversal, the fact of the phase reversal is securely established, and can therefore be used for further theory construction.

8.3.2 Free Moving Rat

8.3.2.1 Neuronal Models of Theta Activity in the Free Moving Rat

Leung (1984c) presented a compartmental model, similar in concept to that of Holsheimer *et al.* (1982) discussed above. This simulated some of the features of the laminar phase profiles found in the free-moving rat, in particular the gradual phase transition in stratum radiatum of CA1, with no definite null point. The construction of the cellular compartments of the model was similar to that adopted by Holsheimer *et al.* but the spatial and temporal arrangement of the theta-modulated inputs was different. Two synaptic inputs were considered. The first was an inhibitory input to the cell body and proximal dendrites. The second was an excitatory input directed at the distal dendrites. The construction of the model is shown in Fig. 34. If the first dipole alone was active, theta reversal was quite sharp (complete within 100 μm on the apical side of the soma), a result too abrupt to match the observations. If only the distal excitatory dipole was active, phase change of 150 degrees was produced over 150 μm of stratum radiatum and reversal was complete within 250 μm. This, though somewhat more gradual than in the first case, was still more abrupt than found experimentally. Using a single dipole, it was not possible to produce a more gradual phase shift than this. Using combinations of two inputs at any location within CA1 or the dentate gyrus Leung found that it was impossible to simulate the observed changes in stratum radiatum, if the two dipoles were in phase with one another. The reason for this was that, with such simultaneous dipoles, the calculated phase gradient across stratum radiatum was always more abrupt than observed, and there was always a zone of approximately zero amplitude, unlike the experimentally observed potentials.

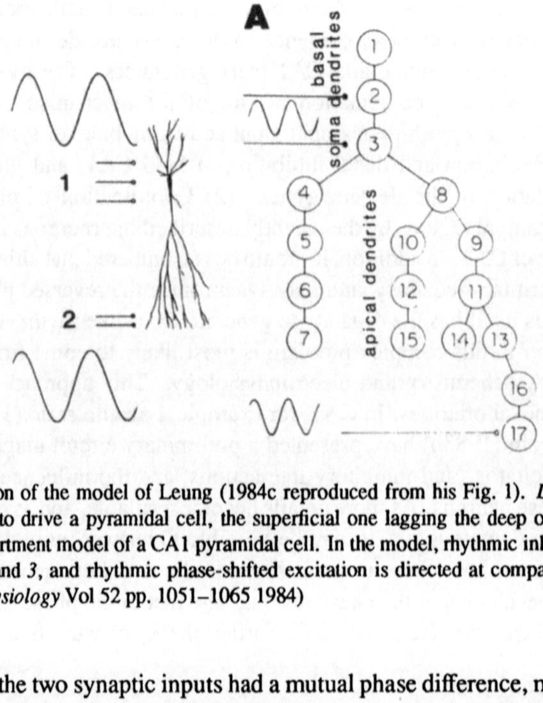

Fig. 34. Construction of the model of Leung (1984c reproduced from his Fig. 1). *Left* Two rhythmic inputs are assumed to drive a pyramidal cell, the superficial one lagging the deep one by 45 degrees. *Right* The 17-compartment model of a CA1 pyramidal cell. In the model, rhythmic inhibition is directed at compartments 2 and *3*, and rhythmic phase-shifted excitation is directed at compartment *17*. (From *Journal of Neurophysiology* Vol 52 pp. 1051–1065 1984)

If, however, the two synaptic inputs had a mutual phase difference, more interesting results were computed, which depended critically on the phase relation between the two theta-modulated inputs. The most interesting case was when the distal excitatory input was 45 degrees phase-advanced compared with the proximal inhibitory input. (By this it is meant that the peak of the depolarizing influence distally occurs 45 degrees before the peak hyperpolarizing influence proximally. In other words, since IPSP's are 180 degrees phase reversed compared with EPSP's, the membrane potential changes at the distal site were actually 180 + 45 = 225 degrees ahead of those at the proximal site). This is illustrated in Fig. 35. In this case, phase shift occurred over a span of about 400 μm of stratum radiatum, with the minimum of amplitude in this region being less prominent than in either of the two simpler cases. In both respects, these calculated profiles correspond fairly well with the experimental observations in this region. The phase gradient over space could be made more abrupt if the relative amplitude of the deep dipole was decreased, and more gradual if the relative amplitude of the superficial dipole was decreased. The phase gradient over space could also be changed by altering the phase relations of the two dipoles. If the phase lead of the deep dipole was increased to 60 degrees, phase change was accomplished over 500 μm. If it was decreased to 30 degrees, it was accomplished in 300 μm.

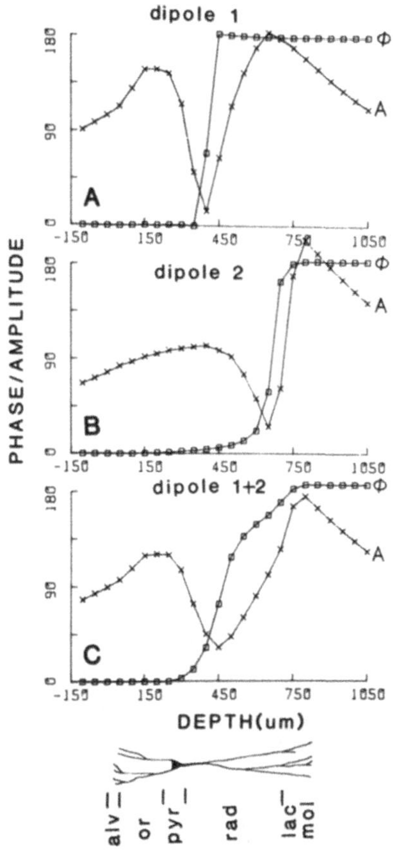

Fig. 35. Phase profiles in Leung's model for dipole 1 *(A)*, dipole 2 *(B)* and both together with 45 degree phase advance for dipole 2 *(C)*. Absolute amplitude (*crosses*) and phase (*squares*) are plotted against depth in a track, with typical pyramidal cell shown below along the depth axis. (From *Journal of Neurophysiology* Vol 52 pp. 1051–1065 1984)

The only respect in which Leung's simulation fails to represent reality is in the *direction* of the calculated phase shift. Thus, in Leung's illustration (his Fig. 3, reproduced in Fig. 36), the peak negativity at the deep site occurs slightly (45 degrees) *before* the peak positivity in the superficial site. Referring to potentials of the same sign, the deep potentials are about 225 degrees in advance of the superficial ones or equivalently, *lag* the superficial site by about 135 degrees. In experimental data, however, the potentials at the deep site are about 135 degrees *in advance* of the superficial site (Leung *et al.* 1982; Buzsaki *et al.* 1986).

In terms of functional interpretation of the potentials in the hippocampus, this point is crucial. However, we need not be hindered by the discrepancy between Leung's simulation and the observed results. Presumably the calculations in Leung's model would give a laminar phase profile exactly the mirror image of that published, if the phase relation of the superficial inhibition and the deep excitation were reversed. This could be

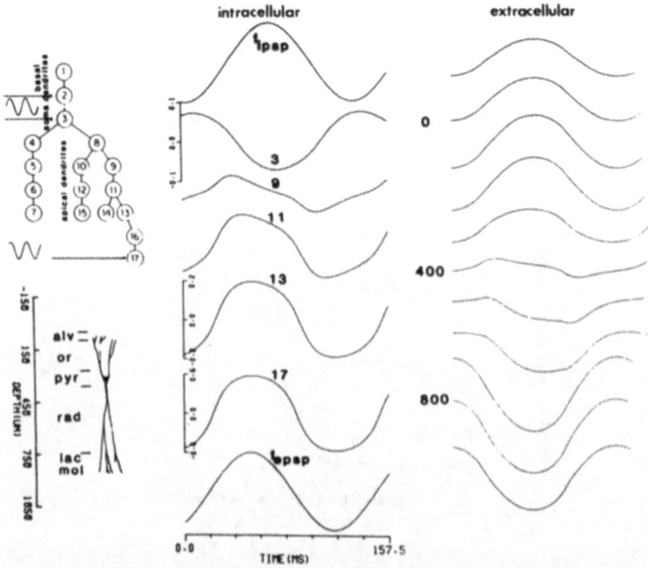

Fig. 36. Computations of laminar wave forms in a hypothetical track through CA1, with the assumptions used in Fig. 34. (Leung 1984c Fig. 3). *Centre* Highest and lowest traces represent the rhythmic synaptic drive delivered to the compartments shown in Fig. 35. Between them are shown the resulting intracellular potentials at various levels of the dendritic tree (equivalent compartments in the model shown in Fig. 35 indicated by *numbers*). *Right* Corresponding extracellular potential profiles through the depth of CA1. *Figures* represent microns as indicated in *inset, lower left*. Compare with Fig. 30, track P31. (From *Journal of Neurophysiology* Vol 52 pp. 1051–1065 1984)

accomplished by a combination of EPSP's and IPSP's at proximal and distal dendritic sites different from that suggested by Leung (viz. a proximal IPSP and a distal EPSP). For instance, the mirror image could be achieved with a 45 degree lead of a superficial EPSP over a deep IPSP. In this case, Leung's computation would give laminar phase shifts in the same direction as well as the same magnitude as are observed experimentally.

No attempt has been made to simulate the laminar profile of theta potentials recorded in the dentate gyrus. However, the model of Leung (1984c) is also instructive in helping us to understand this aspect of the evidence. The results of Buzsaki *et al.* (1986) should be recalled (Fig. 30). In the outer parts of the molecular layer theta activity was phase-coherent with that just above the hippocampal fissure, i.e., about 135–180 degrees in advance of the superficial theta rhythm. At about the junction of the inner and middle third of the molecular layer, the minimum amplitude was recorded, and here the wave shape was asymmetrical and complex, with a double peak sometimes seen, as if representing a harmonic of the fundamental frequency. Deep to this, in the granule cell layer and the hilus, theta activity was again coherent with that recorded around the hippocampal fissure.

The interpretation of this is not certain, but the following remarks are offered, as a development of the evidence previously discussed. Since the entorhinal input to the distal dendrites of CA1 has a laminar origin similar to that to the outer two-thirds of the dentate molecular layer (i.e., from laminae II/III of the entorhinal cortex), it is likely to be phase-coherent with it. This explains the coherence of the potentials recorded in the respective regions on either side of the hippocampal fissure. However, a theta-modulated excitatory input to the outer two thirds of the molecular layer from the entorhinal cortex might be expected to generate a region of extracellular positivity at some distance more proximally along the dendrites from the extracellular negativity resulting from the synaptic excitation. However, no such obvious site of reversal is seen in the proximal granule cell dendrites. There are several possible reasons why this is not observed. Firstly, as argued by Holsheimer *et al.* (1982) a contribution to the fields may be made by the distal dendrites of the hilar polymorph cells in this region. Secondly, it is possible that there is a second rhythmically-active synaptic population acting on the proximal dendrites of the granule cells, just as it appears to do on the CA1 pyramidal cells. The crucial difference from CA1, however, would be that the proximal potentials would have to be phase-coherent with the distal ones, rather than the approximate reverse of the distally generated potentials (as in CA1). In this scheme, almost the whole of the soma and dendritic tree of the granule cells, with the exception of a band between inner and middle thirds of the molecular layer, would be oscillating in synchrony, without the inverse relation between proximal and distal sites commonly seen in neuronally generated field potentials. Thirdly, it is possible that the above informal analysis of potentials may give an incorrect impression: This possibility cannot be rigorously evaluated without reconstruction of neuronal models, similar to that of Leung (1984c) in field CA1.

This interpretation may be continued, again informally, with respect to Buzsaki *et al.*'s (1986) results on CA2 and CA3 theta activity. It will be recalled that there is a phase reversal in stratum radiatum of the deep part of CA3 just above the pyramidal cell layer [track P38, Fig. 1 (Fig. 30 of the present work)], and a smaller, and gradual phase shift just deep to the superficial pyramidal cell layer in this track. In between these two regions is the maximum amplitude theta activity. The deep potentials on either side of their zone of phase reversal are about 45 degrees ahead of the superficial ones at the pyramidal cell layer. Within the deep layer the potential profile appears to be consistent with a source of excitatory input to the deep part of CA3 in stratum radiatum just above the reversal point. The somata and basal dendrites in the deep layer of CA3 pyramidal cells reflect this excitation, with reversed sign, due to their distance. Likewise the superficial potentials could be generated by excitation just beneath the region of phase change in stratum radiatum, reflected in potentials of reversed sign in superficial pyramidal cell somata and basal dendrites.

Very tentatively it could be suggested that the postulated input to the deep layer corresponds to EPSP's arising from the perforant path input to CA3, with a possible contribution from the mossy fibres. The postulated input to the superficial layer may correspond to EPSP's generated by the Schaffer collaterals. These identifications are by no means forced by the available evidence, which is quite scanty. They are mentioned here as a prelude to the discussion below, and the definitive theory formulation in the next chapter.

In summary the theoretical analysis of Leung, together with Buzsaki's experimental data, points to the conclusion that, in the free moving rat, theta potentials below the pyramidal cell layer of CA1 have a phase advance of about 130 degrees compared with those in that layer. Field CA3 and the dentate hilus and granule cells appear to generate theta potentials of significant amplitude, in contrast to the urethanized rat.

8.3.2.2 Evidence Concerning Phase Relations Between Unit Firing and Local Theta Activity

As for the urethanized rat, evidence on phase relations between unit firing and local theta potentials must also be considered in the free-moving rat, since they constrain the conclusions that can be reached. Three papers give relevant detail: Fox *et al.* (1983, 1986) report on phase relations of CA1 cells identified as either pyramidal (complex-spike) cells or interneurons (theta cells), during walking in rats. In contrast to the situation in urethane anaesthetized rats, both classes of cells fire on the positive peak of the dentate theta wave, that is during the negative phase of local theta. In addition, the precision of phase locking of both classes of cells in CA1 was looser than in the urethanized preparation. The other study to address this issue is that of Buzsaki *et al.* (1983), in which phase relations of unit firing throughout the hippocampus were assessed by spike-triggered averaging in rats in a running wheel. In CA1 it was found that pyramidal cells fired on the negative phase of local theta (in agreement with Fox *et al.* 1983, 1986). The CA1 interneurons fired on the positive phase of local theta, a finding in disagreement with Fox *et al.* However, Fox *et al.* point out that the method of spike-triggered averaging would produce an artefactual phase advancement of firing of theta cells, on account of their producing more than one spike per sweep. The data may therefore not be in serious conflict. It should be noted that the interneurons identified in these three studies were likely to be those in strata oriens and pyramidale, rather than those with rather different properties in stratum lacunosum/moleculare.

Fox *et al.* (1986) also identified granule cells in the dentate gyrus, and these fired predominantly after peak local negativity, midway on the rising phase towards peak positivity of the local theta rhythm (precisely at 296 degrees with respect to peak positivity). Buzsaki *et al.* (1983) also report on cells in the dentate gyrus. In agreement with Fox et al., the granule cells fire predominantly on the rising phase after peak negativity in the local theta rhythm (actually at 270 degrees after peak positivity). These workers also report on dentate gyrus interneurons, which they found also to fire mainly on the rising phase after peak negativity. However, these cells had a more widely dispersed phase relation than the granule cells.

In summary, there seems to be a significant difference in the phase of firing of CA1 interneurons between the urethanized and the free moving preparation. In the former preparation they could feed forwards inhibitory influences to the pyramidal cells. In the latter they have the same phase relations to local theta potentials as the pyramidal cells, and so would seem to have some other relationship to them.

8.3.2.3 Free Moving Rat: Discussion and Conclusions

How do we pull these data together? The following hypothesis is presented, representing a very different mode of operation of the hippocampus in the free moving rat compared with the urethanized rat. Firstly, it is necessary to recall the evidence that during type 1 behaviour, such as is displayed by the free moving rat in a running wheel, there is an important component of type 1 theta that is dependent on the integrity of the link between the neocortex and the hippocampus. From the anatomy it is known that the principal pathway by which the neocortex can influence the hippocampus is via the entorhinal cortex, the perforant pathway and the dentate gyrus. Thus, it is plausible to suggest that the primary signal generating theta activity in the free moving rat is the perforant pathway to the outer two-thirds of the dentate molecular layer. We have several reasons to believe that in the free moving rat, control of theta activity in CA1 is achieved by a mechanism different from that in the urethanized rat. These reasons are as follows. *Firstly*, according to the data of Leung (1984b) the implicit relation between theta activity in CA1 and that in the dentate gyrus is one of constant delay, with phase difference changing with frequency. This is a contrast with the situation in the urethanized rat where the relation is one of constant phase reversal, regardless of frequency. Such a relation cannot be explained by any of the processes postulated for the urethanized rat. *Secondly*, it may be remembered that in the urethanized rat there is an inverse phase relation between firing of interneurons and pyramidal cells in CA1. However, this relation appears not to hold for the free moving rat: In so far as interneuronal firing is phase-modulated, it has phase relations with the local theta rhythm in CA1 which are similar to rather than the inverse of the pyramidal cells. In addition, interneurons appear to be more loosely linked to local theta activity than in the urethanised preparation. These data would seem to imply that in the free moving rat, CA1 interneurons have lost their control over the pyramidal cells. Since theta activity is, however, still prominent in CA1 of free moving rats, some other pathway than the septal route via inhibitory interneurons must be in operation. A *third* difference has been documented by Rudell *et al.* (1980) and Rudell and Fox (1984): the phase of theta at which extracellular field potentials (evoked by stimulation of afferent pathways) is greatest differs significantly between the free moving and urethanized rat. However, since there are some problems of interpretation of these data they are not described in detail.

We are therefore within reach of the central tenet of our hypothesis, namely, that *in the free moving rat, theta-modulated signals from the entorhinal cortex and dentate gyrus control theta activity in field CA1*. The pathways by which this occurs are not certain in every respect, but include well known afferents to the hippocampal complex: Most obvious is the transverse path from the granule cells, via mossy fibres to CA3 pyramidal cells, and thence, via Schaffer collaterals to the fields CA2 and CA1. In addition, however, more direct links from the entorhinal cortex to stratum radiatum of CA3 and CA1 appear to play a part. If this is so, several predictions can be made. *Firstly*, in fields CA3 and CA4 (the hilus), which in urethanized rats show no theta activity, there should be theta activity detectable in the free moving rat. This appears to be true, but on the basis of only one paper at present (Buzsaki *et al.* 1986). There, as was described above, CA3 did appear to have a capacity for generating its own theta activity, and the hilus represented a distinct, though low amplitude source of theta potentials. *Secondly*, there should be a gradual phase transition along the successive stages of the transverse pathway. In other words, whereas the dentate gyrus has a 130–180 degree (30–50 ms) lead over CA1, CA3 and CA2 should have intermediate values for the phase lead, CA3 being closer to the dentate in phase, and CA2 being closer to CA1. Buzsaki *et al.* (1986) do not present any analysis of phase shift on traversing CA2 and CA3, so the second prediction remains to be tested. However in the vertical track (P38, Fig. 1: Buzsaki *et al.* 1986; Fig. 30 of the present work) through CA2 and into CA3 the phase lead of deep over superficial dipoles appears to be less than in vertical tracks from CA1 into the dentate gyrus (e.g., tack P31, Fig. 1: Buzsaki *et al.* 1986). The same point is made more clearly in another presentation of these data (Buzsaki 1985). In Fig. 10 of that work (reproduced here as Fig. 37) the relationship is depicted between RSA recorded in CA1 and the phase of preferred firing of neurons in dentate gyrus, CA3 and CA1. The phase of maximum firing is gradually later as one passes from dentate gyrus to CA3 to CA1. Quantitatively, the delay over this pathway is 30–35 ms, a value compatible with the estimate of the constant delay made by Leung (1984b). A *third* prediction is that lesions of CA3 should abolish or greatly modify theta potentials in CA1 in the free moving rat. Such lesions have been made, but their effects have so far been tested only in ether-anaesthetized preparations, where they had no influence on CA1 theta activity. The prediction therefore remains to be tested experimentally, in free moving rats.

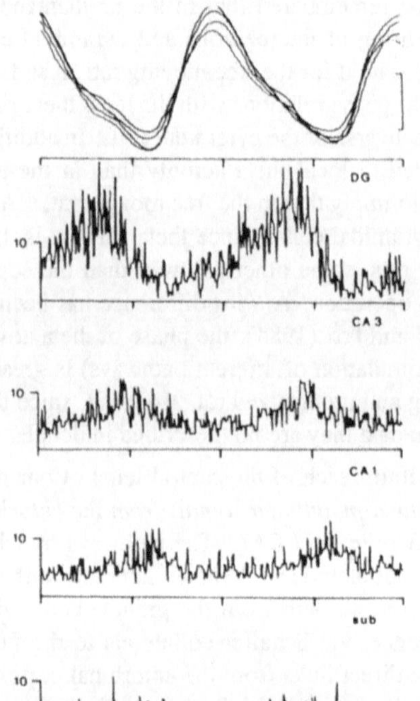

Fig. 37. Relationship between theta activity (*upper trace*) recorded from CA1 with a fixed electrode, and multiple unit discharge recorded in other regions of the hippocampus (*lower traces*), in three penetrations of a roving electrode during running in a rat. Each *trace* represents the average of 256 theta waveforms or concurrent impulse discharge. Note that the preferred phase of firing is gradually later as one passes from the dentate gyrus (*DG*) to CA3 and to CA1. *Sub* subiculum. (Buzsaki 1985, Fig. 10) (From *Electrical activity of the archicortex* Akademiai Kiaido, Budapest, pp. 143–164 1985)

One paper presents this hypothesis with some problems. Winson and Abzug (1978), using unanaesthetized rats, studied transmission along the transverse pathway from dentate gyrus to CA3 to CA1, by observing the evoked fields in each region following a stimulus applied to the angular bundle. This was carried out during slow wave sleep, rapid eye movement sleep (with theta activity), in a still alert condition (characterized by low amplitude, high frequency activity in the hippocampal EEG) or in the awake condition during locomotion where rats were showing theta activity. Transmission through the whole pathway to CA1 was best in slow wave sleep, and worst in the still alert condition. In the awake state or during REM sleep with theta activity, transmission through the dentate gyrus to CA3 was intermediate between the two, though transmission further to CA1 was no better than in the still alert condition. [This is corroborated by the result of Herreras *et al.* (1988a) in which it was shown that transmission from Schaffer collaterals to CA1 was reduced when theta activity was initiated.] In awake or REM theta states responses were also more variable, and correlated, trial by trial, between dentate gyrus and CA3. In order to reconcile these data with the idea that, in free moving rats, theta activity in the dentate gyrus controls that in CA1, the suggestion is made that in slow wave sleep, transmission occurs non-selectively. In the awake or REM sleep stages, when theta activity is present, transmission is a function of the instantaneous phase of the theta rhythm, and so is more variable.

The second difficulty presented by this paper is that the latency between the response in the dentate gyrus and that in CA1 appears to be rather shorter according to the results of Winson and Abzug (i.e., about 15 ms) than would be demanded if the dentate gyrus controlled field CA1 with a lag of 50 ms (i.e., a phase lag of about 135 degrees at 7 Hz). This may be explained in terms of the artefactual nature of electrical stimulation, compared with naturally initiated impulses. With a single intense electrical pulse, there would be synchronous activation of axons, particularly the large, low threshold, short latency axons. Such synchronous activation would be quite unknown in the natural firing of these axons. Therefore it is not surprising that electrical stimulation can produce an evoked field potential with substantially shorter latency than is achieved with naturally evoked impulses. The former depends principally on the connections with high conduction velocity, the latter on the full natural complement of axons, of the full range of conduction velocities. Its effective latency would be longer than that measured using electrically evoked stimuli.

One can only speculate at this stage on the circuitry by which control of theta activity in field CA1 can be changed from the septal input in urethanized animals to the entorhinal/dentate input in free moving ones. It will be noted that in both circumstances the synaptic mechanisms in CA1 favoured in the above discusssion were a combination of proximal EPSP's and distal IPSP's. In one or other (or both) of these components there is likely to be a switch from one controlling mechanism to an alternative. In this context it is relevant to mention the paper by Lacaille *et al.* (1987), in which two classes of interneurons were identified by intracellular recording in field CA1 of hippocampal slices. Both were capable of feeding forwards inhibition to the CA1 pyramidal cells, and one of these types of interneurons was also capable of inhibiting the other. The two types had different laminar distribution, and there were suggestions that they had some differences in their afferent controlling fibres. It is therefore possible that they represent alternative, mutually antagonistic ways of controlling theta-modulated signals in the CA1 pyramidal cells. The role of the interneurons in stratum lacunosum/moleculare, described by Lacaille and Schwartzkroin (1988a,b) is also of great interest in this regard, but will not be unravelled until they can be identified in *in vivo* preparations.

8.3.3 Unanaesthetized Rabbits

So far we have contrasted the urethanized rat with the free moving rat. There exist a number of other reports of experiments conducted on the curarized or free moving rabbit. As discussed above (Sect. 7.3.1.6), such preparations have been said to resemble the urethanized rat more than the free moving one. However, as mentioned in Chap. 7, the evidence is by no means good enough to make a strong point about this, since no full laminar profiles of theta activity in unanaesthetized rabbits have yet been published. We should therefore be cautious in equating these preparations with the urethanized rat, because they are, after all, conscious rather than anaesthetized. In the paragraphs below evidence on these preparations is described, including a few points that suggest similarities with free moving rat preparations.

There are three papers using unanaesthetized rabbits which investigate the phase relation between unit discharges and local theta activity, all from Bland's group. Sinclair *et al.* (1982) studied theta cells in the dentate gyrus and CA1 of freely moving rabbits, during type 1 behaviour and type 2 behaviour. In both regions, cells fired on the negative phase of the local theta activity. CA1 theta cells fired just prior to the negative peak (that is: at 162 degrees during type 1 behaviour, and at 154 degrees during type 2 behaviour, with respect to 0 degrees as the positive peak). Dentate gyrus theta cells fired just after the peak negativity (that is: at 210 degrees during type 1 behaviour, and at 220 degrees during type 2 behaviour). Dentate gyrus cells were more tightly coupled to local theta activity than CA1 cells. Bland *et al.* (1983) confirmed this result as far as the dentate theta units go. Dentate theta units fired maximally, during both type 1 and type 2 behaviour at 252 degrees of the phase cycle, i.e., in the latter half of the negativity phase of local theta wave (peak negativity at 180 degrees).

Bland *et al.* refer to their cells as "theta cells". However, it is somewhat difficult to equate them with theta cells according to the criterion used by Ranck and others. Nevertheless, many of them appear to have quite high firing rates, and so probably are interneurons such as would have been called "theta cells" by Ranck. The fact that they tend to fire on the negative part of the local theta potentials then makes them similar to theta cells recorded in free moving rats, rather than to those in urethanized rats.

The fact that the dentate cells tended to fire after the peak local negativity, and the CA1 cells before the peak suggests that their control is not identical with that of the theta potentials. By analogy with the rat, this could reflect a "conflict" between septo-hippocampal influences which determine theta activity in CA1 under type 2 conditions, and signals relayed from the dentate gyrus which, in rat, may control CA1 theta activity in type 1 conditions.

Bland *et al.* (1984) also presented data on phase relations between theta unit firing in CA1 and slow waves recorded simultaneously in the dentate gyrus in rabbits during walking (type 1 theta) and during sensory stimulation in the immobile animal (type 2 theta). Type 1 theta activity was accompanied by the firing of CA1 theta cells maximally at 9–42 degrees (mean of 23 degrees for three cells) after peak positivity of dentate gyrus theta. For type 2 theta activity, CA1 theta cells fired maximally at 12–46 degrees (mean for three cells: 27 degrees) after peak dentate positivity. Given the above result that CA1 cells fire just *before* peak local negativity, one can draw the conclusion that peak dentate positivity and peak negativity in CA1 are not synchronized in this preparation. This would seem to be a tentative indication that phase difference of theta potentials between superficial and deep theta generators departs from an accurate phase-reversal in this preparation, as they do in the free moving rat, but not in the urethanized rat. This possiblility must, of course, await more definite confirmatory evidence, including full laminar profiles of theta potentials.

An additional source of information about the unanaesthetized rabbit is two studies of the relation between intracellular potentials and extracellular theta activity, both using paralyzed preparations. Fujita and Sato (1964) recorded from cells in CA1, CA2 and CA4 (but apparently not in CA3), tentatively identifiable as pyramidal cells, on the basis of a wide spike (see their Fig. 1B), their location and their response to Schaffer collateral stimulation. In these cells there was consistent phase locking of intracellular potential oscillations with the extracellular theta rhythm. The intracellular phase of hyperpolarization was locked to the extracellular positivity. Spike bursts were sometimes present on the peak of the intervening depolarizations. The second paper, that of Artemenko (1973) studied cells in CA1 and CA2. Most hippocampal neurons fired in synchrony with the local theta waves. Cells were divided into two groups: In group 1, spontaneous activity of these neurons was low, and many were silent. Therefore this type were probably non-theta cells. They were excited during the positive phase (virtually at its peak) of the spontaneous theta rhythm recorded locally or at the hippocampal surface of CA1. In intracellular records hyperpolarization coincided with extracellular negativity. The second group of cells were less numerous. They had higher spontaneous levels of firing than the first group, and were probably theta cells, according to Ranck's criteria. They discharged mainly on the negative phase of the local theta rhythm, especially before the peak negativity.

These two papers are in conflict. The paper of Fujita and Sato (1964) is in accord with the evidence obtained with other experimental approaches: pyramidal cells seemed to show a depolarization in time with the extracellular negativity, which other evidence suggests they generate. This paper is, however, apparently in conflict with the other study mentioned above, which is difficult to explain.

8.4 General Conclusions on Hippocampal Sources Generating Theta Rhythms

8.4.1 Conclusions Concerning the Hippocampal Complex

What can we conclude from this survey of the complex evidence on the generators of theta activity within the hippocampus? First, there are several primary sources of theta activity. Usually theta sources are divided into two groups: (1) The superficial layers of CA1, (2) The deep layers of CA1 and the dentate gyrus. However it is possible that the reality may be more complex than this: both in CA1 and in the dentate gyrus there may be more than one synaptic population responsible for generation of theta potentials. Whatever the truth on this issue, the distinction between sources of theta potentials does not correlate with the distinction drawn in Chapter 6 concerning two modes of theta activity which have different behavioural correlates, frequencies and pharmacological characteristics. Indeed, it seems that all of the groups of hippocampal theta generators contribute to both modes of theta activity, but their mutual relationship is different in the two theta modes.

In the urethanized animal, the preparation which displays most purely the atropine-sensitive, type 2 theta, there seems to be an inverse phase relation between the superficial and deep generators. This can be explained in terms of a single source of theta modulated signals, namely the pathway from medial septal and diagonal band neurons, but *the links*

between this and superficial CA1 pyramidal cell layer introduce a systematic phase inversion compared with those to the deep part of CA1 and the dentate gyrus. The mechanism generating this inversion is uncertain. From the discussion in Section 8.3.1.5, the most likely possibilities are: (1) A phase coherent septo-hippocampal input activates pairs of synaptic populations, with proximal excitation and distal inhibition in field CA1, and proximal inhibition and distal excitation in the dentate gyrus. (2) Coordination of phase between the different hippocampal regions is achieved by the recently described interneurons in stratum lacunosum/moleculare of CA1. A combination of both mechanisms may exist. In the urethanized preparation, the fields CA3 and CA2, which are intermediate stages in the anatomical pathways between the two principal generators, appear not to participate actively in theta generation.

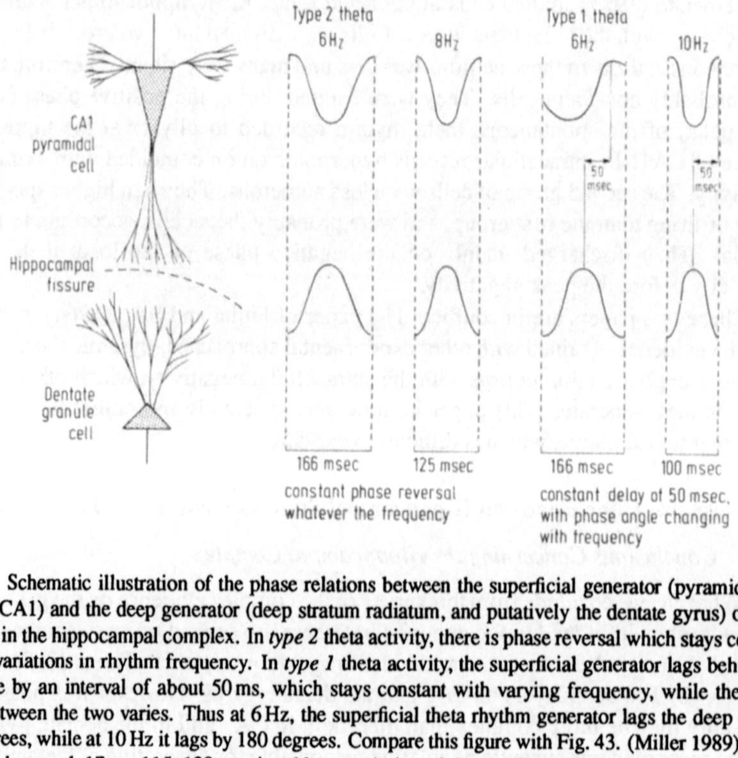

Fig. 38. Schematic illustration of the phase relations between the superficial generator (pyramidal cell layer of CA1) and the deep generator (deep stratum radiatum, and putatively the dentate gyrus) of theta rhythms in the hippocampal complex. In *type 2* theta activity, there is phase reversal which stays constant despite variations in rhythm frequency. In *type 1* theta activity, the superficial generator lags behind the deep one by an interval of about 50 ms, which stays constant with varying frequency, while the phase angle between the two varies. Thus at 6 Hz, the superficial theta rhythm generator lags the deep one by 108 degrees, while at 10 Hz it lags by 180 degrees. Compare this figure with Fig. 43. (Miller 1989) (From *Psychbiology*, vol. 17 pp. 115–128, reprinted by permission of the Psychonomic Society, Inc.)

In the free moving rat, however, theta activity in the superficial part of field CA1, though often approximately phase reversed compared with that in the deeper part of CA1 or the dentate gyrus, actually appears to be related to this by a constant delay. A schematic representation of this relationship is presented in Fig. 38. The evidence is consistent with the idea of theta-modulated signals reaching the dentate gyrus and distal apical dendrites of CA1 along the perforant path, these signals having a phase advance (constant duration of lead) over the signals generated in the superficial part of CA1. In this condition there appears to be theta activity also generated in fields CA3 and CA2, which lie on the transverse pathway between the dentate gyrus and field CA1. *It is therefore plausible that in the free moving preparations the entorhinal cortical input controls theta activity in the dentate gyrus which then controls CA1 theta activity.*

These are the safest minimum conclusions which can be drawn from evidence reviewed. While it would be desirable to construct a full neuronal model, this is not yet possible. Much of the evidence is conflicting, or unintegrated into a satisfactory overall picture. The evidence is especially difficult regarding the biophysical effects of acetylcholine on neurons in field CA1. Satisfactory intracellular studies of neurons in the dentate gyrus are also lacking. Other avenues of evidence which could contribute significantly in fact do not do so, because of difficulties of interpretation which have not so far been resolved. These difficult areas include correlations of evoked potentials with phase of local theta, and correlations of septal unit firing or EEG activity with hippocampal theta activity.

8.4.2 Discussion of Hippocampo-Cortical Interaction

The hypothesis that the entorhinal cortex and dentate gyrus are the principal controllers of CA1 theta activity in the free moving animal leaves open a very big question. Given that the entorhinal cortex is a controller of hippocampal theta activity in the free moving animal, what can be the pacemaker of this type of theta activity?

In Chapter 6, evidence obtained from lesions of the entorhinal cortex, or disconnecting this cortex from the bulk of the isocortex led to the idea that there be some interplay between hippocampus and isocortex, especially in preparations such as the free moving rat. This seemed more likely than that there be some extra pacemaker for theta rhythms other than the septum. The interplay envisaged between isocortex and hippocampus could not initiate hippocampal theta activity by itself, but provided the septo-hippocampal system had already set such activity going, the cortico-hippocampal interplay reinforced or amplified it.

It was also suggested in Chapter 6 that reciprocal *resonant circuits* exist between cortex and hippocampus, which may play a part in the interplay between the two, dependent on the prior rhythmic activity of the hippocampus. This idea begs two obvious questions: how can resonating circuits be selected from the vast repertoire of reciprocal connections between hippocampus and isocortex?; why is it that theta activity is not widely detectable throughout all regions of isocortex connected with the hippocampal formation? These matters will be addressed in the next chapter, after presenting a theory of how resonant circuits can organize themselves. At this stage it is appropriate to summarize the empirical evidence supporting the idea of reciprocal circuits resonating at the theta frequency.

The idea of resonant circuits receives most specific support from the analysis of laminar profiles discussed above. Since the entorhinal cortex and dentate gyrus seem to control theta activity in CA1 in the unanaesthetized preparation, one has evidence that this is the condition in which this part of the resonant circuitry is completed. Further support comes from data that certain areas of cortex most closely connected to the hippocampus can generate electrographic activity (and unit activity coherent with it) at the frequency concurrently shown by the hippocampus. One therefore has at least *some* evidence of the rhythmic activity in the isocortex which would be postulated if there are resonant circuits between hippocampus and isocortex. Additional support comes from data on the amplitude of hippocampal theta potentials. In the free moving rat, the preparation where the interplay is postulated, amplitudes of theta potentials are significantly larger than in the immobile rat (Whishaw and Vanderwolf 1973). Further support for the idea of resonance comes from evidence (cited in Chap. 7) of harmonics in the theta rhythm. If pathways, whose total loop time corresponds to the period of the oscillator, can be set into resonance, so, too, could pathways whose total loop time was a simple fraction of the oscillator period (harmonics of the fundamental frequency). Three results are discrepant with this idea of resonant hippocampo-cortical circuits, suggesting that resonance can occur at the theta frequency even in conditions when it is unlikely that the hippocampus can initiate the rhythmic activity. Azzaroni and Parmeggiani (1967) found that rhythmic driving of theta activity by stimulation of limbic structures could occur even after septal lesions. Destrade and Ott (1982) found that under some circumstances theta activity in the cortex can occur without interplay with the hippocampo-septal rhythms. Borst *et al.* (1987) found that in a minority of rats theta activity could be generated in the cingulate cortex even after medial septal lesions. These data are not a fatal blow to the idea of resonance initiated by the hippocampal rhythm. However, the circumstances under which such autonomous cortical theta rhythms can occur is a matter worth discussing further.

An additional feature of the circuits in either direction on which this resonant relationship depends is that they are sometimes polysynaptic rather than monosynaptic, and in any case long. One may therefore expect substantial delays between theta-modulated signals in the two structures (whichever direction of control is envisaged). Therefore the phase of cortical theta activity is likely to be somewhere intermediate between one beat of the hippocampal rhythm and the next. If this were seen it would be compatible with control in either direction, but is most parsimoniously accounted for by a combination of both, that is by the postulated resonant circuits. Ideas of one-way control of the cortical rhythms by the hippocampal ones, or vice versa, may thus be missing the point. This idea is developed in the next chapter.

Part III

The Theory of Resonant, Self-Organizing Phase-Locked Loops

9 Cortico-Hippocampal Interaction: Theory, Implications and Predictions

9.1 Introduction

The basic idea of relations between hippocampus and isocortex depends on the concept of resonance at the theta frequency. It is envisaged that *Hebbian processes of synaptic strengthening select patterns of loops passing from hipocampus to cortex and back again to the hippocampus. The total conduction delay time round each one of these patterns of loops is envisaged to correspond to the theta period. Such resonance between hippocampus and cortex is envisaged to have an important functional role in registration and retrieval of information in the cortex. Each pattern of resonant loops will raise the activation of specific collections of cells which are widely dispersed across the cortical mantle. Such collections of cells have insufficient mutual interconnections within the cortex to form a cell assembly by the usual Hebbian mechanism described in Chapter 2. Nevertheless, they do represent definite information groupings. In fact, it is envisaged that they represent large-scale aspects of the environment as a whole, those we designate as contexts.*

The essential property of a conventional cell assembly is that the firing of a subset of the neurons it comprises can raise the activation of all the neurons in the assembly. Whether the collections of cells set up by resonant loops has this property is a moot point, discussed below: Therefore such a collection of cells will not yet be identified with the "global cell assembly" referred to in Chapter 2. Instead, a more neutral term will be used, a "resonant group" of cortical cells.

The establishment of such a resonant group allows specific context representations to be available. Consequently, memories latent in the network of cortical connections which were acquired in that context can more easily be reactivated when the matching context representation is reactivated than when it is inactive. The ambiguities inherent in the information structure of the environment (mentioned in Chap. 2), which the cortical network is otherwise unable to interpret, are thus resolved. The hippocampus thus acts in some ways as an "index" to memories, the cortex being, as it were, the "book" to which this index refers.

In Section 9.2 below, the first half of this theory will be addressed, namely: how can these resonant loops be set up? In the following section (9.3) the significance of this complex mechanism for various aspects of cortical information processing will be considered. In both 9.2 and 9.3 predicted consequences of the theory are pointed out. Sometimes there is available relevant evidence on these predictions, which, generally speaking, supports them. In other instances the predictions stand as pointers for future experimenters. One area of the literature, that on the correlations between theta activity and behaviour during learning is so large and complex that a separate chapter is devoted to it (Chap. 10).

9.2 Self-Organization of Resonant Phase-Locked Loops Between Hippocampus and Cortex

9.2.1 Connectional Basis of Phase-Locked Loops

The *first* premise in this theory is that there are reciprocal connections between the hippocampus, and much of the cortex. The detailed anatomical evidence for this has been considered already (Chap. 3). It requires only to emphasize that *direct* afferent and efferent connections between hippocampus and neocortex appear to involve a rather limited extent of cortex, notably some of the parietotemporal association areas, and also the prefrontal and limbic (e.g., cingulate) areas of the cortex. Nevertheless, in view of the very rich connectivity between isocortical areas, there is the potentiality of theta-modulated signals having a very wide influence over the cortex. Thus *the theta-modulated signals are likely to influence limbic and prefrontal areas, and also (directly or indirectly) other areas of (mainly association) cortex.*

The *second* premise concerns the likely conduction delays inherent in these reciprocal pathways. In general, there is no reason to think that the conduction velocity in the axons connecting hippocampus and cortex to each other is any different from that of any other of the cortico-cortical connections. Conduction delays in cortico-hippocampal pathways can therefore be inferred from the properties of other cortico-cortical axons. Typical populations of cortico-cortical connections include up to 40% of unmyelinated axons (Fleischhauer and Wartenburg 1967; Waxman and Swadlow 1976; Remahl and Hildebrand 1982; Wiggins *et al.* 1983). The median diameter of these unmyelinated axons is very small, around 0.2–0.3 μm in the above studies, the finest axons being under 0.1 μm in diameter.

The corresponding conduction velocities are consequently very slow. Swadlow and Waxman (1976) have argued that conduction velocities below 0.8 m/s are typical of unmyelinated central axons. Lee *et al.* (1986) demonstrate empirically that central axons of 1 μm calibre have a mean conduction velocity of about 0.7 m/s, while those of 0.5 μmcalibre have conduction velocities of about 0.5 m/s. The conduction velocity of the slowest conducting central axons has been estimated at about 0.3 m/s (Willey *et al.* 1975; Swadlow and Waxman 1976; Lee *et al.* 1986). However, the slowest axons may well conduct more slowly than this: Antidromic conduction methods, such as used in the above studies generally ignore the terminal arborization where conduction velocity is slowest; and single unit recording is likely to underepresent the fine calibre axons, since it generally selects larger cell bodies which are likely to have thicker axons. Actually, Ferraya Moyano and Molina (1980) have seen antidromically conducted responses in the rat olfactory peduncle with conduction velocities as slow as 0.15 m/s. Similar values for conduction velocity of fibre volleys have recently been seen in hippocampal slices (Miller, Wickens and Abraham, submitted). This figure may be a more accurate estimate of the conduction velocity of the slowest-conducting cortico-cortical axons.

The length of the axonal pathways linking hippocampus to cortex, of course, varies greatly from one species to another. In the rat the direct length is of the order of 10 mm, in each direction. (This figure may be an underestimate of total axonal length, due to the curvature of the axonal trajectories). Therefore there are likely to be links between hippocampus and cortex with conduction delays of the order of 50 ms or even more. Experimental studies have not yet shown this, but have shown that some axons passing between hippocampus and cortex have antidromic conduction delays approaching 30 ms for single axons passing between CA1 and the prefrontal cortex of rats (Ferino *et al.* 1987). Again the antidromic stimulation method probably underestimates the true delay between soma and synapse in an individual axon, because conduction in the axon orthodromic from the stimulation site (including its axonal arborization, where conduction is slowest) is not taken into account. In addition, histograms based on this method will probably underrepresent the slowest axons, because their somata would usually be small, giving extracellular spikes below or close to the noise level.

Let us estimate the longest plausible loop time in the most-studied species, the rat. Taking a value of 0.2 m/s as a commonly occurring conduction velocity amongst the unmyelinated axonal population, and a distance each way of 10 mm, we have the possibility of a 50 ms (or more) conduction delay from hippocampus to some cortical synapse, another 50 ms from a cortical neuron back to the hippocampus. There are two other conduction delays which must be included in calculating possible loop times: cortico-hipopocampal loops would probably include a link from the direct hippocampally connected zones of cortex with other cortical areas, such as the parietal association cortex. One should therefore add to the above figures the possibility of, say, a further 30 ms delay (in either direction) in intracortical connections. Finally we know, from the data discussed in the previous chapter concerning the free moving rat, that there are delays of 30–50 ms in signal transmission through the transverse hippocampal pathway from perforant path to dentate gyrus to CA3 to CA1. Taking the figure of 0.15 m/s as the conduction velocity of the slowest conducting axons, and a distance of approximately 6 mm around the transverse pathway in the hippocampus, it is just plausible to think of this delay as also resulting mainly from axonal conduction delays. In conclusion (see Fig. 39) the slower conducting components of the available connections in the rat could comprize cortico-hippocampal loops whose total loop time is in the range 120 to 200 ms. It must be admitted that this calculation is based on the slower conducting axonal pathways. It certainly need not be typical of all the links in these

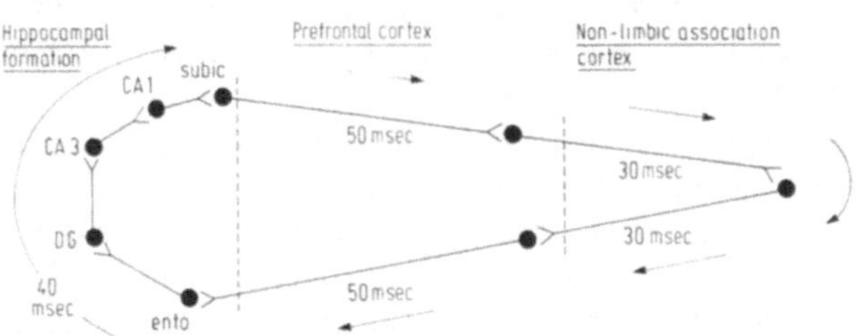

Fig. 39. Schematic representation of plausible delay times in the various parts of a cortico-hippocampal loop. The 40 ms delay within the hippocampus is derived from data considered in Section 7.3.1.4. The other delays are considered in the text (Section 9.2.1). *ento* Entorhinal cortex; *DG* dentate gyrus; *subic* subiculum. Total loop time may be in excess of 200 ms. (Miller 1989) (From *Psychbiology*, vol. 17 pp. 115–128, reprinted by permission of the Psychonomic Society, Inc.)

reciprocal pathways. However, such slowly conducting loops between hippocampus and cortex take on added significance when it is realized that the values given above—120–200 ms—are right in the range of the period for the theta rhythm in the rat (corresponding to frequencies of 5.0–8.3 Hz).

The conduction delays estimated above are likely to be very stable for any single axonal link, but to vary between axons over a considerable range (determined by the histogram of axon calibres). There thus exists a considerable repertoire of different *delay lines* between hippocampus and cortex. There is no necessity that more than a small proportion of the possible recursive loops would accurately match the theta period. However, given the possibilities of selection of connections when Hebbian mechanisms operate, there may be a way of selecting just those loops whose loop-time permits resonance. This possibility is developed below.

9.2.2 Loop Selection by the Hebbian Mechanism

The remaining premises concern the plastic processes which might govern synaptic strengthening at either end of this loop. In Chapter 2 we made the assumption that synaptic strengthening according to Hebbian or quasi-Hebbian rules may operate throughout the neocortex. The abundant literature on long-term potentiation (LTP) indicates that similar processes may occur in all fields of the hippocampus. From study of hippocampal slice preparations (Kelso *et al.* 1986; Levy and Steward 1983; Gustafsson and Wigstrom 1986), we know that the conditions necessary for LTP in the hippocampus include quite close temporal contiguity between cooperating inputs. There appears to be a temporal window of about 20 ms duration, such that if inputs coincide in single neurons within this interval their cooperative effects may lead to synaptic enhancement. Inputs further apart in time than this do not lead to LTP. This degree of temporal precision for the Hebbian mechanism applies to the hippocampal end of the postulated recursive loops. The same may also apply to the isocortical end of the loops, though there is no evidence on this, and probably no theoretical need for it either.

It has been argued elsewhere (Miller 1987) that where a variety of delay lines converge upon single neurons, with a temporally precise Hebbian mechanism at the converging synapses, it is possible for single neurons to come to represent brief temporal patterns. This is related to the degree of temporal convergence at the single neuron level, just as the acquisition of representations of "spatial" (i.e., synchronous) patterns is related to the degree of spatial convergence.

One further condition is required. At the isocortical end of the loop, a subset of the neurons should sometimes have a degree of maintained neural activity, originating from some other synaptic input. This again is plausible: as discussed in Chapter 2, *when an animal is in a particular environment, a collection of widely dispersed neurons throughout the cortical mantle may be set into a tonic state of elevated activation, as a result of signals reaching the cortex via any of its sensory systems.* As yet, however, the activation of these neurons is not coordinated as a "cell assembly", so one cannot speak of this collection as categorically recognizing the environment.

Consider now the relation we find between hippocampus and cortex: There is a series of recursive loops, with some temporal divergence/convergence in the pathway in each direction, and there is the possibility of regular oscillation of neural activity imposed at the hippocampal end. In this situation, *the Hebbian mechanism favours the strengthening of connections which permit resonance.*

The details of operation of the Hebbian process must be spelt out clearly. Imagine that the hippocampal oscillator comes into action at the same time as a group of isocortical cells receives tonic activation from the environment: the cortical neurons will be activated tonically, and will also tend to receive rhythmic signals via connections from the hippocampus. This is a condition in which the Hebbian principal can operate. When it does so the converging synapses relaying the rhythmic activity will be progressively strengthened. As a result of this, the group of cortical cells would show not only tonically elevated activity, driven by environmental stimuli, but, superimposed on this, a rhythmicity phase-locked to the hippocampal rhythm. Pathways returning from these rhythmically active cortical cells to the hippocampal oscillator would therefore come to transmit similar rhythmic signals. Consider those returning connections whose conduction delay allowed these rhythmic signals to approximately synchronize with a succeeding "beat" of the oscillator: For the hippocampal synapses of these connections (and of these only) one has a coincidence of pre- and postsynaptic activity at a certain phase of the theta cycle. One therefore has again the conditions necessary for strengthening these connections by the Hebbian mechanism. Thus the subset of the anatomical connections which is selected for strengthening is that in which a signal transmitted from a hippocampal population of synchronously rhythmic neurons might return to its starting point in phase with the active part of the next (or a subsequent) cycle. In other words, the selected loops are those whose total conduction time approximates to the theta period (or a simple fraction or multiple of it). Resonance can then occur, and synapses at both ends of the loops will progressively strengthen towards maximum strength. In short, the anatomical loops between cortex and hippocampus will self-organize into patterns of resonant circuits.

The process of self-organization depends on the time relations between neural activity originating in the hippocampus and the reflection of it returning from an activated cortical locus. However, it is not necessary that there be perfect synchronization of these. For a start we know that the Hebbian process has a temporal resolution of only about 20 ms. Thus the theta period of about 150 ms may be regarded as consisting of a small number (perhaps seven or eight) temporal "bins". A synaptic link participating in a resonant circuit is selected on the basis of its habitually firing in the appropriate one of these bins, rather than on the basis of perfect synchronization. In addition, it is likely that hippocampal neurons will have sufficient postsynaptic activation to recruit the Hebbian process over a substantial part of the theta cycle (e.g., most of the depolarized half of the cycle). This consideration perhaps helps to explain the experimental findings that in type 1 theta activity in unanaesthetized rabbit, spike discharge is not accurately locked to the negative peak of the (extracellular) theta rhythm, but may slightly precede it in CA1 (Sinclair *et al.* 1982).

The configuration of each pattern of resonant circuits which comes to exist needs to be discussed. It is simplest to consider ones where the links from hippocampal oscillator to isocortical resonant node is monosynaptic in each direction (see Fig. 40). (The paragraphs above are also most easily understood for monsynaptic connections). For each environment and each corresponding subset of cortical neurons with maintained activity, there will be a different pattern of links between cortex and hippocampus

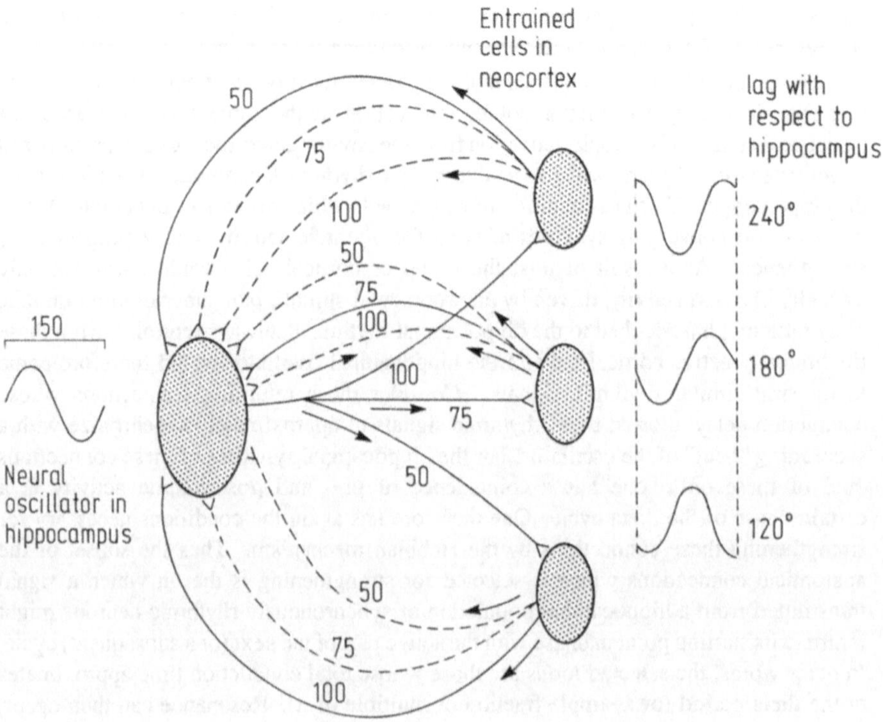

Fig. 40. Scheme for self-organization of phase-locked loops. The master oscillator (hippocampus) has afferent and efferent links with a variety of neocortical loci. All links include lines with a range of delays. [Conduction delays (ms) for each axonal link are shown by *numbers* adjacent to each axon.] By strengthening of selected synaptic connections of axons with appropriate delays (*solid lines*, depicted here as having total loop times of 150 ms), the neocortical loci can become entrained to the hippocampal rhythm. Different neocortical loci will have different phase relations with the hippocampal rhythm (*right*) (Miller 1989) (From *Psychbiology*, vol. 17 pp. 115–128, reprinted by permission of the Psychonomic Society, Inc.)

which resonate. Within each pattern of resonant circuits, the conduction time from hippocampus to cortex of each single axonal link may be different. Likewise the conduction time of the individual links selected to return the resonant signal to the hippocampus may vary from one connection to another. Therefore, within any one pattern, the different resonating cortical nodes will have different phase relations to the "master oscillator" in the hippocampus (as shown in Fig. 40, right). What stays constant, however, is the *sum* of the conduction delays *from* hippocampus to cortex, plus that back from cortex *to* the hippocampus. For all routes which have been selected to make up the pattern of resonance, this summed value should approximate to the theta period at times when resonance is occurring.

In practice it is likely that the axonal pathways between the hippocampal oscillator and the cortical nodes activated by the environment would be polysynaptic reciprocal networks, rather than the simple monosynaptic "arcs" depicted in Fig. 40. A schematic representation of the temporal aspects of the strengthened lines of polysynaptic connectivity which might be set up from hippocampus to isocortex is depicted in Fig. 41. For such polysynaptic pathways to participate in self-organization of phase-locked loops,

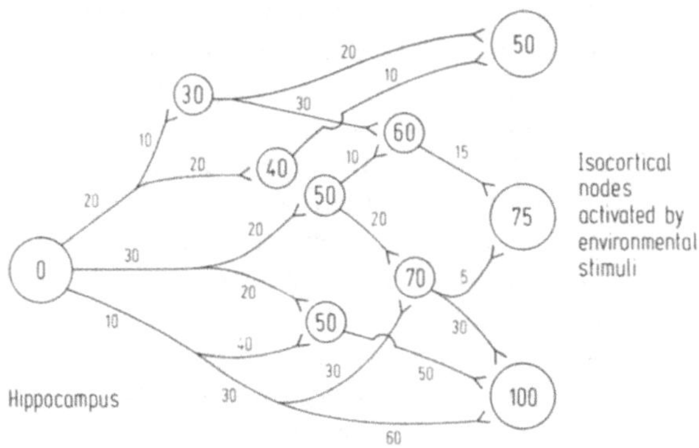

Fig. 41. Schematic representation of the temporal relations between hippocampus and isocortex which might develop as phase-locked polysynaptic loops become established. Only one limb of a loop is depicted. Conduction delays are shown alongside each segment of the connections (ms). On the *left* is the master hippocampal oscillator as in Fig. 40. On the *right* are three isocortical neurons, whose neural activity is likely to be raised when the animal is in a specific environment. All routes from the master oscillator to a single one of these cortical neurons entail the same conduction delay. Different isocortical neurons, however, are connected to the master oscillator with different delays, as in Fig. 40

additional properties of the intermediate cortical local circuits are required: these circuits should transmit to their efferents a rhythm imposed by the afferents, without obliteration by averaging of the several afferent signals which might converge on the local circuit. There is no experimental data on this, but from what we know of cortical networks, it is an entirely plausible conjecture, even if the rhythm transmitted to the efferents exists only in a stochastic sense, requiring averaging techniques for its detection. Provided some trace of the rhythm is transmitted to the cortical node activated by the environment, self-organization will tend to occur. Once it starts to occur, strengthening of the links on which it depends will accelerate, and will occur not only in the principal components of the resonant loops but also in the intermediate synapses between these components.

We can thus envisage the hippocampus operating in two quite different modes, which we can call the *baseline oscillator* and the *cortical entrainment* modes. The first of these is closely related to the innate, species-specific correlations with behaviour discussed in Chapter 4. In experimental analysis of theta mechanisms, this variety of theta activity is seen in its purest form as the type 2 theta activity, solely dependent on cholinergic and other direct inputs from the medial septum/diagonal band neurons. The hippocampus merely generates its theta rhythm without entraining other structures in that rhythm. Here the two generators of theta activity in CA1 and the dentate gyrus are controlled from a common source, in a phase-reversed manner, but do not influence one another. This mode of operation is not influenced by cortical disconnection.

In the *entrainment* mode hippocampus and cortex establish resonant activity in reciprocal loops between the two, the conditions being produced for completion of these loops. Theta activity is then not dominated by cholinergic mechanisms, but is partly dependent on cortical links with the hippocampus, since it is attenuated by disconnection from the cortex. This variety of theta activity is abolished by urethane and other drugs, (leaving only type 2 theta activity). In this mode the dentate gyrus theta rhythms appear to control those in CA1: as described earlier, there is then a standard delay within the hippocampus, one which we can now identify as mainly a conduction delay.

This being so, the chance of resonant loops forming depends very much on theta frequency. If it is too low, the signal from dentate gyrus to CA1 will arrive at CA1 out of phase with theta activity generated there from septal inputs. At higher frequencies, the transverse pathway can be entrained by strengthening of connections. It is therefore of great interest that, in the rat at least, type 1 theta has a characteristic frequency range, which is significantly greater than that of the type 2 (baseline) theta.

The distinction between the "baseline oscillator" and the "entrainment" modes of hippocampal operation also has an important bearing on the data reviewed in Chapter 4. In that chapter a variety of correlates of theta activity were described, differing somewhat between species, and also with many exceptions, which were often related to learning situations. In Chapter 5 (Sect. 5.6.2), it was suggested that most of these species-specific correlations were coordinated in the brainstem, representing the innate control mechanisms for triggering septal rhythmic activity. It was also suggested that the information processing function of the theta rhythm was not to be revealed from these species-specific brainstem-mediated correlations, but rather from the interaction of the theta rhythm at higher levels of the neuraxis. *We can now equate the brainstem-mediated species-specific correlations of theta activity with the "baseline oscillator" mode of hippocampal function, leaving the large amount of data which is exceptional to these correlations as a function of the "entrainment" mode.* One point requires to be added, however: the "type 1" behaviours are *both* species-specific correlates of theta activity, *and* epochs when entrainment of the cortex takes place.

9.2.3 Further Implications of Self-Organizing Phase-Locked Loops

A number of other observations contained in the literature are also consistent with the idea of resonance outlined above, though they are not strong enough to establish the case in the first place.

Firstly, amongst the literature reviewed above is evidence that electrographic waves recorded in the hippocampus may include activity at the second or third harmonic of the fundamental theta frequency (see for instance Feenstra and Holsheimer 1979; Leung *et al.* 1982; Buzsaki *et al.* 1986). This is a likely eventuality in the scheme proposed. Loops having loop times of roughly half or a third of the theta period might become strengthened by the same mechanisms as establish those with loop times of the fundamental period. This would only be expected to occur during the entrainment mode of theta activity, and indeed Leung *et al.* (1982) find that harmonics of the fundamental frequency are more pronounced during type 1 behaviour.

The *second* type of corroboratory evidence relates to the situation of atropine-resistant theta activity, in which the remaining theta activity is dependent on cortical connections as well as on non-cholinergic inputs from the septum. Under atropine, the theta rhythm in CA1 may continue for brief intervals while interneurons in CA1 cease firing (Buzsaki *et al.* 1983). This suggests that for atropine-resistant theta activity an influence other than the cholinergic pathway from the septum can take over complete control of the CA1 theta rhythm for short periods. The present hypothesis suggests that this other source of control is the dentate gyrus, under further resonant control from the cortex.

Thirdly, it has occasionally been found that theta activity can occur when the medial septal pacemaker is incapable of function (see discussion of the results of Azzaroni and Parmeggiani 1967, Destrade and Ott 1982 and Borst *et al.* 1987 in Sect. 8.4.2) These observations suggest that when resonant circuits have been strongly established, the temporal relations so represented between the parts of the circuits may be firm enough to produce some sort of rhythmic theta activity even when the primary pacemaker is out of action.

Following this last idea, it is possible that, once resonant loops between hippocampus and cortex are established, many other links throughout the cortex can become entrained to the rhythm transmitted by these loops. This will tend to set up a large scale structure of strengthened connections, which determines temporal as well as spatial relations between the firing of cortical neurons. This spatio-temporal structure does not need the rhythmic activity of the medial septum to control its operation. In other words, the development of cortico-hippocampal circuits plays a priming role in *establishing* the spatial and temporal structure of cortical activity generally, but is not always needed for *expression* of that structure.

Finally, the conclusion we can carry forwards to a later section concerns the cortical end of the self-organizing phase-locked loops. When resonant circuits have been established between hippocampus and association cortex, there will be a population of cortical neurons, in association or limbic cortex, which are receiving some degree of maintained neural input from elsewhere in the forebrain, together with a pulsatile input from the hippocampus. While the resonance lasts, these neurons will tend to be more excitable than other neurons in the same region which do not participate in the resonant circuits. Reciprocal inhibition is a widespread phenomenon in cortical local circuitry, and so one may envisage that when a population of association cortical neurons is set at a higher level of excitability, due to resonance, the remainder of neurons in that region will undergo a corresponding depression of their excitability. Thus, at the cortical ends of a pattern of resonant loops, a specific group of neurons will be selected to be set at a higher level of activity, simultaneously suppressing the activity of other neighbouring neurons.

9.2.4 A Prediction: Phase-locking of Cortical Theta Rhythms to Hippocampal Theta Rhythms

If phase-locked loops are self-organized between the isocortex and hippocampus, one would expect that widespread loci in the isocortex should show theta rhythms when the hippocampus has these rhythms. The evidence reviewed in Chapter 7 shows that this is true of the isocortical regions most closely linked to the hippocampal formaion. There are additional predictions which can also be made: that any theta activity recorded in the cortex during hippocampal theta activity should have the same frequency as the hippocampal rhythm, a consistent phase relation to it at any one frequency, and constant time delay with respect to the hippocampal rhythm if frequency changes. This subject is discussed briefly below.

9.2.4.1 Phase Relation Between Theta Potentials in Entorhinal Cortex and the Hippocampus

Two studies of theta activity in the entorhinal cortex give evidence of phase/time relations between the entorhinal cortex and the hippocampus. In one of these (Alonso and Garcia-Austt, 1987a,b), the superficial layers of the entorhinal cortex are found to be approximately in phase with the deep hippocampal theta source, around, and deeper than the hippocampal fissure (rather than with that originating superficially in CA1). Control of hippocampal theta activity by the entorhinal cortex is advocated by the authors. This sounds promising since it would be in accord with the ideas developed above (Chaps. 6, 8, and the present chapter). However, these authors used curarized rats under the influence of physostigmine, a preparation which, as regards profiles of theta activity is more like the urethanized rat. Moreover it is difficult to relate this idea to any specific neuronal model since the medial septal input, envisaged as the origin of the entorhinal theta rhythm, innervates both superficial and deep layers of this cortex, and the superficial layers, whose neurons project to the hippocampus also contain dendrites of neurons in the deeper layers.

In the second study (Mitchell and Ranck 1980) theta activity was recorded in the entorhinal cortex of free-moving rats, a preparation in which we have stronger reasons to suspect that cortico-hippocampal resonance may be occurring. Theta activity in the deeper layers of the entorhinal cortex was approximately in phase with that in CA1. Since CA1 projects to the deeper laminae of the entorhinal cortex, this is compatible with the idea that CA1 controls the deep entorhinal layers via the transverse pathway through the hippocampus. The deep layers of entorhinal cortex were in approximate antiphase with that recorded in the superficial layers of the entorhinal cortex and the

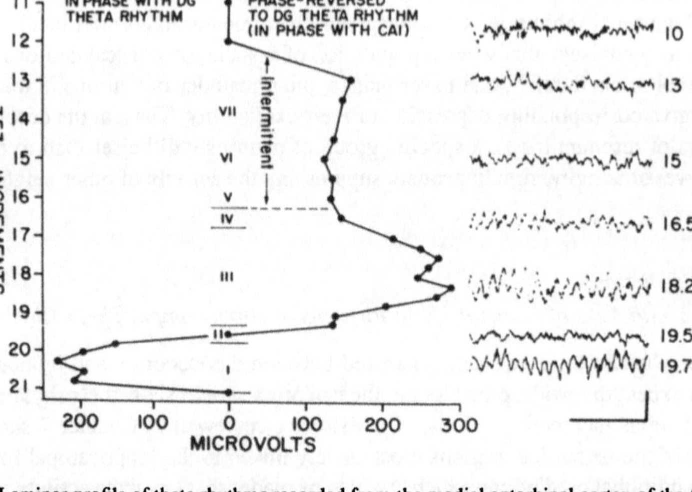

Fig. 42. Laminar profile of theta rhythms recorded from the medial entorhinal cortex of the free moving rat. *Left* Amplitude profile at various points on an electrode track through the medial entorhinal cortex (external layers I and II below). *Right* Samples of theta activity at sites on the track indicated by *numbers* (corresponding to the vertical axis in the graph on the *left*). Calibrations: 1 s, 500 μV. (Mitchell and Ranck, 1980) (From *Brain Research* Vol 189 pp. 49–60 1980)

dentate gyrus (see Fig. 42). Specifically, the dentate gyrus usually led the inverted theta activity recorded from the deep entorhinal cortex by 13–43 degrees. (The frequency of theta activity in these experiments ranges between 6.49 and 7.35 Hz). These values seem to contradict the idea that the dentate gyrus and Ammon's horn control rhythmic activity in a loop which includes the entorhinal cortex, since the delays are much smaller than those apparent just within the hippocampus. However, it may be remembered that in Section 6.3.2.1, a suggestion was made that there may be resonance not only between hippocampus and isocortex, but also by shorter pathways between hippocampus and entorhinal cortex. There is a possible way in which resonance can occur, suitable for pathways shorter than those to the main mass of isocortex: we know that potentials produced at a distance along dendrites from a group of active synapses may be the opposite sign of those produced at the location of those synapses. Therefore phase locking could take place to an inverted rhythm with conduction delays corresponding to just one half of the oscillator cycle. There is some suggestion in Mitchell and Ranck's study that in the entorhinal cortex theta activity locked to the inverted half-cycle is present. For this to occur, the phase difference from the dentate gyrus to the entorhinal cortex would be 180 degrees plus 13–43 degrees, i.e., 193–223 degrees. At the observed frequency of theta, this would correspond to a delay of about 84 ms. Mitchell and Ranck do not give data on the frequency of theta activity for each measurement of phase shift. It is predicted that, following Leung (1984b), that this delay between dentate gyrus theta activity and the inverted deep entorhinal theta should remain constant as frequency changes.

The idea of resonance locked to the inverted half cycle can also help to make sense of the data of Brown (1968) discussed in earlier chapters. In Chapter 7 (Fig. 31/2A) analysis of these results showed that at high frequencies (> 4.3 Hz) the rhythm in the parahippocampal gyrus led that in the dorsal hippocampus. Around 4.3 Hz the negative phases of the rhythms in the two recording sites were approximately in register. Below this frequency the dorsal hippocampus appeared to lead the parahippocampal gyrus. These data make no sense in terms of a constant delay between *equivalent* phases of the two rhythms. However, they make better sense if there is a constant delay between the positive phase of the rhythm in the parahippocampal gyrus and the negative phase of that in the dorsal hippocampus. The data of Brown are of theta periods for each complete cycle rather than the *half-cycle*, the precise site of recording is not stated, and the amount of data presented is small. Hence a proper statistical analysis is of these data is not called for. Nevertheless, qualitative support is given from these data for the hypothesis (represented diagramatically in Fig. 43).

These suggestions are at best tentative, and a detailed neuronal model cannot yet be guessed. In particular, in the results of Mitchell and Ranck (1980) it is unclear whether the deeper entorhinal theta activity is locally generated, or merely volume conducted from CA1 stratum oriens, with whose theta it is in phase. Although it is reasonable to argue that theta potentials in the superficial layers of the entorhinal cortex have the right phase to be involved in control of dentate theta activity, it is not clear how those theta potentials could control firing of the superficial neurons which project to the dentate gyrus. After all, the hypothesis suggests that the superficial theta potentials are generated by dendrites of neurons whose somata are located in the deep layers, where inverted

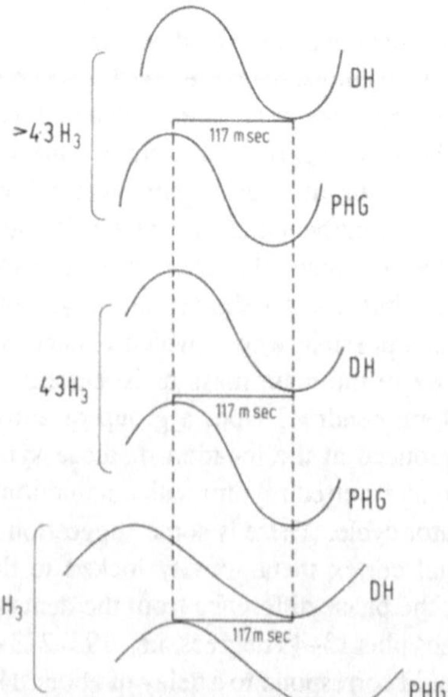

Fig. 43. Schematic representation of the possible relation between theta rhythms recorded in the parahippocampal gyrus (*PHG*) and the dorsal hippocampus (*DH*) of the free moving cat. This diagram is to be used to interpret the data of Brown (1968) presented in Fig. 32. The relation depicted is one of constant delay between the peak positivity of the wave in the parahippocampal gyrus and the peak negativity of the following wave in the dorsal hippocampus. At a frequency of 4.3 Hz the theoretical delay should be 117 ms.

potentials occur. These superficial dendrites of *deep* cells have no obvious means of controlling *projection neurons* with somata in the superficial layers[8].

9.2.4.2 Phase Relation Between Theta Potentials in the Hippocampus and the Isocortex Generally

Three papers have attempted to establish consistent phase relationships between rhythmic EEG activity at specified frequencies in hippocampus, and in specific loci of the cortex (Basar *et al.* 1979a,b; Aleksanov *et al.* 1986). In neither of the first two did coherence reach acceptable levels of significance. In the third, during the course of alimentary conditioning, and at the time of presentation of the CS, the coherence between hippocampus and prefrontal cortex increased, especially in the theta and delta frequency ranges. For the theta activity recorded from the cingulate cortex, the evidence available about phase relationships is not adequate to decide whether this rhythmic activity controls, or is controlled by the hippocampus. These are, however, merely first

[8]However, Krnjevic *et al.* (1986) have shown that the synchronous firing of pyramidal cells in Ammon's horn, produced by antidromic volleys, produces large enough extracellular fields to affect significantly, by ephaptic transmission, the firing of neurons which are below threshold for direct antidromic activation. Conceivably, the fields produced by synchronous firing of deep entorhinal cortex neurons during theta rhythm generation there could similarly influence the firing of superficial entorhinal neurons, which are not themselves the initiators of this rhythm.

attempts to find evidence for a hypothetical phenomenon which is, by its very nature, rather elusive. Failure to observe significant coherence might be due to inadequate spatial resolution at the cortical recording site, use of only a single cortical recording site, or selection of records at times when the hippocampus was not displaying theta activity.

9.2.4.3 Summary of Evidence, and Discussion

It is clear from the above two subsections that we are at a very early stage of investigation of the phase relationships between hippocampal and cortical theta activity. Although the evidence suggests a number of intriguing possibilities for resonance, there is as yet no clear evidence that the more distantly related areas of association cortex generate theta potentials reflecting the hippocampal ones. There is as yet therefore no evidence of phase locking between isocortex and hippocampus.

Despite the lack of evidence at present, the central prediction is still made *that there be cortical nodes with consistent phase relations to hippocampal theta activity*, both as leaders and followers of that activity. It is not surprising that the evidence available is slight, since experimentation has consisted of description (sophisticated but arbitrary) with no hypothesis as a guide. With the present hypothesis in mind, it may be possible to design better experiments to test its central prediction. It is suggested that the reason why evidence for locally generated theta rhythms has not so far been obtained from the isocortex is that the different but closely adjacent loci which generate such rhythms will have different phase relations to the hippocampal rhythm and to each other. In addition, it is likely that the rhythms generated at cortical loci are obscured by noise to a greater extent than in the hippocampus. Since cortical EEGs have usually been studied with quite large electrodes, it is likely that any theta rhythm present there is hidden because of spatial averaging of the rhythm from several different loci, each only weakly rhythmic, and with different phase relations. (This is illustrated in Fig. 40.) This difficulty can easily be overcome by use of smaller-scale electrodes. There is also a requirement for more specialized statistical techniques. Although not yet widely used, the statistical and experimental techniques do exist for revealing consistent phase/frequency relationships (Nunez 1981; Thatcher *et al.* 1986) even though they may be hidden by close proximity of phase-dispersed theta generators.

9.3 Role of Resonant Phase-Locked Loops in Information Processing

9.3.1 Recapitulation

To understand the role of self-organizing phase-locked loops between hippocampus and association/limbic cortex in the overall strategies of information processing in the forebrain, it is necessary to recapitulate the conclusions of Chapters 1 and 2.

1. In early post-Renaissance philosophy it was thought that, given valid premises, the validity of the conclusions derived from them must follow unconditionally. Immanuel Kant, however, argued that for all the more substantial examples of deductive inference, this was not so. There had to be given in addition *the context* for logical operations.

2. In early twentieth century learning theory many psychologists believed that pairing of two stimuli, and if necessary the existence of reinforcement which is regularly

contingent upon the specific response evoked by a specific stimulus, is all that is required to establish a functional link between a stimulus and a response. Given these consistent conjunctions to strengthen specific connections, input was assumed to dictate output unconditionally. Later in the twentieth century, however, it was realized that this idealistic approximation did not match the reality of animal learning and behaviour. Usually, it was found that, if input was to dictate output, other conditions had to be fulfilled. These other conditions were variously described as "contexts", or "dimensions", or the conditional nature of input-output relations might be described in terms of "selective attention". Thus there is an inherent ambiguity in the relations between stimulus and response, just as there is between premise and conclusion. This ambiguity requires further matching patterns of information for its resolution, patterns to which we can give the general term "contexts".

3. From a consideration of the information structure of the environment, and the manner of its representation in the cortex, a biological counterpart of this psychological and philosophical ambiguity can be found: small scale features of the environment may be inherently ambiguous. Single images ("stimuli") may belong to more than one object. Single objects may be potential triggers for more than one response. It is clear that in fact these ambiguities *can* be resolved in many cases by taking into account the whole of the environment in which the image or object presents itself. This appears to require the formation of a "global cell assembly" representing the environment as a whole, within which the representation of the single image or object can be given a clear interpretation. However, the usual principles underlying the formation of cell assemblies are inadequate to accomplish this task. This is because it is only in relatively small blocks of cortical tissue that there are sufficient connections between neurons for Hebbian cell assemblies to form. A "global cell assembly" by definition requires coordinated activity in cells scattered throughout large areas of the cortex. Single neurons contain insufficient connections for this to occur in large blocks of cortical tissue. It is therefore difficult to see how a representation of the environment can be used to resolve the ambiguity in the relation between image and object, or between object and response, unless organizing principles in addition to the Hebbian cell assembly are employed.

9.3.2 Phase-Locked Loops and the Resolution of Ambiguity

In Section 9.2.3 the conclusion was reached that collections of widely scattered cortical cells can be maintained in a state of heightened activation by the pattern of resonant phase-locked loops. These cortical cells—a resonant group—represent the environment[9]. In any environment, other groupings of cortical cells will be activated besides the resonant group: small regions of cortex will contain active local cell assemblies, representing images, and objects which commonly occur in that environment. Other things being equal, the relationship of these cell assemblies to each other and to any other

[9]The word "environment" is not, of course, confined to a spatial environment or "place", but is used more generally as a "context" or "dimension" within which other information, whether spatial or not, can be interpreted.

cell assemblies inherent in the total neural network may be ambiguous. For instance, the cell assembly for a single image may be as likely to address a cell assembly for an object found in a different environment as one found only in the current environment. A cell assembly for an object may be as likely to trigger a response appropriate to a different environment as it is to trigger that appropriate to the current environment.

However, the situation is different when a widely spread resonant group of cortical cells is concurrently active. A local cell assembly is more likely to be activated if it contains a few neurons which are also part of the active resonant group. If it contains none of the cells in the resonant group, it is, relatively speaking, likely to be suppressed by the inhibitory influence spread by the members of the resonant group. Therefore the resonant group can coordinate local cell assemblies throughout the cortex. All local cell assemblies which represent objects or events occurring in a particular environment will tend to be activated by the resonant group which represents that environment. Activation of the assembly for a single image plus a particular resonant group can thus activate the cell assembly representing the specific object to which the image belongs in that environment, though it could not be so specific without the resonant group. A single object cell assembly plus a resonant group could activate just the response intention appropriate to that environment, although the object representation may have equally strong connections with other response intentions appropriate to other environments. Thus the activation of resonant groups allows retrieval of particular classes of information from the totality of information represented in the cortical network (see Fig. 44).

9.3.2.1 An Implication: Units in the Hippocampus Representing Environments

The above theory has a bearing on electrophysiological properties of the hippocampus other than those described so far. If the resonating groups of cells in the cortex collectively represent an environment, it would be expected that the cells in the hippocampus at the other end of the resonating loops would also represent the same environment. Moreover, due to the convergence of axonal pathways on the hippocampus, it is likely to be easier to detect such representation in hippocampal neurons than in isocortical ones.

This conjecture is, of course, true: many of the hippocampal neurons, at least in the rat, appear to be "place units". In other words, as the rat explores its environment, they tend to fire consistently when the rat is in particular parts of its environment. O'Keefe has given this idea considerable prominance after his papers of 1971 (O'Keefe and Dostrovsky) and 1976, and in the monograph by O'Keefe and Nadel (1978). This topic is important here, and so deserves detailed description.

The finding that many hippocampal neurons respond when the animal is in a specific part of its environment has been made in several different laboratories (O'Keefe and Dostrovsky 1971; O'Keefe 1976; O'Keefe and Conway 1978; Olton *et al.* 1978; Hill, 1978; O'Keefe, 1979; Best and Ranck 1982; McNaughton *et al.* 1984; O'Keefe and Speakman 1987; Muller *et al.* 1987; Muller and Kubie 1987). Most of this work has been in rats, but some work in rabbits is mentioned by O'Keefe (1979). In an early paper, Ranck (1973) produced apparently opposed results. However, when Best and Ranck (1982) reinvestigated this topic they found a number of clear place cells and comment

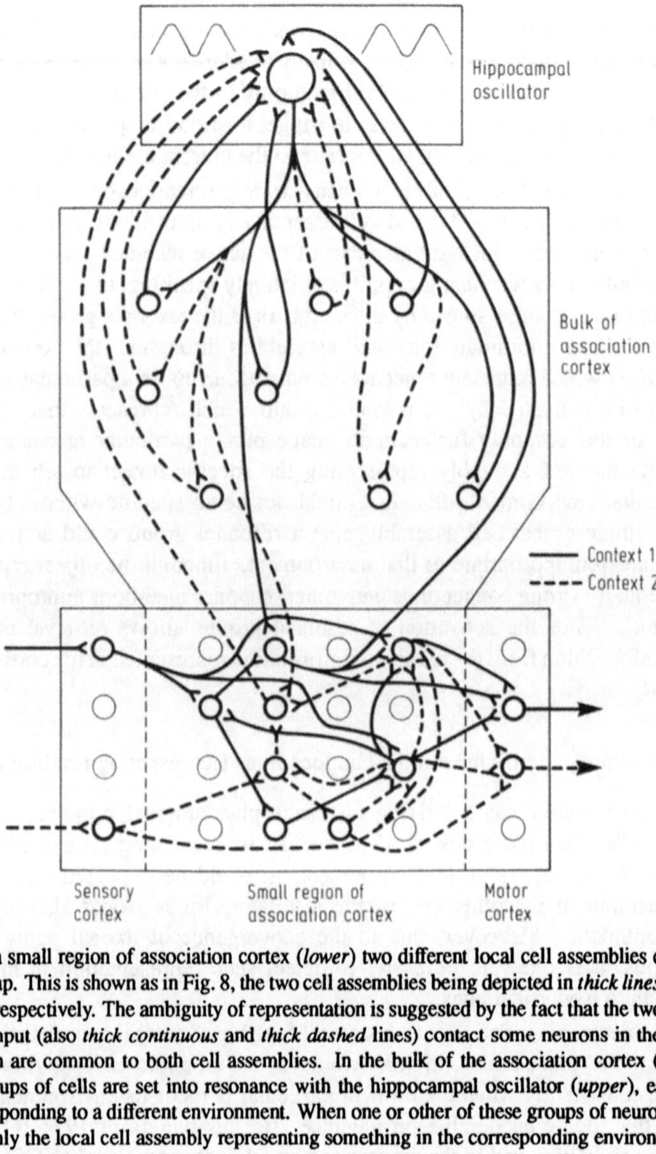

Fig. 44. In a small region of association cortex (*lower*) two different local cell assemblies coexist, with partial overlap. This is shown as in Fig. 8, the two cell assemblies being depicted in *thick lines*, *continuous* and *dashed* respectively. The ambiguity of representation is suggested by the fact that the two active lines of sensory input (also *thick continuous* and *thick dashed* lines) contact some neurons in the association cortex which are common to both cell assemblies. In the bulk of the association cortex (*middle*) two different groups of cells are set into resonance with the hippocampal oscillator (*upper*), each resonant group corresponding to a different environment. When one or other of these groups of neurons is set into resonance only the local cell assembly representing something in the corresponding environment will be activated. Any ambiguity resulting from the overlap of local cell assemblies will thus be resolved, and unambiguous relations between input and output can be established. Compare with Figs. 6 and 8

that in the earlier study Ranck (1973) had seriously underestimated the spatial correlates of unit firing.

Place cells are all complex spike cells, according to Ranck's criteria for rat hippocampus. In contrast, theta cells fire indiscriminately with regard to place whenever the rat is moving (O'Keefe 1976; O'Keefe and Conway 1978; Best and Ranck 1982; McNaughton *et al.* 1984). Place cells are found in fields CA1 and CA3, and in the dentate gyrus and medial entorhinal cortex (O'Keefe 1979). These findings are all to be expected from the present hypothesis, since it is the pyramidal cells rather than the interneurons which participate in resonance. The sensory input on which place selectivity depends is certainly multimodal. Place cells can fire appropriately in a darkened room (O'Keefe 1976). Nevertheless, visual sense appeared to be dominant in the determination of place responses (O'Keefe 1979). According to O'Keefe (1979), it did not matter whether the animal reached its place field passively or actively by its own movement.

Several variations of the "place unit" have been described. Early studies did not see an association between the motivation or reward value of a place and the density of representation by place units. However, O'Keefe and Conway (1978) found that place fields were overrepresented in the goal arm of a maze. A minority of the "place units" require, in addition to the animal being in a specific place, that other easily identifiable specific stimuli or responses of the animal occur. (These behavioural response components of neuronal selectivity are described in a later section). One variety of place cell described by O'Keefe (1976) is the "misplace cell". Such cells require, in addition to specific location of the animal, that it engage in sniffing or exploring, as a response to an unexpected presence or absence of something in the place field. This description tallies with that of Ranck (1973) for "approach-consumate-mismatch" cells.

Various studies have attempted to define the exact nature of the place representation. In particular they have been concerned to evaluate O'Keefe and Nadel's (1978) idea that the hippocampus forms a topographic map of space in Euclidean manner. Early work failed to find a specific stimulus or combination of stimuli which could trigger the firing of typical place cells as efficiently as the abstract global concept of "place" (O'Keefe 1976; Olton *et al.* 1978). However, later studies have suggested that place selectivity can be analyzed as responsiveness to combinations of objects distributed around the environment. This was implicit in some of the earlier studies, where it was noted that some place cells required the animal to face a particular direction as well as being in a particular place (O'Keefe and Dostrovsky 1971; Olton *et al.* 1978). O'Keefe and Conway (1978) and O'Keefe and Speakman (1987) showed definitely that place selectivity of hippocampal cells could be defined in terms of responsiveness to combinations of cues, including those immediately related to the relevant task, other neighbouring cues, as well as stationary background cues.

Muller *et al.* (1987) used a cylindrical enclosure with a single white vertical stripe as the sole cue to place in the environment. They plotted in a systematic way the intensity of firing in each position of the enclosure. They considered the possibility that place unit firing may signal the distance from the centre of the place fields, so that the animal's position in space could then be represented by synthesis of the distance from the centres of several overlapping place fields. However, they reject this hypothesis for several reasons. Some place units have more than one field in a given apparatus, which would introduce unecessary noise into such a representation. Some place fields are non-circular, which is difficult to reconcile with a distance-coding hypothesis. Thirdly, they used information on the firing intensity of various place cells, (whose place fields overlapped) to estimate a calculated position of the animal. This calculated position was in systematic error compared with the actual position of the animal which gave the firing patterns used in the calculation. This argues against the idea that the hippocampus maps Euclidean space.

In a second paper (Muller and Kubie 1987), various manipulations of the cylindrical enclosure were carried out and their effects on place fields studied. Changes in vertical height or width of the white stripe did not markedly affect place fields. Removing the stripe altogether left size, shape and relative radial position unchanged, but caused place fields to rotate to unpredictable angular positions. If the cylinder was changed to one with twice the diameter and height, about one third of the place cells retained the same angular and relative radial positions (scaled up in terms of absolute distances), while a larger proportion of cells (52%) showed a complete change of place-related firing. The same changes were seen in rectangular enclosures of different sizes. However, on transference from a cylindrical enclosure to a rectangular one, there was no trace of similarity between the place fields in the two environments. Placing of small barriers (opaque or transparent) to bisect a previously mapped place field almost always abolished the place field, without altering other neighbouring ones. On one occasion, however, a place field was found to be remapped after it had been bisected, in the vicinitiy of the old place field.

Eichenbaum *et al.* (1987) also argue against the idea of systematic mapping of space. Their argument was based on the fact that place fields were not homogeneously distributed, but were clustered around the location which was to be approached. In addition, place unit firing often consisted of firing in the place-to-be-occupied as much as 2 s in advance of arriving there. This latter observation suggests that a pattern of resonance may serve to recognize a particular environment as a whole and impose a prediction on parts of the representation of that environment, even before that part is actual presented to the animal.

The conclusion to be drawn from all this evidence about place coding by hippocampal complex-spike cells is as follows. There seems to be no topographic mapping of space across the hippocampus, and individual cells do not appear to represent either place or distance in a linear Euclidean manner. It seems more likely that *the responsiveness of each place unit is determined by a combination of a small number of cues in the environment, this combination being a quasi-random selection of the cues available to the animal at each point of space that it occupies.* It also appears that each place unit has access to several such combinations of cue representations. If one set of cues is deleted, other quite different cues may come to control responses of the place units.

The existence of place cells in the isocortex has not been exhaustively studied. The parts of the cortex most closely related to the hippocampus (e.g., the cingulate cortex) have not been explored at all. However, typical place cells appear not to be present in the isocortex. Nevertheless, cells responding to particular stimuli or responses with a spatial aspect are found in various regions of the isocortex (parietal area 7: Mountcastle *et al.* 1975; prefrontal cortex: Niki 1974a,b). The above evidence analyzing the responsiveness of hippocampal place cells suggests that their selective responsiveness results from convergence from several cortical cells with such less-global response characteristics.

A more detailed prediction can be made about the selective responsiveness of hippocampal place cells in relation to the theta rhythm. Either the place fields should be present only when the hippocampus is generating theta activity, or the place field should be more sharply defined in this condition. It is not clear from most of the above evidence whether or not the firing of a place unit when the animal is in the unit's place field depends on concomitant occurrence of theta activity. According to O'Keefe's (1976) illustrations, there appears to be some relation with theta activity but theta activity alone is not a sufficient criterion. However, Muller *et al.* (1987) make a more definite statement, that specific place unit firing could occur in absence of theta activity, but if there *was* theta activity, the firing of the "in-field" units was higher, and that of the "out of field" units was lower than in the absence of theta activity. This statement supports the present theory.

9.3.2.2 Evidence for Acquisition of Context Representations

In Chapter 1, the claim of O'Keefe and Nadel (1978) was mentioned that context representations in the hippocampus were innate, primarily concerned with Euclidean space, this space being topographically laid out in the hippocampus. Evidence against the second of these claims has just been considered. Evidence that the hippocampal representation of contexts is more versatile than a simple topographic representation of space is to be considered below (Sect. 9.3.6.1). The claim that context representations are innate requires attention now, in relation to the discussion of place selectivity of hippocampal neurons.

Most place neurons will respond specifically on first exposure to a new environment (Hill 1978), which appears to support O'Keefe and Nadel's position on this issue. However, there is occasional evidence for acquisition of new response properties on the first one or two exposures. For instance, one misplace cell of O'Keefe (1976) required that the animal engage in exploratory sniffing over a maze arm end as well as that it be in the arm end as a place field. On the first trial in darkness this combination of place and response failed to fire the cell, but on the second and subsequent trials it did. The same was seen for some ordinary place cells on first exposure to darkness. Hill (1978) found 2/12 cells whose place response to a new environment did not emerge until up to 10 min had elapsed in that environment.

O'Keefe and Speakman (1987) provided more definite evidence for an erasable short term memory component of the place unit responsiveness. They used a plus-maze, with six controlled cues just outside the maze. At the start of the trials the controlled cues were present, but they were removed for the second half of the trial. The rat then had to use a working memory of these cues in relation to intra-maze cues and stationary background cues. Ninety percent of the place units maintained their place fields established early in each trial despite this deletion of the cues which were most important in control trials. The longest period for which this working memory was demonstrated was 2 min. In typical trials of this sort the animal was able to locate the goal arm quickly. If, however, at the time of removal of the controlled cues, the animal was forced to choose a non-goal arm, the place unit fields in this arm were still the same as before the cues were removed. Thus, the working memory applied not just to the relation between the start arm and the goal arm, but to the maze as a whole. It seemed that during the early part of each trial (when cues were present) place fields were set up for the entire maze. Since this information was specific to the individual trial (due to variation of the controlled cues between trials), the setting up of the working memory must have been accomplished very quickly (so the term "working memory" is appropriate). A little further evidence for this is mentioned in their paper. Between trials in the plus maze, rats were placed in a rectangular holding platform. Two units were seen to have short-lived place fields with respect to this platform as well.

Thus for most place units the specific responsiveness is established either instantaneously or very rapidly. This suggests that the coding that determined the firing pattern was already present somewhere in the brain, and could be used at short notice independently of the learning that takes place on first introduction to a new environment. This does not mean that this coding was never learned. It is possible that the continual exposure to a variety of spatial environments may give rats a basic repertoire of neural responses to different sorts of place, which can be imposed on each new environment without new learning. This seems to be the implication of the fact that rats can solve detour problems. A few examples of acquisition by place units of their specific responsiveness have been reported. Possibly these might be more common if experiments were conducted in young animals. There does not seem to be an innate map of Euclidean space, and in view of the above evidence, it is unproven that the hippocampal representation, whatever its nature, is innate.

9.3.2.3 Prediction: Memory Retrieval and Performance Dependent on Theta Frequency

The selective retrieval of information is supposed to depend on activation of specific patterns of resonating loops. It may be asked whether the class of information which can be retrieved depends on the frequency of resonance. In one sense this should not be so. At any one frequency of resonance a number of quite different patterns of resonance can be set up, each representing a different context. However, there is a sense in which retrieval of information is frequency-specific. It would be predicted that if the frequency of the theta rhythm at the time when information retrieval was required departed very much from that during the original acquisition of the information, then retrieval would fail. This would be because the appropriate cortical cells could not participate in their specific pattern of resonance and so could not be maintained in an activated state. The prediction from this argument is that there should exist a form of amnesia dependent on the non-matching of theta frequency between learning and retrieval.

There is a quite old set of observations which is compatible with this prediction. Yerkes and Dodson (1908) claimed that the relation between "arousal" and performance level was that of an inverted U. In other words, at very high arousal levels, and at very low arousal levels, performance was poor, with an optimum arousal level in between. More recently, this subject has been reviewed by Korchin (1964) and Eysenck (1982). Commonly the Yerkes-Dodson law, as it is called, has been atributed to an influence of "arousal" on the selectivity of "attention", though these two terms are not at all clearly defined or distinguished. A major drawback of this topic is that no very plausible biological basis has been suggested for why performance should deteriorate at very high arousal levels. From Chapter 5 it may be suggested that increasing arousal leads to an increase in frequency of the theta rhythm. It can then be conjectured that at low and high frequencies of theta activity, there is poor matching of the frequency during performance of a learned task with that obtaining during learning. The level of arousal at which performance is optimal would correspond to the intermediate theta frequency at which resonance was established during original learning.

Another possible example of the predicted amnesia is for dream material available to consciousness during REM sleep. It is common experience that this is often not well recollected in the awake state. According to Leung's (1984b) paper, in both REM sleep and in the awake state, there is a regular relation between theta frequency and the phase difference between deep and superficial generators. However, for a given frequency, the phase relation was systematically different between REM sleep and the awake state (5–30 degrees closer to full phase reversal in REM sleep than in the awake state). This seems to imply that for any one frequency, different systems of phase-locked loops are in operation during REM sleep than in the awake state. Whatever information is represented by the set of loops active in one state may not be accessible in the other state.

9.3.3 Categorical Recognition of an Environment

The question was raised at the start of this chapter whether the collection of cortical cells activated by a pattern of resonance had the essential property of a cell assembly. In other words, one may ask "does activation of a subset of such a collection tend to activate the whole of the collection, as it would in a Hebbian cell assembly?". Before considering theoretical possibilities there is some empirical evidence from behavioural studies which suggests that the answer to this question is "yes".

First, there are hints in the literature on place units which suggest that an environment can be recognized as a whole. As mentioned above, Eichenbaum *et al.* (1987) found that some place units responded predictively. That is, they were activated in an animal moving towards its place field, but actually up 1 or 2 s before it had reached the place field. Whatever the mechanism underlying this, it seems that the animal recognized the environment as a whole.

Secondly, there are behavioural experiments which point in the same direction. Restle (1957) addressed the issue of precisely what sensory cues are involved in maze learning. He cites earlier experiments of Hunter (1929, 1930) and Honzik (1936), in which the effect of destruction of specific senses was studied on either maze learning, or performance of an already learned task. Simple alternation mazes could be learned on the basis of kinesthetic information alone. However, more complex mazes could not be learned without the integrity of visual, olfactory or auditory cues. This suggests that a global representation using the distance senses is required for the complex mazes, whereas individual items of somatic sensory information can mediate simple maze learning in the absense of a global representation. Restle's own experiments showed that fewer cues are needed for *performance of a previously learned complex maze than for its initial acqisition.*

This single paper using the rat is concordant with very common human experiences. Imagine the situation of finding one's way about a darkened room or interior of a building. Provided we have become familiar with that interior in good lighting conditions, we are surprisingly good at navigating in near total darkness. Putting this in more general terms, it seems that acquisition of a complex maze task requires far more information about the environment than performance of it after its efficient acquisition. After acquisition, a very limited range of stimuli can take the place of all the information required during learning.

In both rat and man these navigation tasks are just the type which is supposed to require representation of the environment as a whole. One must conclude that *once a representation of the environment has been laid down, that environment can be recognized on the basis of quite a small portion of the information in that representation.* In other words animals and people do recognize familiar environments categorically, as wholes or Gestalts. As explained in Chapter 2, this feature is typical of pattern recognition on a smaller scale, and is probably accounted for in terms of the essential property of a cell assembly.

This raises another theoretical issue. At the cortical end of the phase-locked loops, the cells of a resonant group are too widely dispersed to be able to form a Hebbian cell assembly. To overcome this hurdle it would not be adequate to have a single functionally central area of the cortex (viz. the hippocampus) giving connections to and receiving them from all other areas of the cortex. Although this would have the convergence required for global representations, it would by the same token have cells whose firing is closely dependent on each other. Most projection neurons would then be so non-specific that they could not be coordinated to form environment-specific cell assemblies. By themselves collections of such neurons would then have little ability to specify one context rather than another. However, by arranging that the hippocampus is also a central oscillator, which can entrain resonant loops between it and specific nodes in the cortex, specificity of representation of a global pattern ("context") is retained, *but not as a Gestalt-like cell assembly.* Yet the data just reviewed seems to necessitate such an assembly to represent environments as wholes.

The resolution of this issue probably depends on the fact that the route between hippocampal oscillator and specific neural nodes of resonance in the association cortex involves the intervening stages of isocortex such as the cingulate cortex and the ento-, and peri-rhinal areas of cortex. Here the totality of all the connections which resonate in a particular context may congregate within a small enough area of cortex such that the omniconnection principle (Miller 1981) still obtains. The contiguities between all the resonant circuits active in a given situation may then be encoded. Therefore conventional cell assemblies would tend to develop in these regions, forming functional links between cells which tend to be activated together. According to the hypothesis developed above, such co-activation of cells would depend on two concurrent events: the cells should be all part of the same pattern of resonant circuits (i.e., activated by the same environment); and they should have the same phase relationship to the master oscillator in the hippocampus. Thus one might expect that in these cortical regions Hebbian cell assemblies may form, defined by the eliciting environment and the phase relation with respect to hippocampal theta rhythm. For such cell assemblies it would be legitimate to use the term "global cell assemblies".

9.3.4 Selective Attention

Closely related to the ability to recognize an environment or context categorically is the fact animals and people can generally only use one context representation at once. This is very obvious in introspective terms, and is also supported by the animal literature (Sutherland and MacKintosh 1971). The narrow selectivity which applies to the use of context representations is generally designated by the term "selective attention".

Clearly the selectiveness must imply that only one global cell assembly can be activated at once. This in turn implies that each global cell assembly tends to inhibit all others. At what site in the brain such reciprocal inhibition occurs is at present uncertain. However, the conjecture that the different global cell assemblies do tend systematically to inhibit one another, rather than merge into one another (as sometimes happens with local cell assemblies) is not surprising if one looks again at the information structure of the environment. Environments are usually sharply distinct from one another, and the occasions on which basic aspects of several environments become mixed together are rare. Therefore, there will be few occasions on which conditions are present in which the cell assemblies representing different environments might tend to coalesce.

Nevertheless in principle, an environment could be constructed which contained features of several previously familiar environments. There is no reason why a context representation, once acquired, should be immutable. One might therefore expect that resonances appertaining to the components of each environment would then be set into operation together. Provided that all resonances were at roughly similar frequencies, coalescence of the respective global cell assemblies could gradually occur. If, however, the frequency of resonance which specifies one of the global cell assemblies departs significantly from the other(s), it could not be incorporated into a single unifying global cell assembly. This prediction is accessible to experimental test in animals.

9.3.4.1 State-Dependence of Information Retrieval

These arguments are particularly relevant to the contexts underlying human conceptual systems. Such context representations may have been acquired in quite different circumstances, and yet have enough in common for them to be integrated. The ease with which such integration can occur is well known to vary greatly from person to person and integration of related but separate conceptual systems may sometimes be strongly resisted. The capacity for information retrieval may also vary from state to state. For instance, information acquired in states of high arousal or anxiety may be difficult to retrieve in more relaxed circumstances (or vice versa). There is here a potential biological basis for such well-attested psychological facts as state-dependent amnesia and repression of anxiety-provoking material. The conjecture is offered that for such failures of mental integration, the categories concerned are associated with substantially different frequencies of theta activity.

9.3.4.2 Electrophysiological Correlates of Selective Attention

Some years ago John *et al.* (1969) performed an experiment which appeared to identify an electrophysiological correlate of selective attention. Cats were trained to respond in different ways to somewhat different visual stimuli (circle or oval). After training, the animals were presented with ambiguous stimuli, whose shape was intermediate between those used in training. In these circumstances the animals would respond variably, sometimes as if they interpreted the shape as a circle, sometimes as if they saw it as the original oval. Records of mass electrical activity of the cortex were obtained at the presentation of each stimulus, both in the non-ambiguous and in the ambiguous conditions. In the non-ambiguous condition systematic differences were seen in the evoked waveform at the time of decision, which reflected the physical shape of the stimulus. More significantly, in the ambiguous condition, there were systematic differences correlating with the animal's *interpretation* of the stimulus, although it was physically the same in all cases. The different patterns of electrical activity in tests of the ambiguous stimuli tended to match the patterns produced by the non-ambiguous stimulus corresponding to the cat's interpretation.

In the context of the present theory, it is suggested that the ability of the intact cat to produce different patterns of electrical activity to a single ambiguous stimulus, depending on the evident meaning of that stimulus to the animal, is a reflection of the global cell assembly in operation. It is predicted that in animals with hippocampal ablation, or with septal lesions to prevent occurrence of theta activity, the electrical waveforms produced by the ambiguous stimulus should vary much less from trial to trial. So should the evident interpretation of the stimulus. There should be no systematic relation between the shape of the waveform and the manner of evident interpretation of the stimulus.

9.3.5 *Improved Economy of Information Storage Resulting from Cortico-Hippocampal Interplay*

In Chapter 2 the design problem for a cerebral cortex ultimately focussed on the fact that individual neurons had insufficient connections to perform the large-scale associative operations which were commonly required. Having presented the central hypothesis for cortico-hippocampal relations, the question remains: "How does this scheme improve the economy of information storage while still using the same anatomical building blocks?"

In the theory proposed, it is envisaged that a *neocortical* projection neuron involved in a resonant loop might use just *one* of its available collaterals, specified by its conduction delay, and strengthened by Hebbian mechanisms. This collateral provides the neuron with its functional link to the hippocampus. However, at the *hippocampal* end of the loop the situation would be different. A projection neuron there having a variety of axon collaterals each with a different conduction delay could use them *all* for different purposes (see schematic diagram in Fig. 40). Each one would take part in specifying different environments, according to its time relations with isocortical neurons activated in each environment. The hippocampus thus should have a potentially more economical way of using its available connections than the isocortex, although its cellular building blocks are not very different from those in the cortex. Some of the evidence about hippocampal place units suggests just such an economical representation:

Up to 90% of complex spike cells in the hippocampus show place fields (Olton *et al.* 1978; O'Keefe 1979). Some place cells have more than one field in a particular environment (O'Keefe 1976; Olton *et al.* 1978; Muller *et al.* 1987). In the study of Muller and Kubie (1987) change of the apparatus in which the rats were tested to one of the same shape but twice the size resulted in over half the place cells showing a complete change of place-related firing. When individual hippocampal cells are tested in more than one environment it is found that a large proportion (40% in O'Keefe and Conway 1978) of place cells has a field in each environment. The respective place fields bear little obvious relation to each other (O'Keefe and Conway 1978). This curious observation helps to explain why such a large proportion of complex spike cells has a field in a single environment. Neighbouring hippocampal neurons do not regularly have contiguous place fields (O'Keefe 1976).

In short, it seems that each hippocampal projection cell takes part in several (perhaps many) different representations of the environment, and few hippocampal neurons can be found which do not participate in the representation of any one environment. Isocortical neurons probably also can encode more than one association, but this multiple function is not so obvious as in the hippocampus. In addition, the redundancy typical of isocortical representation (for instance in the columnar organization of the visual cortex) seems to be quite foreign to the organization of the hippocampus.

9.3.6 Cortico-Hippocampal Interplay in Relation to the Striatum

The theory so far developed is easiest to understand in relation to learning defined by the contiguities of signals within the sensory input stimuli. The most important type of stimulus-defined learning is pattern learning. Global cell assemblies help to make a single ambiguous image of an object address the representation of the object to which it is likely to belong in the prevailing environment. Responses and response intentions have also been mentioned periodically. Modifiability of response patterns is more associated with instrumental learning—that mediated by reward and punishment—than with learning on the basis of contiguity in the sensory input. As described briefly in Chapter 2 (Sect. 2.3.2) such learning is seen as acquired initially in the striatum. However, the pathways from the striatum are envisaged to convey signals to the cortex representing the responses acquired by instrumental conditioning. By Hebbian processes acting in the cortex, the contiguities between these motor signals and other concurrent neural activity can be encoded (see also Fig. 2). In addition the cerebellum feeds signals related to motor responses to the cortex. These are more to do with detailed execution of a response than with its goal or intention. However, the cerebellar input to the cortex is also a source of information to be integrated with other information represented in the cortex by the Hebbian mechanism. We may therefore depict a more complete model of the cerebral cortex as in Fig. 45.

In this model of the cortex, the signals from striatum and cerebellum are the immediate determinant of response patterns efferent from the motor cortex. However, the contiguities encoded in the cortex between these signals and other concurrent events will come to influence motor output. For instance, as explained elsewhere (Miller 1988), in situations of extinction or reversal, the input-output pathways formed through the cortical network can override the immediate contingencies of reward and punishment which influence the patterns of response conveyed from the striatum.

Such ideas were initially conceived in terms of stimulus-response theory. So far no attempt has been made to express this hypothesis of cortico-striatal relations in a more up-to-date conceptual language, the language where cell assemblies represent the important objects and intentions. However, this is necessary if we are going to understand the rôle of contexts and global cell assemblies in instrumental responses.

The input to the cortex from striatum and cerebellum is directed not only to the motor cortex, but to the premotor areas, and the association areas of the parietal cortex (areas 5 and 7). It is very likely therefore that the *local* cell assemblies which form in these regions represent contiguities not only within the sensory input, but between sensory input and motor signals in the cortex. The *global* cell assemblies which form to represent the environment, partly in these areas, would be expected to represent not only the sensory aspects of the environment, but also its relation to motor output. In particular,

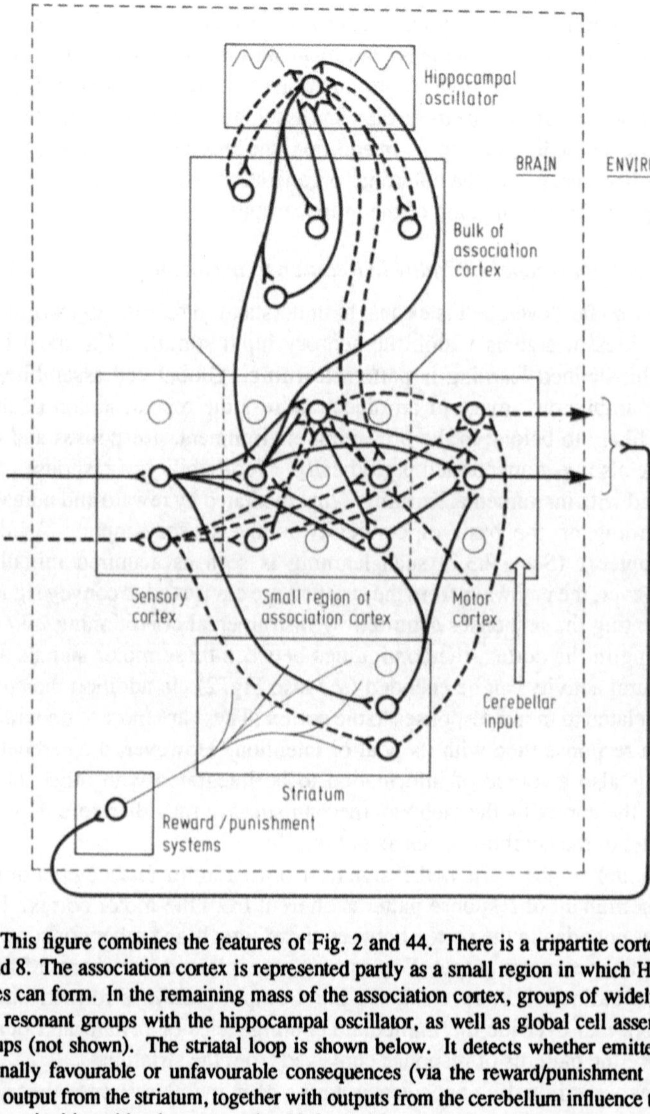

Fig. 45. This figure combines the features of Fig. 2 and 44. There is a tripartite cortex, as in Figs. 3, 4, 5, 6 and 8. The association cortex is represented partly as a small region in which Hebbian local cell assemblies can form. In the remaining mass of the association cortex, groups of widely dispersed cells can form resonant groups with the hippocampal oscillator, as well as global cell assemblies accessing such groups (not shown). The striatal loop is shown below. It detects whether emitted behaviour has motivationally favourable or unfavourable consequences (via the reward/punishment systems, *bottom left*). The output from the striatum, together with outputs from the cerebellum influence the motor cortex, and their contiguities with other concomitant information represented in the cortex can become coded

the specific consequences following each response should be included amongst the cues for recognizing which environment obtains (i.e., which global cell assembly to activate), this information being as important as the solely sensory contiguities. This conjecture receives support from the response properties of neurons at the other end of the resonant loops, that is the hippocampal projection cells.

9.3.6.1 Response Selectivity in Hippocampal Projections Cells

In most studies it has been reported that a minority of the hippocampal "place units" required, in addition to the animal being in a specific place, that other easily identifiable specific stimuli or responses occur. For instance, O'Keefe's (1976) "misplace" cells required, in addition to specific location of the animal, that it engage in sniffing or exploring, as a response to an unexpected presence or absence of something in the place field. This description tallies with that of Ranck (1973) for approach-consumate-mismatch cells. A few of the place cells reported in O'Keefe and Conway (1978) and McNaughton *et al.* (1984) required specific direction of movement as well as specific place. Best and Ranck (1982) also found 3/12 place neurons that required a response component. O'Keefe and Speakman (1987) using a plus-maze describe response components of place unit firing as unimportant, and Muller *et al.* (1987), using a cylindrical enclosure with a vertical white stripe as the only place cue found that place unit firing was remarkably independent of other aspects of the animal's behaviour. However, it was acknowledged in both these studies that small degrees of correlation with concurrent behaviour might have been overlooked.

Eichenbaum *et al.* (1987) recognized that there was a significant issue to be resolved here: On the one hand, in free moving animals, hippocampal units tended to fire in specific places, but in other situations specific stimulus or response correlates were more prominent. For instance, in various conditioning tasks clear responses to the conditioned stimuli could be observed (Segal and Olds 1972; Segal 1973; West *et al.* 1981), and in classical conditioning of the eyeblink in slightly restrained rabbits (Berger *et al.* 1983), the response of identified hippocampal pyramidal units correlated with the developing conditioned response. Eichenbaum et al. (1987) suggested that this apparent discrepancy was a consequence of the use of different behavioural requirements in the different studies. They therefore conducted experiments which had aspects of several of these types of study. Rats performed a successive odour discrimination task. The cage in which this occurred was divided into sections for analysis of place responsivness. Each trial consisted of approach to a sampling port, investigatory sniffing of the odour cue, followed by orientation and approach to a different location for receiving the reward, and finally retrieving the reward (water).

Out of 86 complex spike cells, 17 were "cue-sampling" cells, similar to Ranck's "approach-consummate-mismatch" cells, or O'Keefe's "misplace" cells. They argued, however, that the "mismatch aspect" is not the fundamental correlate of these cells' firing, but just a common concomitant aspect when the animal is engaging in investigatory sniffing. In addition to these cells, 69 out of the 86 were "goal-approach" cells, firing on arrival at either the odour-sampling port, or the reward location. In the apparatus used by Eichenbaum *et al.* both their major classes of cells could also have place correlates, but this aspect represented only part of the conditions necessary for firing of the units. Time-locking to onset of approach movements was often better than the place correlation. Moreover, in addition to the cells showing correlations with combination of place and response, there were more cells with only a response correlate than there were with only a place correlate. Further, some cells fired in a particular place only for one of the two possible approach responses (i.e., to either the sampling port or the reward cup, but not both).

Eichenbaum *et al.* discuss whether the twin classes of cells they describe are sharply distinct, or merely represent different sensitivities within a single population of cells. Although they do not reach a definite conclusion on this issue, their illustrations favour the second of these alternatives: cells which responded prominently either as "cue-samplers" or as "goal approach cells" nevertheless showed significant responses above baseline, during the other part of the task. When discussing place cells above, it was concluded that the responsiveness of each place unit is determined by a combination of a small number of cues, selected from the environment in a quasi-random manner from those available at each location. The same idea can be extended: in those behavioural situations where specific behaviours are required in specific places, behavioural response may be amongst the items from which a hippocampal projection neuron makes its quasi-random selection of things-to-be-represented. In any event response features definitely do contribute to hippocampal place representations.

The consequences of this conclusion are clear: global cell assemblies are partly defined by response information. These global cell assemblies therefore permit specific interpretations of local cell assemblies, which interpretations are determined in part by motor aspects of the environment (especially the relation between motor response and its consequences in each environment).

9.3.6.2 Habits Versus Memories

Mishkin and Petri (1984) have distinguished between two types of learned information, which they call ***habits*** and ***memories***. The former correspond roughly to the tasks studied by stimulus-response theorists. They are learned by appropriate repeated pairings of stimuli and responses, with reinforcement if necessary. They survive hippocampal damage. The latter may be acquired in a single trial, are vulnerable to manipulation of contextual background, and they are disrupted by hippocampal damage.

In the present theory, "habits" correspond to those tasks which can be acquired without the need for global cell assemblies. With sufficient repetition of very bold stimuli, which create very strong representation in the cortex, single image cell assemblies may be able to address uniquely the appropriate object cell assemblies without use of global cell assemblies. Likewise, object cell assemblies may sometimes have specific access to the appropriate response intentions without considering the context. In typical experiments of the stimulus-response type, contextual cues may be so uniform throughout that the animal shows no evidence in its behaviour of its reliance on context representations.

For more subtle information groupings, however, or in experiments where contextual background is one of the variables, impairments in memory retrieval may occur which indicate that contextual representations are important. In terms of the present hypothesis, global cell assemblies are required to open the pathways between input and output, or to allow such pathways through the cortical network to form in the first place. Such learned information corresponds to Mishkin and Petri's true "memories". For them to be acquired in a single trial presumably depends largely on the fact that exactly the same context is established on acquisition as on testing.

9.4 Summary

The major conclusions of this chapter are as follows: Four very plausible premises are required to build the theory: (1) A hippocampal oscillator is required (the generator of the theta rhythm). (2) Reciprocal connections between the hippocampus and the isocortex are needed which contain a rich repertoire of delay lines including those with total loop-times roughly matching the period of rhythms within the theta range. (3) Widespread synaptic modification by Hebbian or quasi-Hebbian rules must occur in both isocortex and hippocampus. (4) Groups of widely dispersed cells in the isocortex must be set into a state of elevated activation by the particular environment in which the animal finds itself. Given these premises, self-organization of resonant phase-locked loops between hippocampus and cortex is likely to occur.

There is as yet little evidence of the resonant activity in the isocortex, possibly because of technical inadequacy in experiments conducted so far. However, specific predictions are made, together with suggestions of the technical requirements for these experiments.

Resonant phase-locked loops are seen as playing a most important part in information processing in the forebrain. In particular, they are envisaged to resolve the ambiguity of cortical representation discussed in philosophical terms and with respect to learning theory in Chapter 1, and with respect to connectivity ratios of cortical neurons in Chapter 2. It is envisaged that ambiguity is resolved by representation of the environment or context as a whole, the latter being possible only by means of the patterned activity in these resonant circuits. Such resolution allows selection of specific categories of information whose representation is latent in the cortex. This is equivalent to the process of selective attention. Some predictions are made of the circumstances in which memory retrieval will fail due to failure of frequency matching of the theta rhythm between acquisition and retrieval of the memory.

Evidence that the hypothetical resonant circuits do in fact contain environmental representations comes from the study of place-selective neurons in the hippocampus, seen especially when the latter structure displays the theta rhythm. It is unproven that these context representations are innate, and in fact there are a few accounts, supporting the present theory, of place representations being acquired during an experiment.

Although the basic theory of self-organization of resonant loops does not predict that the environment be recognized categorically, as are smaller-scale Gestalts, other experimental evidence suggests that this is probably so. Environments can gain representation as Gestalts by normal Hebbian associative processes operating on collections of resonant neurons. The likely sites where Hebbian cell assemblies form to represent the environment are the regions of isocortex most directly linked to the hippocampus, rather than the bulk of the association cortex. The hypothetical resonance which forms between hippocampus and isocortex should allow the hippocampus to represent information more economically than the isocortex, since several collaterals of a single hippocampal neuron, with different associated conduction delays can particpiate in different groups of resonant loops. Evidence on the manner of representation of place in hippocampal neurons, for free moving animals strongly suggests a more economical representation of information than in typical regions of isocortex. The hippocampus is thus strategically of greater importance in memory coding and retrieval than areas of equivalent size elsewhere in the cortex.

The patterns of resonant circuits and the global cell assemblies which form by cortical-hippocampal interplay represent contexts generally, not just spatial contexts (places in an environment). Thus contexts can be specified in part by responses and their consequences, as well as by far more abstract groupings of information in human conceptual operations.

Two types of information processing are identified in this chapter which can be accomplished *without* the aid of the hippocampal oscillator. In Section 9.2.3 it was argued that the spatio-temporal structure of firing patterns in the isocortex is set up by the hippocampal oscillator, but does not always require that oscillator for its expression. Therefore retrieval of context-dependent memories acquired before hippocampal damage may be possible. What is prevented, however, is the registration of new memories in relation to the corresponding context. In this idea there may be a basis for the occurrence of forms of human amnesia (e.g., the Korsakow syndrome) where old memories are retrievable, but new events cannot be registered. In Section 9.3.6.2, it was suggested that a more elementary type of learning could be accomplished without the hippocampal circuitry. This is the broad class of learning tasks studied by the stimulus-response theorists, where information is "stamped in" to the cortical circuitry, dependent only on repetition of contiguous events, without involvement of contexts. In this case the hippocampus is not involved for either registration or retrieval. (In fact the distinction between registration and retrieval disappears for stimulus-response learning.) However, *the most powerful and specific information processing of which the hemisphere is capable requires that the appropriate context representation first be activated, and this involves resonance with the hippocampus.* Amongst such information processing is the ability to reconstruct in memory an event or sequence of events which have occurred only once.

Theta activity can, of course, occur without entraining the isocortex into resonance. Whether or not entrainment occurs, there are inevitably correlations between behaviour and occurrence of theta activity which have nothing to do with the entrainment itself. These are the innate correlations discussed in Chapter 4, determined by the fact that the brainstem mechanisms which initiate certain species-specific forms of behaviour also tend to trigger theta activity in association with these behaviours.

In different mammalian species, the relative size of hippocampus and isocortex varies enormously. In smooth-brained species such as rat or mouse the hippocampus is a significant fraction of the bulk of the whole hemisphere. In the human brain (or even more in the brain of whales) the hippocampus is quite miniscule in relation to the vast expanse of isocortex. This wide variation is a little difficult to comprehend. However, taking the model of information processing presented here in broad view, it is clear that the hippocampus is concerned with *organizing* and *accessing* stored information. However, the main *store* of that information would seem to be the isocortex. Clearly the size of the store can vary very greatly, in relationship to the varied ecological needs for large or small storage capacity in different species; but this does not require corresponding variation in the size of whatever processors organize and access the information in that store. One can therefore begin to understand the large variation in relative size of hippocampus and isocortex, in comparisons between species.

10 Theta Activity and Learning

10.1 Introduction

In Chapter 4, evidence was reviewed of the correlations between behaviour and theta activity in various species of experimental animal. Most of this data concerned correlations between theta activity and behaviours which were apparently innate rather than acquired. However, amongst the information reviewed was a variety of exceptions, which did not appear to fit into the regular correlations. Many of these exceptions occurred during learning situations.

In later chapters a theory was developed of how the hippocampus, in mutual interaction with the cortex, could resolve the ambiguity of representation in the cortex, which must be inherent in that structure, on grounds of its connectional anatomy, and the information structure of the environment. The theory involved resonance between hippocampus and cortex at the frequency of the theta rhythm. The theta rhythm could also occur without entraining resonance with the isocortex, as a "baseline oscillator" mode of function. This mode appeared to be relevant to the many static species-specific behavioural correlates reviewed in Chapter 4. However, in the "entrainment mode", cortical function was not treated in static fashion implicit in these species-specific correlations, but in a dynamic one. In other words, the information whose representation was ambiguous was assumed to be acquired by learning processes which are continually updating and reorganizing the cortical bank of memory. One may then ask whether the theory advanced in the previous chapter can give a detailed account of all the exceptions to the species-specific correlations between behaviour and theta activity.

It is therefore requisite that the original data on behavioural correlations of theta activity be reassessed. The aim of the present chapter is to consider the correlations between theta activity and various stages and forms of learning. It is thus hoped to account for some of the exceptions noted in Chapter 4. In fact, a number of specific predictions can be made about the role of theta rhythms in the interplay between cortex and hippocampus during learning. As phase-locked loops become established, and type II theta activity changes to type I, several changes should occur. Theta activity should stabilize and its frequency should become more restricted. As this happens, the brain regions in which it is found should extend. Phase relationships between theta activity in various regions of the brain should change. Examples of all these predicted developments of theta rhythms during learning are provided below. Some of these predictions require analysis of EEG records beyond mere visual examination. Fortunately a number of studies of theta activity during learning have conducted such analyses. Presentation of much of this evidence is organized with respect to the type of learning situation. Later sections deal with the relation between electrically driven theta activity and learning, and the regional spread and interregional phase relations of theta activity during learning.

Theta activity is not the only aspect of hippocampal electrophysiology which might be correlated with learning processes. There is also a substantial literature on single-unit firing in various stages and in various types of learning (e.g., Segal 1973; Segal and Olds, 1973; Deadwyler *et al.* 1979; Berger *et al.* 1980; Deadwyler *et al.* 1982; Berger *et al.* 1983; Hirano 1984). However, this literature will not be dealt with here because it does not generally document whether or not theta activity occurs at each stage of learning, and also (with the exception of Berger *et al.* 1983) generally fails to identify the cell type (projection cell or interneuron) from which recording is made.

10.2 Theta Activity in Response to Novelty

A single paper (Irmis *et al.* 1970) using repeated acoustic stimuli supports the early idea (Green and Arduini 1954) that *discrete, novel stimuli* can elicit theta activity on first presentation, with progressive habituation of this response thereafter. A larger number of studies investigate responses in a *novel environment*. Some of them (Porter *et al.* 1964; Routtenberg 1968; Bennett 1970) also indicate that theta activity may be triggered when an animal is first exposed to a totally new environment. These data seem to imply that theta activity is an indicator of the novelty of the environment. However, other studies indicate a rather different process, the production of regular theta waves only when a new environment is becoming familiar. Radulovacki and Adey (1965) found, in cats, that during the first introduction of the animal to a test situation in a T-maze when there was no orienting, the hippocampal EEG showed a wide spectrum of activity (3–7 Hz) in the pyramidal cell layer, with low voltage fast activity in basal parts of the dentate gyrus and in the subiculum. At a later stage, when there was orienting, there was more regular theta activity at 4–5 Hz. Brown (1968), again using cats, found that theta activity was not present on initial exposure to a novel environment, but developed only a little while later as visual attention/search became prominent. Occurrence of theta activity then declined as the animal adapted to the new environment, and what theta activity there was had a lower frequency than during visual search in early orienting. Thus, from the small amount of evidence available, the early idea that theta activity is a response to total novelty meets some difficulties. There is, however, agreement that when the process of familiarization to a new environment is completed, theta activity declines markedly, and not simply because locomotor activity declines too.

It is possible that this initial response to apparent novelty of a discrete stimulus occurred because the sensory stimulus had already acquired a signal value even before the experiment was started, due to previous exposure to the same or a similar stimulus. An experiment of Karmos *et al.* (1965) attempts to investigate this possibility. To do this cats were kept in a quite new environment for some weeks, with totally controlled environmental stimuli. Then, a strange single non-intense stimulus was given. It was reasoned that, though this stimulus or something resembling it may have been experienced before, the occurrence of that stimulus in the controlled environment was probably a totally new stimulus complex. In these conditions, the single non-intense stimulus did not produce theta activity, until it had been repeated enough to become somewhat familiar. On the other hand, intense stimuli could trigger theta activity on first presentation in the controlled environment. In addition the appearance of the experimenter (a stimulus to which the animal had become familiar in the controlled environment) provoked continuous theta activity. They concluded that novelty was not a sufficient condition for elicitation of theta activity or orienting.

The earlier chapters of this book provide two alternative ways of interpreting the above evidence. *On the one hand*, the hippocampal response to apparent novelty could be a simple consequence of control of the hippocampus from the septum and brainstem. In this context it may be mentioned that there are neurons in the brainstem which respond to novelty of a stimulus, with rapid habituation as the stimulus is repeated (Bell *et al.* 1964; Schiebel and Schiebel 1965; Brown and Buchwald 1973). In addition the single unit studies by Vinogradova (1975) can be mentioned. She found that (unidentified) hippocampal neurons responded to the quality of novelty, and argued that this selectivity was a function of the ascending input to these neurons from the septum. Thus one has a basis for results such as those of Porter *et al.*, Irmis *et al.*, Routtenberg, and Bennett, and in terms of *innate* mechanisms controlling theta activity.

On the other hand, some of the results mentioned suggest that the response to novelty is *acquired* over the first few occasions of exposure to the novel item or environment. The result of Radulovacki and Adey in particular suggests a learning process compatible with the theory of resonance presented in the previous chapter, since an EEG which initially had a wide frequency spectrum rapidly focussed to a narrower frequency band. The behavioural response of orienting correlated with the occurrence of this narrow theta frequency band.

There is no need to choose between these two alternatives: Both probably apply together.

10.3 Theta Activity During Learning

10.3.1 Theta Activity During Classical Conditioning

Schwartzbaum (1975) presented rats with a CS in a conditioned emotional response paradigm. This brought about an increase in theta activity, accompanied by the behavioural response of "freezing". The effect on hippocampal EEG was seen even after just a single pairing. In rabbits, Powell and Joseph (1974) used a classical conditioning procedure and found that theta activity tended to increase throughout training, but did not correlate well with gross movement. Buzsaki *et al.* (1981) studied cats undergoing conditioning to an unavoidable shock, signalled by a tone. Theta frequency increased from 3–4 Hz to 5–6 Hz in response to the CS and developed over trials. This EEG response was apparently acquired in relation to the signal value of the CS rather than in relation to a response component, since it occurred despite immobility.

Buzsaki *et al.* (1979a) have reported on some related experiments using rats. They used a classical conditioning task using a water-lick response as the UCS, with spatial separation of CS source from UCS and goal. Data were expressed in terms of the ratio of 9–10 Hz theta activity to 7–8 Hz theta activity. No data are presented on the evolution of theta responses in the very first trials. However, in the first *session* of conditioning (32 trials) theta activity at 9–10 Hz was high throughout each trial, but was very variable between animals. Also in the first few sessions, when orienting responses to the CS were usually produced, *high-frequency theta activity was locked to the initiation of the tone CS, and could be averaged over trials to produce a synchronous rhythmic trace*. By the 10th session the proportion of 9–10 Hz theta had become far less variable between

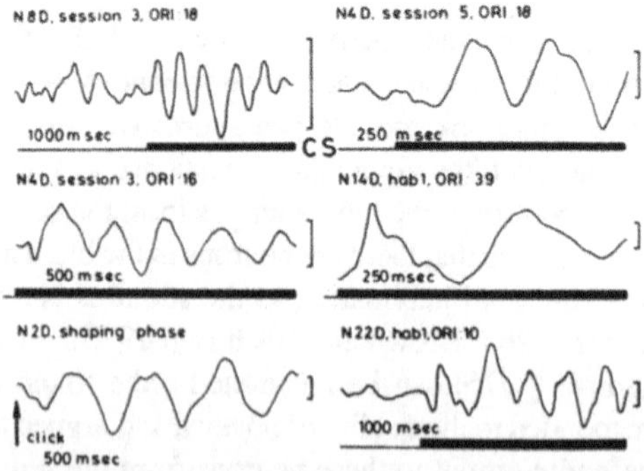

Fig. 46. Averaged hippocampal activity recorded during a single session (32 trials) of classical conditioning in rats. *Each EEG trace* represents a different animal. *Lower trace* is event marker, with *heavy lines* indicating duration of CS. Note that time-base varies between traces (total duration of trace in ms is indicated). Note presence of theta waves phase locked to CS in averaged traces. (Buzsaki *et al.* 1979a) (From *Electroencephalography and Clinical Neurophysiology* Vol 47 pp. 64–74 1979)

animals, had fallen in the prestimulus and goal-response epochs, but remained high in the first half of the CS. At this stage of training the electrographic response was less well coordinated than earlier. The onset of the tone tended to produce an immediate evoked non-rhythmic potential. This was followed by theta activity which was not locked to CS onset and could not be averaged to a rhythmic trace. At this stage, orienting scores were much lower than earlier in training. Thus in this learning task a high frequency rhythmic response of great consistency appears in early conditioning. This response is specific to the time of CS, emerges as the signal value of the CS becomes established, and declines later in training. Fig. 46 shows one of the illustrations from this study. However, in the same paper, using a CS in classical pairing with an unavoidable shock to the paw, theta activity declined as conditioning proceeded.

10.3.2 Theta Activity During Appetitive Learning

Grastyan *et al.* (1959, 1966) reported that in cats during alimentary conditioning tasks theta rhythms did not occur on first presentation of an entirely unknown stimulus, but only a few trials later, at a time when orienting was strongly shown, and before an appropriate response had been elaborated (see also Lissak and Grastyan 1960). Continuing this work, Grastyan and Vereckei (1974) used an apparatus consisting of either a straight runway or T-maze, with reinforcement delivered on every trial. The CS tone originated from a separate spatial location from the goal box. It was thus possible to assess the separate correlations of the CS–directed behaviour and goal-directed behaviour in the acquired theta response. In the first 50 or so trials, the CS merely elicited desynchronization. Between trials 50 and 100, theta activity appeared, correlated with the development of a conditioned orienting response to the CS. This orienting response sometimes interfered with the goal-directed response. During the conditioned orienting response theta activity at 4–5 Hz occurred, significantly different from that which could occur during approach to the goal (6–7 Hz).

Adey *et al.* (1960) produced partially congruent results. They trained cats in approach learning. On the first presentation of the CS, irregular 4–7 Hz activity appeared which became more regular at a frequency of about 5 Hz in both dorsal and ventral hippocampus over the next few trials. By the time of the 20th CS this had become even more regular, but no longer occurred in the ventral hippocampus.

Bennett (1970), in experiments on cats, scheduled that a clicker signalled delivery of milk, which the animal could then approach and drink. On the first presentations of the clicker, no theta activity was elicited. Within a very few presentations, however, theta activity appeared at a time when orienting also appeared. However, when the clicker had gained sufficient control over responding to elicit an immediate approach response without orienting, theta activity was no longer produced by the clicker.

Coleman and Lindsley (1977) studied continuously reinforced lever pressing for water in cats. Theta activity became prominent at an early stage of acquisition, as the correlation between lever pressing and reward was first noticed. As performance improved, the hippocampal EEG became more irregular.

10.3.3 Theta Activity During Active Avoidance Conditioning

Grastyan *et al.* (1959; 1966) and Lissak and Grastyan (1960) obtained results with active avoidance conditioning which were essentially similar to those obtained during alimentary conditioning. Support for Grastyan *et al.*'s findings also comes from a recent paper by Gralewicz (1981), in which cats were trained in a signalled two-way avoidance task. Stable 6 Hz theta activity developed in the course of training, occurring at the time of crossing the barrier between the two compartments. Pickenhain and Klingberg (1967) trained rats in a signalled avoidance task. The CS did not evoke theta activity on its first presentation, even if it was a nociceptive stimulus. However, in well-trained animals, the CS did regularly evoke theta activity, in the epoch of each trial before any avoidance or other motor response occurred.

10.3.4 Theta Activity During Discrimination Learning

Discrimination learning may be scheduled by using both a CS+ and a CS− from the outset of training, or the CS− may be introduced after earlier non-differential training to a (positive) CS. Evidence using the first of the two is described below. The latter schedule, which involves disconfirmation of expectancies, is described later, in Section 10.5.1, in relation to extinction.

Adey and Walter (1963) and Porter *et al.* (1964) carried out studies using a T-maze visual discrimination task in cats, with a daily session consisting of forty trials. Six Hz theta activity appeared at the start of all trials, including the first. However, it appeared most strongly a little while after the start of training, at the phase of most rapid learning, when performance was correct on 50–80% of trials. In addition, in this phase of learning the theta rhythm showed a synchrony similar to that described above in Buzsaki *et al.*'s (1979) work on rats: *the rhythmic activity was phase-locked to the onset of the tone which started each trial, so that a highly coherent rhythmic trace could be produced by averaging over trials.* These results are shown in Fig. 47. When performance level had improved to about 90% correct, this synchrony disappeared or diminished, though asynchronous theta activity at 6 Hz persisted. On reversal of the discrimination, at a time of low performance level, very high amplitude theta activity quickly emerged at a frequency of 6 Hz, this rhythm being again synchronous over trials. After the first day of reversal, synchrony of the 6 Hz rhythm decreased, though asynchronous 6 Hz activity

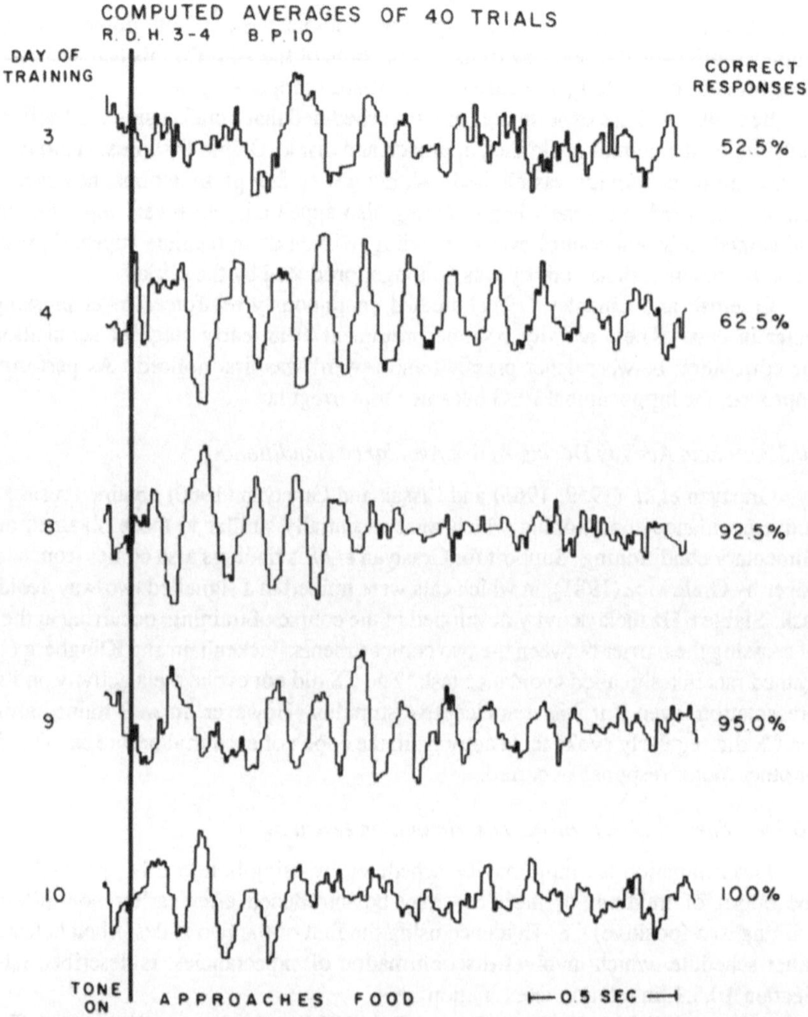

Fig. 47. Averaged EEG traces obtained from cats learning a discrimination task, and recorded from the dorsal hippocampus (CA1/2). During the phase of most rapid learning (*days 4–8*) the tone CS produced synchronous theta rhythms, which could be averaged over trials to produce a coherent rhythmic trace. By *day 10*, when performance was 100% correct, the rhythmic averaged potentials were smaller in amplitude and had a shorter duration from tone onset. (Porter *et al.* 1964) (From *Experimental Neurology* Vol 10 pp. 216–235 1964)

was still present during the approach phase. As performance built up again above 50% correct, a high degree of synchrony reappeared.

Elazar and Adey (1967a) trained cats in a light-dark discrimination in a T-maze. They found that at an advanced stage of training (80% correct or more, after 70 daily 30-trial runs) there was a regular sequence of changes of theta frequency through a trial. In the prestimulus epoch, frequency was about 4 Hz. In the stimulus epoch it increased to 5 Hz. During approach it was 6 Hz. After approach it fell back to 4–5 Hz. The 6 Hz spectral peaks during approach were sharper at this stage than at any earlier stage of training. Thus there seemed to be entrainment of rhythmic activity as a result of training, and suggestions of characteristic frequencies for each part of the task. At a later stage in training when performance was 100% correct, this sequence of EEG events and the sharpness of the 6 Hz peak during approach tended to disappear.

10.3.5 Theta During GO-NO GO Discrimination

In an early study by Dalton and Black (1968) of GO-NO GO training in dogs, results were explicable entirely as innate correlations with movement such as discussed in Chapter 4, rather than as correlates of learning processes. In two other studies, however, theta activity correlates better with the stage of learning than with concomitant movement. Crowne *et al.* (1972) studied the EEG activity of monkeys by spectral analysis, during GO-NO GO discrimination. As training progressed, activity in the 3–5 Hz band became more prominent in the first 1.5 s of a 5 s NO GO signal, but declined on correct GO trials. It should be noted that NO GO performance is very difficult for monkeys, who have to pay continuous attention, while the GO response becomes easily automated. Buzsaki *et al.* (1981) studied cats undergoing appetitive GO-NO GO training with the GO components (during which lever presses would produce a reward) signalled by a tone. An increase in theta frequency from 3–4 Hz to 5–6 Hz occurred in response to this CS and developed over trials. This EEG response was apparently acquired in relation to the signal value of the CS rather than to the response, since it did not occur so much during lever pressing.

10.3.6 Acquired Theta Activity which Correlates with Response Components

In most of the learning tasks considered so far, there has been a definite, discrete stimulus either directly prompting the response, or indicating the schedule which is in operation. Some of the available literature, however, concerns learned responses where the signal is far less definite. In these cases it appears that theta activity may develop in correlation with the acquisition of the response, rather than to recognition of the stimulus. Pickenhain and Klingberg (1967) found for an unsignalled avoidance task, that theta activity did not occur in early trials when there were merely chaotic unsuccessful attempts to escape, but developed with first successful avoidance responses. Coleman and Lindsley (1977), used a schedule of unsignalled alternation between periods of reinforcement by lever pressing, and non-reinforcement. They found that, during training, theta activity developed in the non-reinforcement period, as if the recognition of the response-reinforcement contingency, rather than that of the signal value of an explicit CS, was related to presence or absence of the theta response. In this, case the theta response correlated with the acquisition of inhibition of lever pressing, rather than that of an active response. Worth mention here are also the early results of Holmes and Adey (1960). Theta activity was recorded from the entorhinal cortex of cats, during the performance of delayed-response choice between two bridges to gain access to food. During the bridge-walking, theta activity in the 4–6 Hz range increased, in comparison to the delay period which occurred before this response could be made. This applied both whether the animal chose the correct arm or the incorrect arm. However, in intertrial bridge walks (i.e., non-purposeful responding) there was no increase in theta activity.

Thus, theta activity seemed to be a necessary correlate of responding, provided that it was performed purposively.

10.3.7 Reward and Theta

When discussing the relation between theta activity and learning, an issue which must be considered is whether theta activity has any consistent relationship to reinforcement. Paxinos and Bindra (1970) noted that rats could show both locomotion and theta activity during rewarding brain stimulation. However, when they were shaped to be immobile during self-stimulation, they did not show theta activity. The conclusion was that theta activity correlates with movement, rather than reward. Other evidence confirms that reward and theta are not related: driving theta activity by brainstem stimulation is no more rewarding than blocking it (Ito 1966; Routtenberg and Kramis 1968; Ball and Gray 1971). In addition, Pond and Schwartzbaum (1972) showed that in rats performing on a fixed ratio-16 lever-pressing schedule, hippocampal synchrony was minimal during reinforcement. It must be admitted that a little evidence shows that under some circumstances, theta activity may occur *after* completion of a successful rewarded conditioned response (Grastyan *et al.* 1966; Lopes da Silva and Kamp 1969; Pond and Schwartzbaum 1972). This observation is, however, susceptible to an interpretation other than a post-reward effect (see below).

10.3.8 Summary and Discussion So Far

In many of the types of learning reviewed above, it is found that a theta response is not present right at the start of training but develops during the course of learning. This has been documented in classical conditioning, in non-differential appetitive and avoidance training, and in discriminative learning (including GO-NO GO discrimination). The amount of training required before the theta response appears varies a good deal between the various experimental reports. However, most of the data are consistent with the idea that the theta response develops at the stage of training when performance is improving most rapidly, and/or when the details of a stimulus are important in determining the timing of a response. In some of these studies there is evidence that frequency of theta activity shifts during learning, and becomes focussed on a narrower frequency band or phase-locked to the onset of important stimuli.

This evidence is therefore suggestive of the development of resonant circuits. It is compatible with the theory advanced in the previous chapter that the resonances which develop represent the overall situation and its more detailed contingencies. However, there are difficulties with some parts of the evidence. For instance, in classical conditioning using aversive stimuli, theta activity is sometimes acquired and sometimes suppressed. Also some instances of the occurrence of theta activity *after completion* of a goal-directed task are difficult to understand in terms of the present hypothesis.

One more detailed conclusion is also suggested: in cats, the synchronous rhythm indicative of resonance probably has a preferred frequency below the maximum frequency of the innately triggered rhythm. Therefore innate rhythms (for instance in relation to locomotion) can appear at higher frequencies than the acquired resonance. This is a situation different from that in rats, where the resonant frequency tends to be at the top end of the frequency range for theta activity. This tentative conclusion is easily compatible with the fact that the rat brain is smaller than the cat brain, so the connections on which resonance depends are shorter, and the period of the resonant theta activity consequently higher.

10.4 Decline or Persistance of Acquired Theta Activity as the Response Becomes Automatic

10.4.1 Examples of Decline of Theta Activity as the Behavioural Response Becomes Automatic

A variety of evidence shows that the theta activity displayed during the course of learning declines or disappears when learning has run its full course, and the learned response has stabilized. Grastyan *et al.* (1959, 1966) showed this for the acquisition of appetitive and avoidance responses. The decline was shown both over the course of acquisition of the initial conditioned response, and over that of the later differential conditioned response. It also applied as an extinguished response stabilized in extinction. Grastyan and Vereckei (1974), in their experiments with spatial separation of CS and goal box, showed that the specific response of 4–5 Hz theta produced to the CS earlier in training was, at a very late stage, replaced by desynchronized activity on many trials. On other trials 6–7 Hz theta was produced, the same frequency as shown on approach to the goal on both early and late trials. This was probably therefore a theta rhythm related to a response component rather than to the CS. It may be commented that at this late stage of training there was no orienting, and the approach response was immediate. However, both the 4–5 Hz theta response and the 6–7 Hz response were definitely acquired ones. Some of these observations have been confirmed by Bennett (1970) for an appetitive task.

From Adey's laboratory come a number of similar demonstrations. Holmes and Adey (1960) studied theta activity in the entorhinal cortex of cats during delayed response discrimination of two bridges giving access to goal boxes. Theta activity occurring in the delay interval was more prominent in early than in later training. Adey and Walter (1963) and Porter *et al.* (1964) showed that the prevalent 6 Hz theta at the time of most rapid learning declined as performance on the T-maze reached 80–90%. Elazar and Adey (1967a) made similar observations during a light/dark discrimination task and during relearning the task after extinction. The theta response acquired during GO-NO GO discrimination disappears when the behavioural "GO" response becomes automatic, according to Crowne *et al.* (1972). Several papers have also shown that well trained bar pressing is not associated with theta activity, though it might be present at earlier stages of training (dogs: Lopez da Silva and Kamp 1969; rats: Pond and Schwartzbaum 1972; Feder and Ranck 1973; Whishaw and Vanderwolf 1973; Coleman and Lindsley 1977).

Gralewicz (1981), using cats in a shuttlebox, found that, in the criterion phase of learning, theta activity was present during performance, but not if the response was of unusually short latency. Thus, on a trial-by-trial basis, the automatization of a response appears also to relate to absence of theta activity.

10.4.2 Examples of Persistence of Theta Activity in Well-Trained Tasks

In contrast to these papers, there are a number of results indicating that theta activity may *persist without degradation* during performance of some very well trained tasks. In some cases, the persistent theta activity is apparently response-related. For instance, in Adey *et al.*'s (1960) study, once theta responses in approach learning had stabilized, they were then shown without adaptation for the rest of training (i.e., for up to 400–500 trials). Likewise, Grastyan and Vereckei (1974) found that in a very late stage of training a proportion of trials still showed 6 Hz theta activity which was non-habituating, and apparently related to an approach response. Adey and Walter (1963) and Porter *et al.* (1964) observed that, in performance of a visual T-maze discrimination or its reversal, the theta activity, which had declined after the period of most rapid learning, tended to build up again when performance was getting up to near-perfect levels. Some of these examples may be no different from the innate correlations with movement emphasized in Chapter 4.

In others cases, theta activity persisting in overtrained animals appears to be related to a stimulus cue requiring vigilance. A number of studies have used the DRL ("differential reinforcement of low rates") schedule of responding where a cue is used to indicate the change from one stage of the schedule to the next. According to Bennett *et al.* (1973) and Frederickson *et al.* (1980), using cats, and Bennett *et al.* (1978) in rats, there was theta activity between presses at all stages of acquisition and performance. This did not occur in the non-cued DRL task (Bennett *et al.* 1973). A related finding is that of Pond and Schwartzbaum (1972), who studied rats performing on a well-trained fixed-ratio-16 lever pressing schedule. Theta activity was greater in the first half of each fixed ratio segment than in the second, although response rates were the same in both halves. In terms of behaviour it was as if the previous reinforcement acted as a signal for the start of the next segment of the schedule.

If acquired theta activity persists when vigilance is required towards stimuli acting as cues, it would also be expected to persist when cues are used to signify stages in other complex schedules of learning. An experiment of Bennett (1970) is relevant here. He established differential responding on an approach response with stimuli presented as S+ and S−. As described earlier, in the later stages of conditioning, S+ never produced theta activity despite stable responding. If a separate alerting signal was now scheduled 3 s before the S+/S−, it gradually acquired the capacity to produce theta activity. When the lever was moved to the other side of the cage, the theta response to the alerting signal declined, but built up again as discriminative responding recovered.

There is a little evidence that in relearning after a reversal of a discrimination task, theta activity may show a greater tendency to persist in comparison with the learning of the original task. Adey and Walter (1963) found that as the reversal task was acquired to a performance level of 50–80%, theta activity in the 5 Hz band was stronger than at the same stage of acquisition in the original task. As reversed performance approached 100%, theta activity persisted, whereas it declined at this stage of the original task. This evidence, though unconfirmed, is also compatible with a role for theta activity in relationship to a cue which acquires a role in signifying the lower-order schedule which is in operation: a discrimination which is subject to reversal requires closer attention to detail to spot which of the alternative versions is currently in operation.

When a task can elicit theta activity even when very well trained, there is some evidence that the occurrence of theta activity correlates, on a trial-by-trial basis with performance accuracy. Yoshii *et al.* (1967), using dogs, claimed that there are differences in hippocampal EEG between correct and incorrect choices, in a delayed response task, but evidence presented is inconclusive. Elazar and Adey (1967a) showed that trials of a light/dark discrimination task which were incorrect lacked the theta activity present on correct performances. Holmes and Beckman (1969) trained cats using a 10 s auditory signal to indicate whether the animal should GO or STAY and the hippocampal EEG was monitored by telemetry. The presence of 4–7 Hz RSA from the hippocampus in response to the tone predicted very significantly whether the animal would go or stay. Crowne *et al.* (1972), using monkeys performing a left/right alternation task, found that hippocampal activity in the 3–5 Hz band increased only on the correct trials. Vanderwolf and Cooley (1974) studied hippocampal EEG in a task in which rats repeated a jump to exhaustion. When the rats were fatigued, failure to jump correlated, on a trial-by-trial basis, with either absence of theta activity altogether, or failure of theta frequency to rise just before take-off. Grastyan and Vereckei (1974) found in their experiments that when there was a conflict between orientation and approach, the approach response failed at the same time as theta activity failed. Arnolds *et al.* (1979c), using dogs in a spatial sound discrimination task with two levers, found that the frequency, amplitude and synchrony of theta activity was greater on correct lever presses than on "pressings in between" signals. They speculated that this could have been due to minor topographical differences which they observed in the nature of lever press in the two cases, but this conjecture is unproven.

10.4.3 Discussion

In trying to summarize the data reviewed in the above two subsections, one cannot but be struck by the teleological nature of the disappearance or persistance of theta activity in a well-trained task: when a learned response has given evidence of automatic performance, by either close-to-perfect performance rates, or minimal-latency responses, theta activity declines. When the performance of a task is liable to fail, theta activity is maintained. If it cannot be performed accurately without maintained close attention to the nature or timing of a cue, then acquired theta response tends to persist. In the latter cases there is evidence for correlation, on a trial-by-trial basis, between presence of the theta response and correct performance. One is therefore tempted to suggest that *the theta response is serving some important function in the instances when it does persist.* The theory presented in the previous chapter gives an obvious explanation of what this function is. Resonant circuits, representing the situation as a whole may be required to resolve the ambiguities of representation of the smaller details of that situation. Tasks whose representation in the cortex is very strong do not need such resonances to resolve their details.

However, an important question is still begged: what *causes* the acquired theta activity to be present when it is needed to be present, and to disappear when representation is apparently strong and unambiguous? To answer this question it is necessary to refer to papers of Dalton (1969), Black *et al.* (1970) and Black (1975) indicating, both for rats and dogs, that occurrence of the hippocampal theta rhythm could be influenced by instrumental schedules of reward. If contingencies of reward can be used to determine the presence or absence of a theta response, this could be very significant in "priming" the nervous system so that it can analyze information in a way allowing successful behavioural performance.

Consider a learned task which is either inherently difficult, or difficult because it is only partially learned. Unless the overall state of nervous activity in the forebrain is "standardized" by resonance of the appropriate loops, performance of the task will be vulnerable to failure. If, however, the same set of resonant circuits is activated on each trial, the representation which is "stamped in" by the contingencies of response and reward will be reproduced faithfully on each trial, and performance will thereby be more robust. Suppose that, at the same time as behavioural performance is reinforced, the septal mechanisms initiating theta activity are also reinforced. Then theta activity will accompany each trial in this phase of learning. Initially this will not involve cortical entrainment, but entrainment is likely to develop of any resonant circuits appropriate to the environment the animal is in. Only on trials where this happens will accurate performance be probable. On just these trials, the reinforcement mechanism will be triggered, strengthening not only the representation of the task, but also the link between the experimental situation and the initiation of theta activity. Thus for such difficult or partially learned tasks, there will be differential reinforcement of the theta rhythm in the environment in which the experiment is conducted on the trials where performance is successful.

If, however, the tasks is so strongly represented that a context representation is not needed, resonance of circuits will offer no advantage in accurate performance. There will thus be no differential reinforcement of theta rhythms on correct trials. Moreover, when it happens that the response is performed successfully even without resonance at the theta rhythm, there may be reinforcement of states of the nervous system other than (or even incompatible with) one of theta resonance. Thus when a task become automatized, its link with theta rhythms is likely to decline. This is the effect described in much of the evidence reviewed above.

The circumstances in which persistance of the theta rhythm occurs even in well established tasks requires a little further comment. This happens when a stimulus is used as a cue indicating exactly *when* to respond (e.g., in the cued DRL experiments described above), or when the animal has to distinguish between more than one version of a task (as in discrimination where reversals are possible). In these situations it is plausible to envisage that to detect or analyse the stimulus, a standardized state of the nervous system is required such as is hypothetically provided by a pattern of resonant circuits. The other class of occasions on which theta activity persists during performance of well learned tasks is in relation to response aspects of the task. It would be parsimonious to account for these cases as also requiring close attention, and a standardization of the state of the nervous system to accomplish the necessary information processing. However, the evidence available does not allow one to make this generalization. The significance of acquired theta activity persisting in relation to a well-learned behavioural response is thus currently unresolved.

10.5 The Hippocampal EEG During Extinction, or Other Examples of Response Omission

10.5.1 The Evidence

Amongst the experiments described by Grastyan *et al.* (1959) are some observations on extinction. It was found that theta activity developed early in extinction but disappeared again when the extinguished response had stabilized (see also Lissak and

Grastyan 1960). Holmes and Adey (1960) also found that theta activity occurred during early but not later extinction of a delayed response bridge walking habit.

There are at least five possible interpretations of these observations. Firstly, theta activity in extinction could have a relation to "unlearning" similar to that to learning (i.e., in both cases a *change* of associations). A second possibility is that in this circumstance theta activity is related to some aspect of the developing response inhibition. Thirdly it could be that theta activity occurs when expectancies previously built up are disconfirmed. Fourthly, reward omission *per se* could define the occurrence of theta activity. Fifthly, any apparently distinctive EEG response to reward omission may be a more direct correlate of concurrent behaviour in the epoch after omission. The first of these alternatives ("unlearning equals learning") receives indirect support from much of the other evidence reviewed so far in Chapter 10. However, before it can be accepted, other correlations must be excluded.

Grastyan *et al.* (1959) provide some evidence concerning the second alternative (response inhibition), which has also been investigated specifically in experiments by Bennett (1970) and Bennett and Gottfried (1970). In Grastyan *et al.*'s (1959) study, when the conditioned response had stabilized, theta activity declined, as discussed above. If from this point differential conditioning was scheduled, no theta activity occurred in the first few trials when reward omission was paired with the S−, but soon theta activity did occur, in response to the S− only, this time accompanied by absence of the behavioural response. Bennett (1970) obtained a similar result. Initially a non-differential approach response was established in cats to the sound of a clicker. Then the cats were trained on a schedule consisting of alternate periods of reinforcement (20 s), signalled by light or tone, when paddle pressing would deliver food, and intervening periods without the stimulus (20, 40 or 60 s, randomized) when it would not. The S+ never produced theta activity (just as before the differential schedule was in force). In the absence of the S+ a theta response occurred periodically, provided that, correctly, no behavioural response was emitted. However, this theta response never occurred at the moment of incorrect behavioural responses. Bennett writes: "The development of this EEG response correlated with the acquisition of the stimulus discrimination task, as measured by the animal's ability to inhibit responses when the positive stimulus was off".

These results are compatible with either the idea that "unlearning" is equivalent to learning, or with the "response inhibition" hypothesis. However, they are incompatible with the third and fourth hypotheses ("disconfirmation of expectancies" and "reward omission") since according to these hyptheses, the theta response should have occurred on the first occasion that the S− was introduced.

To decide between the first two alternatives Bennett and Gottfried (1970) trained cats on a continuously reinforced schedule, and then shifted them to a DRL schedule with progressive increases in the delay when reward from bar pressing was not available. This delay was unsignalled, so the animal had to rely on internal time cues rather than exteroceptive ones for accurate performance. Low amplitude fast activity dominated the hippocampal EEG throughout training, with no theta activity evident except occasionally when there was orienting. This result does not support a role of theta activity during response inhibition. It is corroborated by the paper of Frederickson *et al.* (1980) who found that theta activity during a cued DRL schedule occurred only in the first half of each interval, while in the second half, when responses were equally well inhibited, theta responses were not found.

These results may be contrasted with that of Bennett *et al.* (1973), in which the end of the delay was signalled. In this circumstance theta activity was abundant during the no-reinforcement period, except in those delays where the animal responded incorrectly (i.e., before the signal of the end of delay). In this experiment theta activity was apparently related to acquired signal value rather than response inhibition (discussed above). As far as extinction goes, the evidence fails to support the "response inhibition" correlate.

Data relevant to the fourth possibility (reward omission) was obtained by Kimsey *et al.* (1974) in rats trained on a straight alley with single alternation. They found a consistently higher frequency of theta activity on non-reward trials (8 Hz), than on reward trials (7 Hz), regardless of the stage of learning. Coleman and Lindsley (1977) performed a similar experiment, using cats and a lever press response, with alternating periods of reinforcement and non-reinforcement. Theta activity was greater in non-reinforcement, and tended to increase as the animal learned to withhold a response. This finding applied both when non-reinforcement was signalled and when it was not. Since theta activity was low in the early instances of reward ommission, but increased for later non-reward periods, this result suggests that theta activity was gradually acquired by the influence of reward ommission (in other words it occurred as a result of "unlearning") rather than being an inevitable consequence of that omission regardless of learning.

The fifth possible explanation of Grastyan *et al.*'s finding (correlation with behaviour emitted in relation to extinction) was suggested by the results of Morris and Black (1978) on rats jumping. Detailed analysis of the topography of behaviour after reward and after reward omission was made, and compared with the EEG profile seen on other occasions when the same behavioural components were elicited. The hippocampal EEG reaction to non-reward in rats could be predicted accurately on the basis of the concurrent behaviour, without postulating a distinctive EEG response to non-reward.

10.5.2 Summary and Discussion of the Evidence

Thus, in summary, the "response inhibition" and the "disconfirmation of expectancies" hypotheses are not supported by the evidence, and evidence in favour of the "reward omission" hypothesis is susceptible to an alternative interpretation. On the whole, the evidence favours a combination of the "unlearning equals learning" hypothesis and the simpler notion of a correlation with concurrent unlearned behaviour. The latter correlation is easy to deal with in terms of the innate control mechanisms of theta activity discussed in Chapter 4. The result of Holmes and Adey (1960) cited above is still difficult, but may reflect continued control by environmental cues, which have not yet lost their signal value, despite several reinforcement failures.

The idea that theta activity occurring during extinction is a function of "unlearning", which may be subsumed as another example of learning, deserves a few comments in relation to the theory of resonant circuits advanced in Chapter 9. It seems as if some aspect of the extinction process is treated in the same manner as the learning process. The most likely feature by which this could occur seems to be a form of avoidance learning. Although extinction is complex, part of it is probably a failure of an expected reward which is equivalent to introduction of punishment (the evidence for this view cannot be discussed here, but the interested reader is referred to Leitenberg 1965; MacKintosh 1974, p. 262). Given this, it is likely that the first failures of the behavioural response when extinction is scheduled will tend to punish that response. By withholding the response subsequently, the animal avoids this punishment. Hence the representation in the brain which led to the withholding of the behavioural response will be reinforced. In a manner similar to that suggested in Section 10.4, if a theta response occurs in one of these trials, it also will tend to be reinforced. This will set up resonances which will standardize the state of the nervous system in subsequent extinction trials, just as it apparently does during learning. This will aid the formation of the representation of the extinguished behavioural response. When, however, the behavioural reponse *stabilizes* in extinction, this representation will be so strong that a theta response is not necessary to ensure withholding of the response, and its occurrence will not be differentially reinforced. The theta response therefore declines when extinction stabilizes.

10.6 Theta Activity in Records Made Outside the Task Performance Epochs in Relation to Rate of Subsequent Learning

A small number of papers show, in between-animal comparisons, that the prominence of theta activity in periods other than the learning experiences themselves may correlate with the rate of learning when specific tasks are scheduled. Landfield *et al.* (1972) and Landfield (1976) trained rats in single trial passive avoidance and other learning tasks, and the EEG, recorded in the cortex, was subjected to spectral analysis. Electroconvulsive shock or drugs such as strychnine or pentylenetetrazol were used to modify the EEG. It was found in both papers that retention of the learning was closely linked to the abundance of theta activity immediately after the original learning trial. In fact, it was more closely linked to this than to the presence or absence of electroconvulsive shock or drug treatment per se. In these two studies it is uncertain whether the activity in the theta range was of cortical origin, or was volume-conducted from the underlying hippocampus. However, they support the idea that theta activity (of whatever origin) is positively correlated with learning processes.

A single study (Berry and Thompson 1978), using rabbits, established a correlation between animals on a measure of learning and theta activity in the hippocampus itself. Nictitating membrane conditioning was conducted over several days, with each trial preceded by a 2 min recording session of the hippocampal EEG. The greater the activity in the 2–8 Hz range, the fewer trials taken to reach the criterion of learning. Fig. 48 shows some of Berry and Thompson's data.

10.7 Electrically Driven Theta and Learning

10.7.1 Introduction

The fact that it is possible to drive the hippocampal theta rhythm by electrical stimuli to the medial septal nucleus and elsewhere has prompted a number of attempts to determine the relation between learning processes and experimentally manipulated theta activity. There are two main thrusts to this work. On the one hand, there are several papers which show that theta driving before, during or after a learning experience can *improve learning and retention*. On the other hand, a body of work suggests that theta driving can potentiate or even mimic the effects of *non-reward* (as seen in extinction or

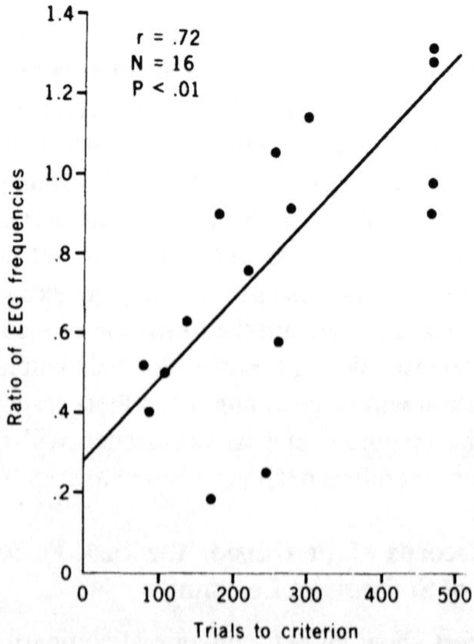

Fig. 48. Diagram showing correlation between low frequency EEG activity, and trials to criterion in nictitating membrane conditioning. The *vertical axis* represents the ratio of 8–22 Hz activity (percentage) divided by 2–8 Hz activity (percentage). *Line* is the best-fitting regression line between the two variables. (Berry and Thompson 1978) (From *Science* Vol 200 pp. 1298–1300 1978) (© 1978 by the AAAS)

partial reinforcement). The latter line of research, particularly associated with the name of J. A. Gray, started earlier than the former. However, since it is *prima facie* following a rather different theoretical basis from that developed here, it will be dealt with after considering the evidence relating theta driving to learning.

10.7.2 *Theta Driving and Improvement of Acquisition/Retention*

Wetzel *et al.* (1977) studied brightness discrimination in rats. Medial septal stimulation was given at 7 Hz for the period between 5 and 15 min after a learning session, and this was found to improve retention scores 25 h later. This effect was not shown if stimuli were given at 100 Hz, a frequency which would block rather than drive hippocampal theta rhythms. In addition, if the low frequency stimulation was given at 4 h after the learning, no improvement was seen. Across all animals there was a correlation between the amount of hippocampal theta activity seen in the 5–15 min period after learning, and the 25-h retention scores. This correlation was mainly a function of the stimulation/no stimulation condition, but within the no-stimulation control group there was also a trend in the same direction.

Landfield (1977) trained rats in a one-way active avoidance task in two trials on successive days. Medial septal stimuli were delivered for 10 min immediately after each trial. In a different task rats were trained in single trial passive avoidance learning (which also contained an incidental active component when the animal escaped from the punishing shock). Septal stimuli were given for 20 min immediately after the trial. Compared with controls, which had electrodes implanted but which were not stimulated, rats given septal stimuli at 7.7 Hz improved retention 2 days later. This applied for both tasks, despite their opposite response requirements. It was concluded that driving of the theta rhythm improves memory.

Destrade (1982) allowed mice 15 min, without any shaping, to acquire lever pressing for food. Stimuli delivered to the medial forebrain bundle elicited theta activity at 7–9 Hz, and if given within 5 min of learning would enhance retention 24 h later. Dorsomedial hypothalamic stimuli elicited theta activity at 9–12 Hz, and could enhance retention if given as much as 30 min after learning. Galey *et al.* (1983) gave mice sufficient training for partial acquisition of a continuously reinforced lever pressing task (i.e., 15 min in the apparatus without shaping). The mice were then given medial septal stimuli, either 30 s or 15 min after this session, with non-stimulated animals serving as controls. Twenty four hours later, the 30 s group had improved, the 15 min group were unaffected and the implanted controls were impaired. The lesion effect of implantation presumably caused the impairment, while some correlate of septal stimulation could improve retention. However, theta records made over 80 s periods before, during and after the stimulation period showed that stimulation did not increase theta activity while it was in progress, and there was a fall off in theta activity immediately afterwards. Thus, although septal stimulation seemed to be involved, it apparently was involved without producing an increase in theta activity. However, this latter result was inadequately controlled: the statement is based on comparison of stimulated animals before during and after stimulation, not on a comparison of stimulated versus non-stimulated animals. The latter animals might have had a sharper fall in theta after sham stimulation than the former did after real stimulation.

All the above studies are concerned with the effects of theta driving after the learning episode. A single study (Deupree *et al.* 1982) investigates the effect of theta driving applied before the learning takes place. Three groups of rats were firstly matched for baseline theta activity. Before training, group 1 was subjected to medial septal driving of theta activity at 7.5 Hz. Group 2 experienced 100 Hz stimulation of the medial septal nucleus, while the third group consisted of implanted controls. The hippocampal EEG was again recorded after stimulation. There followed 3 days of light/dark discrimination training in a T-maze. After criterion had been reached, the hippocampal EEG was recorded immediately and again 2 days later. It was found that the theta-driven group learned significantly faster. Overall there was a correlation between the difference in theta count before and after stimulation, and the trials to criterion in the learning sessions which followed. Activity levels did not differ between groups.

10.7.3 Theta Driving and the Effects of Reward Omission

The line of research linking theta driving to the effect of omitted reinforcement started with pharmacological and behavioural studies of the theta rhythm. Gray (1970) and Gray and Ball (1970) were struck by the occurrence of specific theta frequencies during different parts of an alley-running task in rats: on introduction into a novel cage or during exploration in the start box on each trial, 7.5–8.5 Hz was the dominant frequency. During the alley running, frequencies in the 8.5–10 Hz range were dominant. When the rat found and consumed water in the goal box, a mixture of 6–7.5 Hz theta and high frequency activity was found. When it found no reward in the goal box a 7.5–8.5 Hz rhythm reappeared, and the 7.7 Hz frequency seemed to be especially associated with failure of expected reward. Influenced by the then recent work on the frustation effect of non-reward (Amsel 1962), Gray suggested that 7.7 Hz theta activity was particularly involved in processing the information associated with non-reward.

This thesis was elaborated (Gray 1970) with respect to pharmacological studies with amobarbitone. This barbiturate, in small subanaesthetic doses (20 mg/kg), elevated the threshold intensity for driving theta activity from the septum. The effect was specific to the 7.7 Hz frequency of driving. The link between this finding and the occurrence of 7.7 Hz theta activity with non-reinforcement was that amobarbitone given to rats running an alleyway prevented one of the effects of non-reinforcement: when the animal is rewarded on only a portion of the trials, subsequent extinction is more protracted than in continuously reinforced animals (the "partial reinforcement extinction effect"—PREE). However, when partial reinforcement is scheduled for animals running under the influence of amobarbitone, no PREE is seen. The runway speed in extinction declines at least as rapidly as in animals continuously reinforced during acquisition, whether they are drugged or undrugged (Gray 1969). Gray's argument was therefore that amobarbitone suppressed 7.7 Hz theta activity, which was a normal mediator of the effects of ommitted reward. It should be pointed out that although this drug raised the threshold for 7.7 Hz theta driving, it is only a supposition that amobarbitone suppresses 7.7 Hz theta activity produced in spontaneous behaviour.

Glazer (1972) produced some complementary pharmacological evidence, making use of the fact that physostigmine could elicit prolonged theta activity. He trained rats on a fixed-ratio lever pressing schedule, progressively to FR10, using discrete trials (4 per day). Following this there were 15 further days of FR10 acquisition, with physostigmine given in one group for each daily session, undrugged animals being the control group. Subsequently there were 11 days of extinction of lever pressing, with a comparison of animals receiving and not receiving physostigmine during extinction. Physostigmine given *during acquisition* retarded later extinction. If the drug was given *during extinction*, however, it accelerated extinction. Thus theta activity could be implicated as promoting another effect of non-reward, namely extinction.

Gray pursued these ideas with further studies of theta driving in relation to learning. Gray (1972) applied stimuli to the septum to drive theta rhythms at 7.7 Hz, either during acquisition of a runway task, or during extinction of it. Theta driving during acquisition retarded extinction, that given during extinction accelerated it. Comparing animals, it was found that the amount of 7.7 Hz theta recorded during aquisition correlated with resistance to extinction, the amount recorded during extinction correlated with the ease of extinction. In Gray's experiment, the performance level attained after a session of accelerated extinction was maintained the next day, and so presumably was not merely an effect on performance at the time of stimulation. In addition, the animal's behaviour was changed as a result of stimulation during extinction, so that it was topographically more typical of a later stage of extinction than would have been expected just on the basis of number of extinction trials. (The animals systematically avoided the goalbox, but scrutinized all other parts of the runway).

Gray again interprets this result as a potentiation by 7.7 Hz theta activity of the frustrative effect of non-reward. His results corroborate both of Glazer's (1972) results, indicating that pharmacological elevation of theta activity during acquisition retarded extinction, and during extinction accelerated extinction. Gray actually makes a somewhat stronger point with regard to the effect of theta driving during acquisition.

In this experiment, theta driving was performed on only a proportion of continuously rewarded trials. Since this retards extinction (just like partial reinforcement), it was argued that theta driving could not only potentiate the effects of non-reward (viz. the accelerated extinction seen when theta was driven during extinction), but could actually mimic non-reward. However, this was going beyond the evidence, since no controls were run to exclude a similar effect on extinction when theta driving was given on every rewarded acquisition trial.

Further evidence of a role of 7.7 Hz theta activity in extinction processes appeared subsequently. Gray *et al.* (1972) trained rats to run a straight alley on a 50% partial reward schedule. High frequency septal stimulation (which blocked any concurrent theta activity) was delivered when the rats were in the goalbox on non-rewarded trials. This reduced the resistance to subsequent extinction, a result similar to that produced with amobarbitone, but the opposite of that with theta driving. Glazer (1974a) trained rats on an FR10 lever pressing schedule, with electrical stimuli given on 50% of acquisition trials to drive theta activity at 7.7 Hz. This procedure retarded subsequent extinction, by comparison with a control group and a group stimulated at 200 Hz.

Again, it is unclear whether these effects would also have been produced had the theta-blocking stimulation been given on every acquisition trial. That this might be so was suggested by an experiment of Glazer (1974b) showing that in rats operantly conditioned to produce higher than normal theta activity, FR10 responding extinguishes more slowly than in controls or animals conditioned not to produce theta activity. Further evidence that the retardation of extinction produced by theta driving can occur when theta activity is driven on occasions other than the time of non-reinforcement comes from Holt and Gray (1983). Three groups of rats were pretrained to lever press on an FR5 schedule. One group then received 6 s trains of 7.7 Hz theta driving stimuli on a random interval schedule, with food delivered on the fourth second of this train. The second group also received 6 s theta driving stimuli and food delivery, but these events were on independent random interval schedules. The third group received food delivery alone. These schedules were continued for ten daily 10 minute sessions. During extinction of the FR5 schedule, both stimulated groups showed retarded extinction, compared with the control (food only) group. In another experiment, Holt and Gray showed that retarded extinction could be produced even if the theta driving stimulation was given entirely before the 12 days of lever pressing acquisition.

In summary, elevated theta activity during acquisition does appear to correlate with resistance to subsequent extinction, and high theta activity during extinction correlates with accelerated extinction. The idea that theta activity plays a special role in the effects of reward-omission is not well supported by the above evidence, since the effect of theta driving during partial reinforcement is shown also when theta driving is given at times other than non-reinforced trials. This conclusion is also concordant with the view obtained in Section 10.5, that theta activity is not an inevitable consequence of reward omission, but one acquired only during a sequence of several omissions. Exactly this observation is also reported by Gray and Ball (1970) for their alley-running task. Finally, the original observation of Gray that different frequencies of theta activity were characteristic of different phases of an acquired alley-running task is entirely consistent with the theory advanced in Chapter 9. In fact, a tentative example of exactly the same phenomenon in cats has been described above (Elazar and Adey 1967a).

10.7.4 Discussion of the Above Evidence

In comparing the respective themes of Sections 10.7.2 and 10.7.3 one is immediately struck by the fact that different classes of information have been selected as the basis of the two lines of research, without much attempt to integrate them into a unified picture. The first body of information uses data on the effect of theta driving on acquisition and/or retention. The second uses data about its effect on extinction. This split in the evidence arises in part because, for various technical reasons, it is difficult to study the effect of a variable on both acquisition and extinction in the same experiment. To study extinction, one needs to start from an initial high level of performance after acquisition, and one which is uniform between groups; or alternatively an effect can be shown if the extinction curves of the respective groups actually cross over in the course of extinction. If extinction is to start from a high level of performance, it is likely that a ceiling has been reached during acquisition. This may prevent one from detecting concurrent effects on acquisition produced by the variable in question. To study adequately an acceleration of acquisition one has to use a degree of training, which, in control animals, produces only partial learning. Any variable applied during acquisition will affect the degree of learning at the end of acquisition. It is then difficult to compare groups during subsequent extinction.

Another aspect of the evidence reviewed in Section 10.7.3 is that much of it concerned fixed-ratio lever pressing tasks. In the experimental protocol of these experiments, there was always some pretraining in lever pressing, such as a progressive increase in the ratio required to deliver a reward. Only after this had been done were the experimental variables introduced, and the definitive acquisition and extinction phases scheduled. The consequence of this again may be to minimize the observed effects of the variable on acquisition, because the most important part of the acquisition phase may have taken place before the experimental variable was introduced.

Bearing these comments in mind, is there any evidence from the experiments reviewed in Section 10.7.3 for the sort of improvement in acquisition/retention described in Section 10.7.2? There is a little. In Gray's (1972) paper, the runway experiment, where theta driving was carried out on a proportion of acquisition trials, showed a little evidence for an accelerated learning curve. In the first 2 days of stimulation the stimulated animals improved their performance more than the controls (though starting from a poorer baseline level). In fact, theta driving stimuli tended to produce seizures, and so were administered (on 4 out of 12 trials) only in these first 2 days, and later when performance had stabilized at asymptotic levels. For the central part of the learning curve, no stimulation was given to either group. Thus, in the only part of the curve where performance improvement could possibly be produced by the theta driving, a trend in this direction was actually observed (see Fig. 49). From the lever pressing evidence, the only experiment in which theta driving occurred before pretraining was that of Holt and Gray (1983, expt. 2). In this, in the early phase of acquisition of the FR5 schedule, there was actually seen an acceleration of the learning curve in the theta driven group, compared with controls.

The overall effects produced by theta driving in relation to learning are therefore as follows. Theta driving given during acquisition improves learning and retention and it retards extinction, but if given during extinction it accelerates extinction. The acceleration of learning, and the acceleration of extinction by theta driving given while those processes are taking place can be economically subsumed under the title of learning (where learning includes "unlearning"). In terms of the theory developed earlier in this chapter, both processes can be regarded as examples of "embedding" a particular representation of the task in its appropriate context. A procedure which tends to promote theta generation by the hippocampus may reasonably be expected to favour the development of the resonances between cortex and hippocampus, the putative basis for such "embedding".

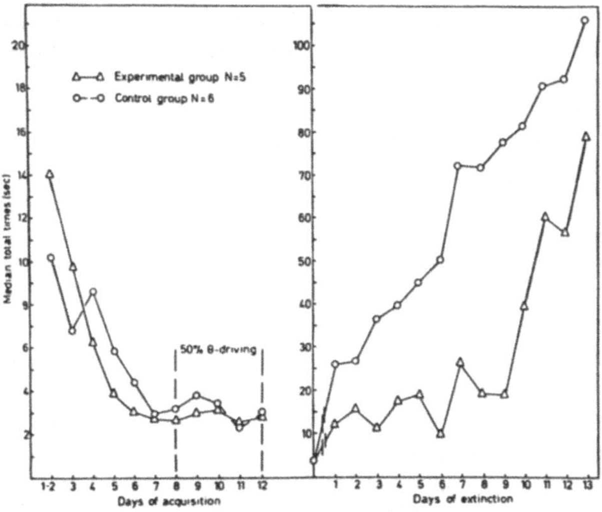

Fig. 49. Learning curves for rats in an alleyway task. *The vertical axis* represents running times during acquisition (*left*) and extinction (*right*). The experimental group received theta driving stimuli on day 1 and 2 during alley running itself, and, on days 8–12 on 50% of trials, when the animal was in the goalbox. (Reprinted with permission from *Physiology and Behaviour* Vol 8 J. A. Gray "Effects of septal driving of the hippocampal theta rhythm on resistance to extinction" Copyright 1972, Pergamon Press PLC)

The retardation of extinction for theta driving during acquisition is more difficult to account for. A small body of psychological literature is relevant here, however, dealing with the relation between the number of acquisition trials and the subsequent resistance to extinction. This is described in MacKintosh (1974, pp. 423–7). In lever pressing tasks, it appears that extinction becomes progressively slower, the larger the number of acquisition trials preceding extinction. In runway tasks, however, the more common result is that increase of number of training trials *increases* subsequent rate of extinction (the so-called overtraining extinction effect: OEE). In neither case is the literature unanimous, there being unaccounted exceptions. If theta driving during acquisition were supposed to improve acquisition, it might have an effect on extinction similar to an increase of number of trials for lever pressing tasks. Most of the examples of retarded extinction described above concerned lever pressing tasks, and so can be explained by this reasoning. That of Gray (1972) is, however, an alley running task, and does not fall within this generalization. The problem would seem to be not so much to explain the retarded extinction after theta driving, but to explain this exception, and also the OEE for runway tasks generally.

This raises the issue of how contexts acquired during training may be used in extinction of the same task. It may be that the changes involved in extinguishing a response require a different context from that involved in the original acquisition, in which case the slope of the initial part of the extinction curve would be limited by the persistence of the original context. Extinction could only then proceed when this context representation had been erased. In this case, overtraining would retard extinction. Alternatively, a well-established context might be able to serve as a facilitator of extinction of the same task as that involved in acquiring the context. Either alternative could be possible, because the extinction situation is partially similar and partially different from the acquisition situation. The psychological literature reviewed by MacKintosh (p. 423–7) makes it plausible to suggest that both processes may occur, with some degree of conflict. The progressive improvement in the rate of extinction shown by animals after successive acquisitions and extinctions (MacKintosh 1974, p. 441) is perhaps an indication of the circumstances in which extinction could be subsumed under the same context as that operating during acquisition. However, under more usual circumstances, differing contexts are probably employed in acquisition and extinction. (For instance, if conditions during acquisition are varied from trial to trial, extinction is retarded, as though there are more associations to break than for acquisition during constant conditions: see MacKintosh, p. 434.) An electrophysiological study which points in the same direction is that of Albino and Caiger (1971). During acquisition of a classically conditioned response there was, over days, a progressive increase in synchrony and frequency of theta activity on CS presentation. During the first day of subsequent extinction, the high frequency on CS presentation acquired by the end of acquisition remained in force, and synchrony increased even further. By the second day of extinction, however, frequency was beginning to fall again. This suggests that in the first day of extinction the context acquired during training remained in force, and was even further strengthened, but as extinction progressed further, either this context was abandoned, or an alternative context might have started forming.

In conclusion, the retardation of extinction produced by theta driving in acquisition can plausibly be accounted for as an increase in the number of associations (between contexts and items) to be broken in the extinction phase. The overtraining extinction effect, however, is an issue of some complexity, and cannot be completely resolved here.

10.8 Changes in Regional Occurrence, or Phase of Theta in Different Generators During Learning, or Between Correct and Incorrect Trials

Before discussing the overall relation between theta activity and learning, a final body of evidence should be dealt with on the varying phase relations between theta activity recorded in different parts of the hippocampal circuitry in different stages of learning, or in correct versus incorrect trials. This is important because it bears directly on the idea of self-organizing phase-locked loops between hippocampus and cortex as a means of retrieval of learned information. Most of the evidence here comes from Adey's laboratory.

Adey *et al.* (1960) studied cats in the course of approach learning. They found that early in training the theta activity in CA2 and CA4 had a phase lead over that in the entorhinal cortex of 20–35 ms. The corresponding frequency (in their Fig. 9) is 5.7 Hz, so that theta rhythms in Ammon's horn had a lead of about 55 degrees over those in the entorhinal cortex. Later in training, the phase relation became inverted: The entorhinal cortex had a lead over that in Ammon's horn by as much as 65 ms. (In their Fig. 10, with theta frequency at about 5.0 Hz, this corresponds to a a phase lead by the entorhinal cortex of 117 degrees.) (see Fig. 50). These data suggest that in early training theta activity in Ammon's horn controls that in the entorhinal cortex, but the situation is reversed in later training. This is compatible with a transition from the baseline state (where hippocampal theta activity has only local effects) to the resonant state (produced by learning, and serving to retrieve the learned information). The situation described by

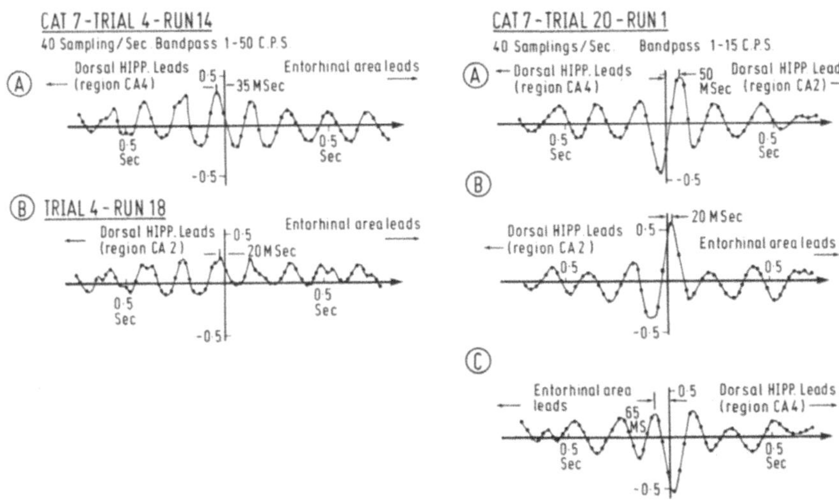

Fig. 50. Cross-correlograms between theta rhythms recorded in two fields of the dorsal hippocampus and in the entorhinal cortex, from cats during the course of approach learning. The basic rhythm has a frequency of about 6 Hz. *Left* early training (Adey *et al.* 1960, Fig. 9). Upper trace, correlogram of dorsal hippocampus (CA4) versus the entorhinal cortex. The entorhinal cortex rhythms lead by 35 ms. Lower trace, correlogram of dorsal hippocampus (CA2) ms. *Right* late training (Adey *et al.* 1960, Fig. 10). Upper trace, correlogram of CA2 versus CA4. CA2 leads by 50 ms. Middle trace, correlogram of CA2 versus entorhinal cortex. Entorhinal cortex leads by 20 ms. Lower trace, correlogram of CA4 versus entorhinal cortex. Entorhinal cortex leads by 64 ms. Note the shift in phase relationships between the early and late training. (From *Archives of Neurology* volume 3, pp. 74–90 1960, copyright 1960, American Medical Association)

Adey *et al.* is actually a little more complex than this: in the highly trained state, not only did the entorhinal cortex theta activity lead that in the dentate gyrus and Ammon's horn, but also field CA2 had a significant phase advance (about 50 ms) over theta activity in CA4 and dentate gyrus, areas traditionally thought to lie in between the entorhinal cortex and CA2, in terms of connections. These additional data are intriguing, but do not offer a fundamental challenge to the theory of self-organizing phase-locked loops. There are possible ways in which Adey *et al.*'s data can be explained. Most likely is the possibility that resonances set up initially through the complete transverse pathway can bypass the dentate gyrus and CA4 at a later stage of evolution of the resonant loops, so that their theta activity is no longer an essential part of the loop.

A related set of results was reported by Adey *et al.* (1961), in which cats performed a visual discrimination in a T-maze. On correct trials, the theta rhythm in the entorhinal cortex led that in Ammon's horn by 150 degrees (about 70 ms at 6 Hz). On incorrect trials, the theta rhythm in Ammon's horn led that in the entorhinal cortex by 90 degrees (about 40 ms at 6 Hz) (Fig. 51)[10]. Once again the implication can be made that correct performance depends on resonance between the hippocampus and the cortex, so that theta in Ammon's horn is controlled by that in the entorhinal cortex. Incorrect trials occur when this resonance fails, and phase relations revert to their baseline state.

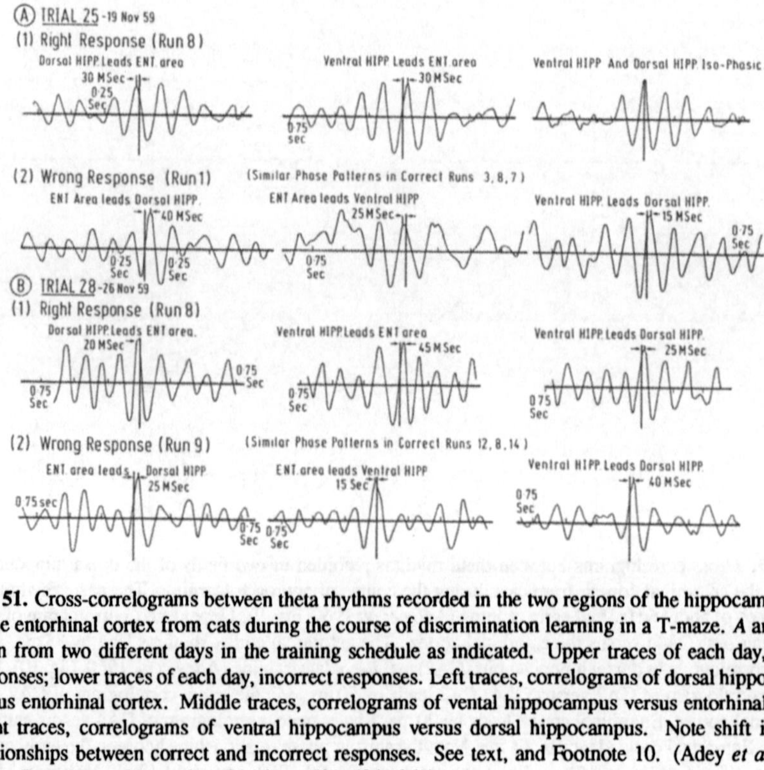

Fig. 51. Cross-correlograms between theta rhythms recorded in the two regions of the hippocampus and in the entorhinal cortex from cats during the course of discrimination learning in a T-maze. *A* and *B* are taken from two different days in the training schedule as indicated. Upper traces of each day, correct responses; lower traces of each day, incorrect responses. Left traces, correlograms of dorsal hippocampus versus entorhinal cortex. Middle traces, correlograms of vental hippocampus versus entorhinal cortex. Right traces, correlograms of ventral hippocampus versus dorsal hippocampus. Note shift in phase relationships between correct and incorrect responses. See text, and Footnote 10. (Adey *et al.* 1961) (From *Experimental Neurology* Vol 3 pp. 501–524 1961)

Radulovacki and Adey (1965), again using cats performing in a visual T-maze discrimination, found that, during performance, theta activity became very regular at 6 Hz, and was now present in dentate gyrus and subiculum as well as Ammon's horn. In contrast, on initial exposure to the maze or in inter-trial intervals at a later stage only the hippocampus showed this frequency. Thus the prediction, which was made at the start of this chapter, that theta activity should progressively involve more brain structures as

[10]Note these are not the values originally reported, but are those calculated in the light of a correction of the original data, as reported in Walter and Adey, 1963, footnote p. 488).

training proceeds, receives support, as do the other predictions documented more fully in this chapter.

10.9 Summary and Discussion in the Light of Earlier Theoretical Notions

In some circumstances a novel stimulus can trigger theta activity, but this appears to be an innate response initiated by the brainstem and septum, rather than dependent on cortico-hippocampal interplay. However, at other times it has been shown that growing familiarity of an environment or of a stimulus in relation to an environment is a determinant of an acquired theta response. In classical, appetitive, avoidance and discriminative conditioning the theta response appears not on first presentation of the task, but grows at the time that the contingencies in the task come to be assimilated (as evidenced by increasing correctness of responding). In GO-NO GO discrimination progressive development of a theta response has also been seen, but there are species differences concerning its preferential occurrence in the GO or the NO GO phase: the theta response is most prominent in that phase of the task which each species finds most difficult. Some experiments where CS– and goal-directed responses are separable have shown that a theta response commonly develops specific to the CS, rather than to the response components, in the above forms of learning. When there is no specially distinct discrete stimulus in a learning task, a theta response can be acquired which has a correlation with response aspects of the task. The occurrence of theta activity in the extinction phase is best explained as a correlation with the process of "unlearning", which, like learning, involves modification of associations. *All these correlations between an acquired theta response and some aspect of learning can be regarded as instances when resonant circuits between the cortex and the hippocampus are activated.* Theta responses do not correlate with the delivery of reward, but may sometimes come to occur after the completion of a behavioural sequence terminating successfully with reward, for reasons which are not immediately obvious.

For many tasks, once a high degree of training has been established the theta response declines or disappears. This has been shown for approach learning, active avoidance learning, lever pressing, and discrimination tasks. However, under some circumstances an acquired theta response may occur during performance of very highly trained tasks. These cases include examples of both acquired response-related and acquired stimulus-related theta activity. Stimulus-related theta responses persist for signals where vigilance is required. The circumstances in which a theta response respectively persists or disappears during the course of training appear to serve a function with respect to learning and performance: once a task has become so strongly represented that it rarely fails, theta activity disappears, and is apparently no longer necessary. *This teleological occurrence of theta activity in relation to learning can be accounted for in terms of instrumental conditioning of the control mechanisms initiating baseline theta activity, concurrent with any conditioning of the behavioural response itself.* (The successful performance of the behavioural response, which controls reinforcement, depends on setting the "state" of the cortical network, a function of the theta rhythm, as well as on activating the appropriate input-output pathways.)

The abundance of theta activity, outside specific periods of learning, appears to correlate, between animals, with the facility those animals have in acquiring a variety of learned responses. Experimental elevation of the abundance of theta activity by administering stimuli to the medial septal nucleus has been reported in a variety of papers. There is an apparent divergence of conclusions, with some results indicating that theta driving improves acquisition/retention (or generally "consolidation" of a memory), while other papers appear to support a role of theta driving in mediating the effects of reward ommission. Amongst the latter effects are both retarded extinction (if theta driving is conducted during acquisition) and accelerated extinction (if it is conducted during extinction). However, the apparent discrepancy between the two bodies of evidence is a result of their tackling rather different aspects of the learning process. There is no real conflict of the evidence, and some indication that the "reward omission" evidence can be subsumed under the umbrella of "improved learning/unlearning". The significance of this putative generalization is that when theta-driving during acquisition retards extinction this may be explained by an increase in the number of associations to be broken during extinction, similar to varying the conditions during acquisition. *These appear to be further instances when resonant circuits between the cortex and the hippocampus are activated.*

There is evidence that early in training, or on incorrect trials for a discrimination task, theta activity in Ammon's horn leads that in the entorhinal cortex, and presumably controls it; but for well trained tasks, and on correct trials for a discrimination task, theta activity in Ammon's horn lags that in the entorhinal cortex, and is presumably controlled by the neural pathways afferent from this region. This is compatible with the idea of resonance between hippocampus and cortex becoming established as learning proceeds.

Many of these occasions when an acquired theta response correlates with some phase of acquisition, extinction or performance of a learned response are also times when the animal produces an overt response (e.g., orienting). However, orienting, or any other behavioural correlate of theta activity, is not its cause, because theta activity can occur in paralyzed animals. As correlations, both orienting and phases of learning can co-exist without rivalry. In fact, a stronger statement than this can be made: we can be certain that learning involves complex, if unspecified internal processes. Moreover, we know that theta activity can sometimes occur in absence of overt behaviour, and so must be a reflection of such an internal process. Therefore it seems plausible to regard the evident correlations between theta activity and phases of learing as more fundamental than those with any external behaviours.

Realization of this is, however, made more difficult because of one further factor: the production of theta activity appears to occur in two separate phases. On the one hand, there is the "baseline" or "ticking over" phase, when septum and hippocampus display theta activity which has not succeeded in capturing a resonance with the cortex. On the other hand, there is the resonant mode of theta activity. Correlations with the former sort of theta activity include many examples which are apparently innate, species-specific electrographic responses, which tend to occur in just those situations where individuals of that species do most of their information gathering, whether or not the mechanism of resonance is in operation. In these cases, it is natural to emphasize the correlation with *behaviour*, because it might have an ecological or evolutionary significance for each species. However, these examples should not be confused with the evidence discussed in the present chapter, which is limited to the second type of theta activity, that involving hippocampo-cortical resonance. In such cases it appears that one can generalize across species, with reference to an internal process related to learning, which cuts across any concurrent correlation with overt behaviour.

11 Synopsis and Epilogue

11.1 Synopsis

In Part 1 of this work, a threefold analogy was drawn, with respect to input-output relations in the behaving organism:

First of all, from a perspective of post-Renaissance philosophy, it was argued that the overt premises for a logical inference do not unconditionally dictate the conclusions which might be derived therefrom. There is required in addition a statement of the general framework of ideas within which premises and conclusions are to be linked, this framework generally being hidden in the statement of a syllogism.

Secondly, in twentieth century learning theory the relation between stimulus and response is not an unconditional one: again one requires an additional category of information, defining the dimension of the stimuli and/or responses which are significant for a particular learning task.

Thirdly, a biological counterpart to these philosophical and psychological ambiguities can be seen in the style of connectivity in the cerebral cortex. Cortical neurons have insufficient connections to perform the associative operations which the animal apparently does perform, if one is limited by organizational principles presently understood. Thus, cell assemblies representing objects cannot be addressed uniquely by individual component stimuli which are images of that object. Likewise, the cell assemblies representing response intentions cannot be addressed uniquely by the objects which should trigger appropriate responses. In both cases the resolution of the ambiguity in representation of information apparently involves representation of the environment as a whole, as a "global cell assembly". However, neurons in the cortex have insufficient connections to form such a global cell assembly.

The "framework of ideas" for deductive inference, the "dimension" for relating input to output in learning, and the "global cell assembly" all refer to the same thing, namely, the information pattern which provides the conditions in which input can be uniquely linked to output. To this entity, the name "context" can be given. To explain how context representations can form, and how they are used for information retrieval and manipulation requires principles of organization beyond the conventional (Hebbian) cell assembly, though that principle is likely to have a most important contributory role. A number of lines of empirical evidence suggest that these additional principles involve the hippocampus, especially in its interaction with the rest of the cerebral cortex.

Part 2 of the monograph deals with empirical evidence which constrains and directs theory formulation. Firstly, details of the connections within the hippocampus, and between it and the isocortex are reviewed. The most important conclusion from this review is that there are a variety of recurrent loops of connections which involve the hippocampus. The size of these loops varies: some exist entirely within Ammon's horn, some between it and the transitional cortical regions of the parahippocampal gyrus, others between hippocampus and limbic and prefrontal regions of isocortex. The latter can also connect with many other regions of isocortex, especially the association areas. While many of these connections are reciprocal and therefore non-polarized, those which involve the perforant path (i.e., the majority of the possible larger recurrent loops) are clearly polarized, since this pathway (rather unusually) is not a reciprocal one. The implication is that whatever neural activity circulates in such axonal loop configurations, the influences within the hippocampal formation must pass *unidirectionally*, that is from entorhinal cortex to dentate gyrus, to Ammon's horn, and then to the subiculum, and isocortical regions. In addition, the area of isocortex most strongly linked with the hippocampal formation is the part which is spatially most remote from it (the prefrontal cortex). This could be regarded as a design feature which ensures the longest possible conduction delays around any axonal loops between isocortex and hippocampus.

With this anatomical information as a background, the regular rhythmic activity of the hippocampus known as the theta rhythm is introduced. The first body of information which is described is the collection of more obvious correlations between overt behaviour and presence of rhythmic activity in the theta frequency range. These correlations vary somewhat between species of experimental animal, but a large part of the complex data available can be subsumed under a single generalization: that theta activity is generated at times when an animal is likely to be gathering information from its environment. Since different species do this in different ways, so the correlations between behaviour and occurrence of theta activity vary between species. In addition to this simplifying principle there are numerous hints that there is more to the hippocampal theta rhythm than just an innate correlation with the species-specific mode of behaviour in which information is gathered. Many of the exceptional cases which allow this statement to be made concern learning tasks as studied in the laboratory. As a whole they leave the impression that the rhythmic activity at the theta frequency may be acquired at particular stages of learning, as well as being produced automatically and innately at times when animals of a particular species are *likely* to be learning. However, since learning is an internal process appropriate to theoretical treatment, it is impossible to develop ideas of this relationship further until we have a theory to test. Before this can be done, a larger data base concerning the theta rhythm needs to be established.

The medial septum/diagonal band complex plays a crucial role as a pacemaker for the hippocampal theta rhythm, (although under unusual circumstances, such as slice preparations, the hippocampus by itself can generate rhythmic activity). Rhythmically active neurons in this region deliver rhythmic neural activity to the hippocampal fields to which they project, and are thus a major influence controlling the occurrence and frequency of the hippocampal rhythm. It is unlikely that the rhythmic signals which pace the hippocampal rhythm are transmitted from any site lower in the neuraxis than the septum. No nuclei caudal to the septal ones are known to generate rhythms in the theta frequency range. However, non-rhythmic neural activity, ascending from various loci in the brainstem controls the rhythmic activity of the septo-hippocampal system. Some brainstem loci are known to promote the theta rhythm, while others are known to desynchronize the hippocampus. It is probable that the innate correlations between species-specific information-gathering behaviour and theta activity are coordinated at the level of these brainstem nuclei, rather than at the septal or hippocampal level.

There is a body of experimental data suggesting that two independent mechanisms for pacing the theta rhythm exist, apparently both at the level of the septal nuclei. The two mechanisms are supposed to differ in frequency range, pharmacological properties and behavioural correlates. Briefly, a high frequency rhythm, occurring during vigorous whole body movement, can occur after cholinergic mechanisms have been blocked, but not after low doses of anaesthetics; and a low frequency component, occurring during smaller body movements or under anaesthetia is blocked by anticholinergic drugs. However, as argued in Chapter 6, it seems more parsimonious to suggest that there is a single system of control of theta activity in the septum, which can cover the entire theta frequency range, this being under control of ascending influence from the brainstem. In addition, a cholinergic mechanism within the septal region determines the mutual recruitment of rhythmically active septal neurons. The evidence which appears to demonstrate two separate control mechanisms can best be explained by the idea that during the pharmacological conditions, or the behavioural epochs which distinguish the two, there is concomittant modification in the ascending drive or the mutual recruitment of septal neurons.

Lesions separating the hippocampus from the isocortex limit the occurrence of the higher frequency component of theta activity, leaving only the atropine-sensitive one. This finding is best explained in terms of some *rhythmic interplay* between hippocampus and isocortex, rather than in terms of an additional pathway controlling theta activity, which reaches the hippocampus by way of the isocortex.

Detailed experimental analysis of theta potentials shows that there is more than one site of generation in the hippocampal complex. The principle sites are located in field CA1 and in the dentate gyrus. In the anaesthetized preparation field CA3 probably does not generate theta rhythms, while in the free moving animal it forms a third source of rhythmic activity. The temporal relation between the source in superficial layers of CA1 and that in the deeper parts of CA1 and in the dentate gyrus is subtly different in the anaesthetized and the free moving rat: in the former case there is fairly accurate phase reversal between the superficial and the deep generator, whatever the theta frequency. In the free moving rat, however, theta activity in the dentate generator and in deep layers of CA1 leads that in superficial layers of CA1 by a constant delay, regardless of frequency, so that angular phase difference between the two increases as frequency increases.

The hippocampal complex is not the only site to generate theta potentials. Although the amount of available information is small, the entorhinal cortex and cingulate cortex can also sometimes show theta rhythms, apparently generated locally. In the rest of the isocortex theta rhythms have been seen also, but as yet it cannot be excluded that they are merely volume-conducted there from deeper sources.

It is clear that the cell types which are mainly responsible for generating the macroscopic theta rhythm are the principle neurons, that is the pyramidal cells of field CA1 and the dentate granule cells. The interpretation of the profiles of theta potentials in the hippocampus is however, complex, and requires separate treatment in the various preparations in which they have been studied.

In the urethane-anaesthetized rat, it is likely that there are rhythmic EPSP's directed at the superficial layers of CA1, and rhythmic IPSP's directed at the deep layers of CA1 (the distal parts of the apical dendrites of pyramidal cells). In the dentate gyrus there may be a converse arrangement, with rhythmic EPSP's directed at the middle region of the granule cell dendrites, and rhythmic IPSP's directed at the level of their somata. If an arrangement such as this applies, and both EPSP's and IPSP's are under direct monosynaptic control by the rhythmic signals from the septal region, the phase reversal between the superficial and the deep theta rhythms would be explained. Some arrangement like this is the most likely explanation of the phase relations between the different generators of theta activity in the hippocampal complex. However, the synaptic relation between the septal rhythmic activity and that in the various fields of the hippocampus appears to be much more complex than a monosynaptic one. Therefore the precise mechanism by which phase reversal between deep and superficial layers is coordinated is at present unresolved.

In the free moving rat, it may also be true that the theta rhythms are set up by combinations of rhythmic EPSP's and rhythmic IPSP's. However, the control mechanism must be very different from the case of the urethanized rat, because there is not an accurate phase reversal between the superficial and the deep generators of theta potentials, but rather a constant time lag of the superficial with respect to the deep rhythms. The deep layers of CA1 and the outer part of the molecular layer of the dentate gyrus receive inputs from a common source, the entorhinal cortex. Moreover, there is evidence that, in the free moving rat, the isocortex can influence the hippocampal rhythms via the entorhinal cortex. In view of these two facts, it is likely that the entorhinal cortex, rather than the septal region is the immediate controller of the phase-advanced rhythms in dentate gyrus and deep part of CA1. The constant delay between the rhythms in the deep generator and that in the superficial part of CA1 suggests that the dentate gyrus controls the superficial rhythms in the free moving rat. This conjecture is supported by the fact that the field which intervenes in the pathway between these two other rhythmically active sites, namely CA3, is also rhythmically active in the free moving rat, though not in the anaesthetized one.

There are several lines of evidence suggesting a rhythmic interaction between the hippocampus and the isocortex in unanaesthetized preparations. These include the following: (1) Rhythmic activity at the frequency simultaneously shown by the hippocampus is found in at least some areas of the isocortex. (2) Lesions disconnecting the hippocampus from the isocortex, attenuate the higher frequency rhythmic activity shown by the hippocampus in rats. (3) Harmonics of the fundamental theta frequency have been seen in several studies, especially those which subjected theta rhythms to spectral analysis.

The last point suggests that the rhythmic interplay between hippocampus and cortex is a form of *resonance*. Support for the idea that resonant circuits are completed in the free-moving rat is to be found in the evidence that there is a constant delay from dentate gyrus to CA1 in this preparation. Thus, the transverse pathway through the hippocampal complex, being a part of this circuit, is completed in this preparation.

In Part 3 these various experimental data are unified by a theory in which loops of axonal connections between hippocampus and isocortex convey circulating neural activity which resonates with defined phase relations to the theta rhythm concomitantly generated by the hippocampus. These loops are capable of organizing themselves, by selection from a larger substrate of axonal connections with the same loop configurations. The selection process involves strengthening of specific connections by rules such as those envisaged by Hebb. Self-organization is possible because the axonal connections from which these loops are selected for strengthening have very long conduction delays around the loop, which can approximate to the period of the hippocampal theta rhythm. The isocortical end of these patterns of loops is envisaged to be activated tonically by component stimuli of the prevailing environment, while the hippocampal end is activated rhythmically as part of the theta rhythm. The strengthening process involves Hebbian or quasi-Hebbian processes of synaptic modification at each end, and, if there are intervening synaptic relays, at these as well. Each pattern of resonant circuitry is specific to the environment which triggers it into activity, and thus represents the environment, when conventional cell assemblies could not. Additional strengthening of connections, possibly in the entorhinal cortex and adjacent regions, may ensure that the environment is recognized categorically as a global Gestalt, rather than merely as a collection of items.

Extensions of this theory helps explain data on phase relations between theta rhythms in the entorhinal cortex and the hippocampus: between these two areas shorter resonant loops may be set up, phase-locked to the inverted half-cycle, rather than to the full theta period.

There is a body of corroboratory evidence which shows that the neurons in the hippocampus do indeed represent environments (or more generally, contexts). In experimental animals, the majority of the principal neurons of the hippocampal complex encode *place* in the environment. More detailed consideration of the neuroethological data available suggests that the specific responsiveness of these neurons is controlled by aspects of the environment in addition to place, including apects of the responses appropriately emitted in each environment. Thus the concept of a context is defined not only by the contiguities in sensory input, but also by those between motor output and its sensory consequences. In human mental life, the concept of a context takes on even greater versatility.

According to this theory, contexts are not innate, and are not simply representations of Euclidian space or any other simple topographic mapping of the physical environment. They are acquired to match the total information structure of the environment, and thus are much more flexible in their scope than spatial representations. Amongst the realms of information-processing which can be coordinated by these hippocampally derived contexts, are the many different dimensions of human thought.

A theory which can account for the encoding and use of contexts for information processing can, in principle, address a central issue in relations between psychological descriptions of behaviour and brain mechanisms. This issue is the problematical concept of *selective attention.* It is suggested that only one context representation can be in operation at once, due to inhibitory processes acting at some part of the cortico-hippocampal loops. Consequently, selective attention is a macroscopic reflection of the competition between the patterns of loops which form context representations.

Thus, even when different patterns of loops resonate at the same theta frequency, recall of a memory may be impossible unless the correct pattern has been activated. There is, however, a variety of different frequencies within the theta range which can entrain resonant loops. The possibilities of recalling memories is even more remote if the frequency obtaining during recall is not the same as that present during acquisition of the memory. This is a very likely explanation of at least some instances of state-dependent amnesia.

The theory of self-organizing, phase-locked loops derived originally from the need for a more economical use of available number of synaptic connections between different cortical areas. Hippocampal place-selective neurons provide some evidence for just such an economical use of connections, in the site in the brain which coordinates representations of the environment.

When the hippocampus is disabled by either a lesion of it, or of the pathways afferent to it which control theta activity, many types of information processing are lost (see O'Keefe and Nadel 1978). Some important functions remain, however. In part, these depend on spatio-temporal structures laid down in the cortex while the hippocampus was still intact, allowing specific memories to be retrieved relating to times before hippocampal damage was sustained. The other main category of information processing retained after hippocampal ablation is learned information, where the signals are strong enough to be unambiguous, even when no context representation is present. This second type of stored information may correspond to what Mishkin and Petri (1984) referred to as *habits* (in contrast to true memories, which are context dependent).

The overall concept of the hippocampal complex is thus as one of the information processors which organizes, coordinates and indexes information stored in the isocortex; but the hippocampus is not the primary repository of that stored information, this being the much larger cortex itself. This concept can accommodate the dramatic fact that the hippocampus shows wide variation amongst mammalian species in its size relative to that of the isocortex.

Given this hypothesis, there are a number of further implications about the dynamic aspects of theta rhythms during the changing situations of specific learning tasks. The circumstances in which theta rhythms initiated in the hippocampus can entrain cortical nodes are those in which learning occurs. When this happens and loops are completed, one should expect that the frequency of the theta rhythm may shift or the rhythm may become more highly focussed on a specific frequency; the phase relations between theta rhythms in different parts of the hippocampal circuitry should change; and the loci from which theta rhythms can be recorded should also extend. The literature on theta rhythms during learning contain examples of all these predictions. In addition, in discrimination tasks there is some evidence for consistent differences in phase relations between different parts of the hippocampal circuitry which correlate with production of correct versus incorrect responses.

Often, the acquired theta response is prominent only during the stage of learning in which performance is improving most rapidly. Once the task can be performed consistently, the acquired theta response declines. This finding suggests that theta rhythms are part of the animal's total response which fulfills the goal of correct performance in the learning tasks. Reinforcement mechanisms can promote the occurrence of theta rhythms as well as of the behavioural response. However, in the well-learned task, correct performance does not require a context representation, and so takes place whether or not a theta rhythm is present. There is then no differential reinforcement of theta rhythms on correct trials, and the theta rhythm associated with the task declines.

In some types of task, however, the theta rhythm persists even when the task is well learned. Generally, these are tasks where vigilance is required to detect a sensory cue which indicates the appropriate time for a response, or the time to change the strategy of response.

Further evidence for a relation between theta rhythms and learning is that the abundance of spontaneously produced theta activity may correlate with ease of learning. In addition, experimental modification of the amount of theta activity in an animal can produce changes in the rate of learning. All these data support the idea that resonance between cortex and hippocampus at the theta frequency plays a part in the registration of information in the cortex during learing.

The literature on the relation between theta rhythms and extinction is quite complex, but is best accounted for by the notion that extinction, like the original learning, involves changes of representation which are facilitated by the presence of resonant theta activity.

Overall, the literature on the relation between theta activity and learning supports the idea that resonant phase-locked loops between hippocampus and cortex mediate the registration, organization and retrieval of learned information. Some of this evidence supports the theory in quite specific detail. Undoubtedly, theta rhythms also have their correlates in terms of overt behaviour, such as orienting. However, the relationship of theta rhythms to the internal processes of learning are more fundamental causal ones, supported by the logic of a hypothetical mechanism, rather than merely the empirical validity of a correlation.

The argument presented in Chapters 1 to 10 has been long and complex. From the welter of experimental scientific literature reviewed doubts may still be raised that the mechanism proposed would actually work as it is envisaged to work. The empirical data can take one so far, but in a field as complex as this they may not relieve one of the burden of proof that the forebrain actually could operate in the way proposed. After all, where much of the evidence can be given alternative interpretations, a mechanism that is logically impossible could still be made to fit the evidence. This sends the tennis ball back to the theoretician's court.

From the theoretical point of view there is a method other than literature-based theory which could allow the ideas put forward here to be formulated more rigorously. The method is simulation. If a computer programme could be written which embodies the central premises of the above theory and also produces the macroscopic behaviour which the theory is supposed to explain, the plausibility of the theory would be enhanced considerably. However, that is a task for the future.

11.2 Epilogue

In an earlier work (Miller 1981 *Introduction*), some comments were made about the general characteristics of brain research at that time. Amongst other things, attention was drawn to the overload of good experimental information in the neurosciences, combined with a dearth of good theoretical ideas with which to integrate that information. Much brain research was seen as motivated by a desire to exploit the available *techniques* rather investigate *ideas* of brain function. It was pointed out that the history of science has often shown periods of intense data collection, followed by phases of rapid conceptual or theoretical rethinking. In the years since 1981, the relationship between experimental and theoretical neurosciences has started to develop in more fruitful ways. Experimental work continues with remarkable intensity, and experimental papers are more sharply focussed in relation to concepts of brain function. Theoretical research has a higher profile now than in 1981, for both library-based informal theorizing and the more technical studies using mathematical or computer simulation methods. There is, however, still a major underutilization of the accumulated knowledge we have in the neurosciences. Like a supersaturated solution, we wait for major syntheses of this information to "crystallize out." For this the most fundamental requirement is a *context*.

Where does the theory presented in the present work fit into the large panorama of accumulated knowledge now available about the mammalian forebrain? The *electrophysiological* knowledge we have about various cortical regions suggests strongly that the cerebral cortex has a general information processing function of registering the association between signals which contiguously activate single neurons. This general function has, of course, to by justified for each cortical region in turn. The precise function of neurons in each area depends on just what is the source of the afferents to that area. For instance, motor functions of the cortex can be incorporated into this picture, since both basal ganglia and cerebellum are afferent to that same cortical mantle. The basic logic revealed by these experimental results is "association on the basis of signal contiguity". Although it is not always valid to translate such a logic from the single cell level to that of the network as a whole, or that of the behaving organism, these results could be taken as supporting a "stimulus-response" version of the psychology of learning.

On the other hand, *psychologists* now deal with a very different set of concepts, a language of their own, not yet grounded in a solid base of biological structure and function. The concepts of this language are those of cognitive science, including of course contexts, dimensions, and above all "selective attention". These concepts require to be integrated into the main framework of scientific knowledge, which started with post-Renaissance physics, and has gradually incorporated chemistry, biochemistry, anatomy, physiology, and many other more recent disciplines. Central to this integration are the questions posed by the term "selective attention." In biological terms, what is selected? On what basis is it selected? What is the mechanism of that selection? These are central questions to be answered if the science of the forebrain is to become a coherent body of theory and experimental evidence. If these questions could be answered, one would have a real hope of integrating thousands of other research papers into a comprehensive account of the forebrain. In other words, the answer to these questions is the context for which contemporary brain research is looking.

Into this conceptual rift are introduced some very obvious anatomical facts, which have been known from the start of the present century: neurons have connections which diverge and converge so widely that neurons as relays are bound to have an ambiguous representation; and yet they have too few connections to perform the associative operations which animals actually do perform in the many examples of context-dependent behaviour. From this paradox, the theory of self-organizing phase-locked loops for representing environments as a whole was developed.

The present work is not the complete theory of the mammalian forebrain. However, its author entertains a hope that the arguments presented herein do, in fact, represent at least part of the context for a large amount of that experimental work. To end with an analogy, and a rhetorical question: *We have innumerable pieces to the jigsaw puzzle that is the forebrain; but do we yet know what that picture is about?*

References

Adey WR, Walter DO (1963) Application of phase detection and averaging techniques in computer analysis of EEG records in the cat. Exp Neurol 7:186–209

Adey WR, Dunlop CW, Hendrix CE (1960) Hippocampal slow waves. Distribution and phase relationships in the course of approach learning. Arch Neurol 3:74–90

Adey WR, Walter DO, Hendrix CE (1961) Computer techniques in correlation and spectral analyses of cerebral slow waves during discriminative behaviour. Exp Neurol 3:501–524

Adey WR, Walter DO, Lindsley DF (1962) Subthalamic lesions: effects on learned behaviour and correlated hippocampal and subcortical slow-wave activity. Arch Neurol 6:194–207

Albino RC, Caiger K (1971) Hippocampal slow rhythms in ongoing behaviour and during classical conditioning. S Afr Med J 45: 135–140

Aleksanov SN, Vainstein II, Preobrashenskaya LA (1986) Relationship between electrical potentials of the hippocampus, amygdala and neocortex during instrumental conditioned reflexes. Neurosci Behav Physiol 16:199–207

Alger BE, Nicoll RA (1980) Epileptiform burst afterhyperpolarization: a calcium-dependent potassium potential in hippocampal CA1 pyramidal cells. Science 210:1122–1124

Alonso A, Kohler C (1982) Evidence for separate projections of hippocampal pyramidal and non-pyramidal neurons to different parts of the septum in the rat brain. Neurosci Lett 31:209–214

Alonso A, Kohler C (1984) A study of the reciprocal connections between the septum and the entorhinal area using anterograde and retrograde axonal transport methods in the rat brain. J Comp Neurol 225:327–434

Alonso A, Garcia-Austt E (1987a) Neuronal sources of theta rhythm in the entorhinal cortex of the rat. I. Laminar distribution of theta field potentials. Exp Brain Res 67:493–501

Alonso A, Garcia-Austt E (1987b) Neuronal sources of theta rhythm in the entorhinal cortex of the rat. II Phase relations between unit discharges and theta field potentials. Exp Brain Res 67:502–509

Amaral DG, Kurz J (1985) An analysis of the origins of the cholinergic and non-cholinergic septal projections to the hippocampal formation of the rat. J Comp Neurol 240:37–59

Amaral DG, Insausti R, Cowan WM (1983) Evidence for a direct projection from the superior temporal gyrus to the entorhinal cortex in the monkey. Brain Res 275:263–277

Amaral DG, Insausti R, Cowan WM (1987) The entorhinal cortex of the monkey. I. Cytoarchitectonic subdivisions. J Comp Neurol 264:346–355

Amsel A (1962) Frustrative non-reward in partial reinforcement and discrimination learning: some recent history and a theoretical extension. Psychol Rev 69:306–328

Anchel H, Lindsley DB (1972) Differentiation of two reticulo-hypothalamic systems regulating hippocampal activity. Electroencephalogr Clin Neurophysiol 32:209–226

Andersen P (1978) General discussion. In: "Functions of the septo-hippocampal system." Ciba Foundation Symposium 58 (New Series) Elliott K, Whelan J (eds), Elsevier-North Holland, Amsterdam, p.312

Andersen P, Bliss TVP, Lomo T, Olson LL, Skrede KK (1969) Lamellar organization of the hippocampal excitatory pathways. Acta Physiol Scand 76:4–5A

Andersen P, Bliss TVP, Skrede K (1971) Lamellar organization of hippocampal excitatory pathways. Exp Brain Res 13:222–238

Anokhin PK (1961) Electroencephalographic analysis of cortico-subcortical relations in positive and negative conditioned reactions. Ann NY Acad Sci 92:899–938

Apostol G, Creutzfeldt OD (1974) Crosscorrelation between the activity of septal units and hippocampal EEG activity during arousal. Brain Res 67:65–75

Arnolds D, Lopes da Silva FH, Kamp A, Aitink W (1975) Motor acts and firing of reticular neurones correlated with operantly reinforced hippocampus theta shifts. Brain Res 85:194–195

Arnolds DEAT, Lopes da Silva FH, Aitink JW, Kamp A (1979a) Hippocampal EEG and behaviour in dog I. Hippocampal EEG correlates of gross motor behaviour. Electroencephalogr Clin Neurophysiol 46:552–570

Arnolds DEAT, Lopes da Silva FH, Aitink JW, Kamp A (1979b) Hippocampal EEG and behaviour in dog II. Hippocampal EEG correlates with elementary motor acts. Electroencephalogr Clin Neurophysiol 46:571–580

Arnolds DEAT, Lopes da Silva FH, Aitink JW, Kamp A (1979c) Hippocampal EEG and behaviour in dog III Hippocampal EEG correlates of stimulus-response tasks and of sexual behaviour. Electroencephalogr Clin Neurophysiol 46:581–591

Arnolds DEAT, Lopes da Silva FH, Aitink JW, Kamp A, Boejinga P (1980) The spectral properties of hippocampal EEG related to behaviour in man. Electroencephalogr Clin Neurophysiol 50:324–328

Arnolds DEAT, Lopes da Silva FH, Boejinga P, Kamp A, Aitink W (1984) Hippocampal EEG and motor activity in the cat: role of eye movements and body acceleration. Behav Brain Res 12:121–135

Artemenko DP (1973) Role of hippocampal neurons in theta-wave generation. Neurophysiology (Neirofiziologia [Kiev]) 4:409–415

Assaf SY, Chung SH (1984) Release of endogenous Zn++ from brain tissue during activity. Nature 308:734–736

Azzaroni A, Parmeggiani PL (1967) Feedback regulation of the hippocampal theta rhythm. Helv Physiol Acta 25:309–321

Baisden RH, Woodruff ML, Hoover DB (1984) Cholinergic and non-cholinergic septo-hippocampal projections: a double-label horseradish peroxidase-acetylcholinesterase study in the rabbit. Brain Res 290:146–151

Baleydier C, Maugiere F (1980) The duality of the cingulate cortex in monkey. Neuroanatomical study and functional hypothesis. Brain 103:525–554

Ball GG, Gray JA (1971) Septal self-stimulation and hippocampal activity. Physiol Behav 6:547–549

Barnett SA (1981) Modern Ethology: the science of animal behaviour. Oxford University Press, New York

Bartel's VI, Urayev YV (1982) Hippocampal electrical activity under instrumental choice conditions. Neurosci Behav Physiol 12:240–243

Basar E, Demir N, Gonder A, Ungan P (1979a) Combined dynamics of EEG and evoked potentials. I. Studies of simultaneously recorded EEG-EPograms in the auditory pathway, reticular formation and hippocampus of the cat brain during the waking stage. Biol Cybern 34:1–9

Basar E, Durusan R, Gonder A, Ungan P (1979b) Combined dynamics of EEG and evoked potentials. II. Studies of simultaneously recorded EEG-EPograms in the auditory pathway, reticular formation and hippocampus of the cat brain during sleep. Biol Cybern 34:21–30

Bell C, Sierra G, Buendia N, Segundo JP (1964) Sensory properties of neurons in the mesencephalic reticular formation. J Neurophsyiol 27:961–987

Ben-Ari Y, Krnjevic K, Reinhardt W, Ropert N (1981) Intracellular observations on the disinhibitory action of acetylcholine in the hippocampus. Neuroscience 6:2475–2484

Bennett TL (1970) Hipocampal EEG correlates of behaviour. Electroencephalogr Clin Neurophysiol 28:17–23

Bennett TL (1975) The electrical activity of the hippocampus and processes of attention. In: Isaacson RI, Pribram KH (eds), The hippocampus, Vol II Plenum Press, New York, London, pp. 71–100

Bennett TL, French J (1977) Electrical activity of the cat hippocampus during the species-typical gape response: evidence against the voluntary movement hypothesis. Behav Biol 21:432–437

Bennett TL, Gottfried J (1970) Hippocampal theta activity and response inhibition. Electroencephalogr Clin Neurophysiol 29:196–200

Bennett TL, Herbert PN, Moss DS (1973) Hippocampal theta activity and the attention component of discrimination learning. Behav Biol 8:173–181

Bennett TL, French J, Burnett KN (1978) Species differences in the behaviour correlates of hippocampal RSA. Behav Biol 22:161–177

Berger TW, Swanson GW, Milner TA, Lynch GS, Thompson RF (1980a) Reciprocal anatomical connections between hippocampus and subiculum in the rabbit: evidence for subicular innervation of regio superior. Brain Res 183:265–276

Berger TW, Milner TA, Swanson GW, Lynch GS, Thompson RF (1980b) Reciprocal anatomical connections between anterior thalamus and cingulate-retrosplenial cortex in the rabbit. Brain Res 201:411–417

Berger TW, Laham RI, Thompson RF (1980c) Hippocampal unit-behaviour correlations during classical conditioning. Brain Res 193:229–248

Berger TW, Rinaldi PC, Weisz DI, Thompson RF (1983) Single unit analysis of different hippocampal cell types during classical conditioning of the nictitating mebrane response. J Neurophysiol 50:1197–1219

Berry SD, Thompson RF (1978) Prediction of learning rate from the hippocampal EEG. Science 200:1298–1300

Best PJ, Ranck JB (1982) Reliability of the relationship between hippocampal unit activity and sensory-behavioural events in the rat. Exp Neurol 75P:652–684

Bienenstock E, Fregnac Y, Thorpe S (1983) Iontophoretic clamp of activity in visual cortex neurones in the cat: a test of Hebb's hypothesis. J Physiol (Lond) 345:123P

Bilkey DK, Goddard GV (1985) Medial septal facilitation of hippocampal granule cell activity is mediated by inhibition of inhibitory interneurones. Brain Res 361:99–106

Bilkey DK, Goddard GV (1987) Septohippocampal and commissural pathways antagonistically control inhibitory interneurons in the dentate gyrus. Brain Res 405:320–325

Black AH (1975) Hippocampal electrical activity and behaviour. In: Isaacson RI, Pribram KH (eds), The hippocampus, Vol II, Plenum, New York, pp. 129–168

Black AH, Young GA (1972) Electrical activity of the hippocampus and cortex in dogs operantly trained to move and to hold still. J Comp Physiol Psychol 79:128–141

Black AH, Young GA, Batenchuk C (1970) Avoidance training of the hippocampal theta waves in flexedilised dogs and its relation to skeletal movement. J Comp Physiol Psychol 70:15–24

Blackstad TW (1956) Commissural connections of the hippocampal region in the rat, with special reference to their mode of termination. J Comp Neurol 105:417–537

Blackstad TW, Brink K, Hem J, Jeune B (1970) Distribution of hippocampal mossy fibres in the rat. An experimental study with silver impregnation methods. J Comp Neurol 138:433–450

Blakemore C, Mitchell DE (1973) Environmental modification of the visual cortex and the neural basis of learning and memory. Nature 241:467–468

Blaker SN, Armstrong DM, Gage FH (1988) Cholinergic neurons within the rat hippocampus: response to fimbria-fornix transection. J Comp Neurol 272:127–138

Bland BH, Vanderwolf CH (1972) Diencephalic and hippocampal mechanisms of motor activity in the rat: Effects of posterior hypothalamic stimulation on behaviour and hippocampal slow wave activity. Brain Res 43:67–88

Bland BH, Whishaw IQ (1976) Generators and topography of hippocampal theta (RSA) in the anaesthetized and freely moving rat. Brain Res 118:259–280

Bland BH, Andersen P, Ganes T (1975) Two generators of hippocampal theta activity in rabbits. Brain Res 94:199–218

Bland BH, Sainsbury RS, Creery BL (1979) Anatomical correlates of rhythmical slow activity (theta) in the hippocampal formation of the cat. Brain Res 161:199–209

Bland BH, Andersen P, Ganes T, Sveen O (1980) Automated analysis of rhythmicity of physiologically identified hippocampal formation neurons. Exp Brain Res 38:205–219

Bland BH, Sainsbury RS, Seto M, Sinclair BR, Whishaw IQ (1981) The use of sodium pentobarbital for the study of immobility-related (type 2) hippocampal theta. Physiol Behav 27:363–368

Bland BH, Seto MG, Rowntree CJ (1983) The relation of multiple hippocampal theta cell discharge rates to slow wave theta frequency. Physiol Behav 31:111–117

Bland BH, Seto MG, Sinclair BR, Fraser SM (1984) The pharmacology of hippocampal theta cells: evidence that the sensory processing correlate is cholinergic. Brain Res 299:121–131

Bland BH, Colom LV, Konopacki J, Roth SH (1988) Intracellular records of carbachol-induced theta rhythm in hippocampal slices. Brain Res 447:364–368

Bland SK, Bland BH (1986) Medial septal modulation of hippocampal theta cell discharges. Brain Res 375:102–116

Borst JGG, Leung L-WS, McFabe DF (1987) Electrical activity of the cingulate cortex II. Cholinergic modulation. Brain Res 407:81–93

Bragin AG, Vinogradova OS (1983) Comparison of neuronal activity in septal and hippocampal grafts developing in the anterior eye chamber of the rat. Dev Brain Res 10 (Brain Res 312): 279–286

Braitenberg V (1978) Cortical architectonics: general and areal. In: Brazier MAB, Petsche H (eds) Architectonics of the cerebral cortex, Raven, New York, pp. 443–465

Brazier MAB (1968) Studies of the EEG activity of limbic structures in man. Electroencephalogr Clin Neurophysiol 25:309–318

Braznik ES, Vinogradova OS, Karanov AM (1984) Control of the neuronal theta-bursts in the septum by cortical and brain stem-diencephalic structures. Zh Vyssh Nerv Deiat 34:71–80

Braznik ES, Vinogradova OS, Karanov AM (1985) Frequency modulation of neuronal theta-bursts in rabbit's septum by low frequency repetitive stimulation of the afferent pathways. Neuroscience 14:501–508

Braznik ES, Vinogradova OS (1987) Action of anticholinergic drugs and their combinations with pentobarbital on theta burst neurons of the rabbit septum. Neurosci Behav Physiol 17:386–394

Braznik ES, Vinogradova OS (1988) Modulation of the afferent input to the septal neurons by cholinergic drugs. Brain Res 451:1–12

Brodal A (1969) Neurological Anatomy in relation to clinical medicine. Oxford University Press, Oxford

Brown BB (1968) Frequency and phase of hippocampal theta activity in the spontaneously behaving cat. Electroencephalogr Clin Neurophysiol 24:53–62

Brown BB, Shryne JE (1964) EEG theta activity and fast activity sleep in cats as related to behavioural traits. Neuropsychologia 2:311–326

Brown KA, Buchwald JS (1973) Acoustic responses and plasticity of limbic units in cats. Exp Neurol 40:608–631

Brucke F, Petsche H, Pillat B, Deisenhammer E (1959a) Die Beeinflussung der "Hippocampus-arousal-Reaktion" beim Kaninchen durch elektrische Reizung im Septum. Pflügers Arch 269:319–338

Brucke F, Petsche H, Pillat B, Deisenhammer E (1959b) Über Veränderungen des Hippocampus-Elektrencephalogrammes beim Kaninchen nach Novocaininjektion in die Septumregion. Naunyn Schmiedebergs Arch Pharmacol 237:276–284

Brugge JF (1965) An electrographic study of the hippocampus and neocortex in unrestrained rats following septal lesions. Electroencephalogr Clin Neurophysiol 18:36–44

Buzsaki G (1985) Electroanatomy of the hippocampal rhythmic slow activity (RSA) in the behaving rat. In: Buzsaki G, Vanderwolf CH (eds) Electrical activity of the archicortex. Akademiai Kiado, Budapest, pp. 143–164

Buzsaki G, Eidelberg E (1983) Phase relations of hippocampal projection cells and interneurons to theta activity in the anaesthetised rat. Brain Res 266:334–339

Buzsaki G, Grastyan E, Tveritskaya IN, Czopf J (1979a) Hippocampal evoked potentials and EEG changes during classical conditioning in the rat. Electroencephalogr Clin Neurophysiol 47: 64–74

Buzsaki G, Grastyan E, Kellenyi L, Czopf J (1979b) Dynamic phase-shifts between theta generators in the rat hippocampus. Acta Physiol Hung 53:41–45

Buzsaki G, Haubenrauser J, Grastyan E, Czopf J, Kellenyi L (1981) Hippocampal slow wave activity during appetitive and aversive conditioning in the cat. Electroencephalogr Clin Neurophysiol 51:276–290

Buzsaki G, Leung L-WS, Vanderwolf CH (1983) Cellular bases of hippocampal EEG in the behaving rat. Brain Res Reviews 6 (Brain Res 267):139–171

Buzsaki G, Rappelsberger P, Kellenyi L (1985) Depth profiles of hippocampal rhythmic slow activity ("Theta rhythm") depend on behaviour. Electroencephalogr Clin Neurophysiol 61:77–88

Buzsaki G, Czopf J, Konakor I, Kellenyi L (1986) Laminar distribution of hippocampal rhythmic slow activity (RSA) in the behaving rat: current source density analysis, effects of urethane and atropine. Brain Res 365:125–137

Cajal SR (1911/1972) Histologie du systeme Nerveux de l'Homme et des Vertebres, Vol II, (translated from Spanish to French by Azoulay L), CSIC, Madrid

Cavada C, Llamas A, Reinoso-Suarez F (1983) Allocortical afferent connections of the prefrontal cortex in the cat. Brain Res 260:117–120

Chandler JP, Crutcher KA (1983) The septo-hippocampal projection in the rat: an electronmicroscopic horseradish peroxidase study. Neuroscience 10:685–696

Chauvin R, Muckensturm-Chauvin B (1977) Behavioural complexities. International Universities Press, New York

Chozick BS (1983) The behavioural effects of lesions of the hippocampus: a review. Int J Neurosci 22:63–80

Christie BR, Bland BH (1989) An investigation of a diencephalic- septal-hippocampal pathway. Paper presented at 7th Australasian Winter conference on Brain Research, Queentown, New Zealand, August 1989

Chronister RB, DeFrance JF (1979) Organization of projection neurons of the hippocampus. Exp Neurol 66:509–532

Chronister RB, Zornetzer SF, Bernstein JJ, White LA (1974) Hippocampal theta rhythm: intrahippocampal formation contributions. Brain Res 65:13–28

Claiborne BJ, Amaral DG, Cowan WM (1986) A light and electronmicroscopic analysis of the mossy fibres of the rat dentate gyrus. J Comp Neurol 246:435–458

Coleman JR, Lindsley DB (1975) Hippocampal electrical correlates of free behaviour and behaviour induced by stimulation of two hypothalamic-hippocampal systems in the cat. Exp Neurol 49:506–528

Coleman JR, Lindsley DB (1977) Behaviour and hippocampal electrical changes during operant learning in cats and effects of stimulating two hypothalamic-hippocampal systems. Electroencephalogr Clin Neurophysiol 42:309–331

Colom LV, Ford RD, Bland BH (1987) Hippocampal formation neurons code the level of activation of the cholinergic septohippocampal pathway. Brain Res 410:12–20

Colom LV, Christie BR, Bland BH (1988) Cingulate cell discharge patterns related to hippocampal EEG and their modulation by muscarinic and nicotinic agents. Brain Res 460:329–338

Costin A, Bergmann F, Chaimowitz M (1967) Influence of labyrinthine stimulation on hippocampal activity. Prog Brain Res 27:183–188

Cragg BG (1965) The efferent connexions of the allocortex. J Anat 99:339–357

Creutzfeldt O, Houchin J (1974) Neuronal basis of EEG waves. In: Creutzfeldt O (ed) Handbook of electroencephalography and clinical neurophysiology. Vol 2 (2C) Elsevier, Amsterdam pp. 3–54

Creutzfeldt O, Ito M (1968) Functional synaptic organization of primary visual cortex neurones in the cat. Exp Brain Res 6:324–352

Creutzfeldt O, Maekawa K, Hosli L (1969) Forms of spontaneous and evoked postsynaptic potentials of cortical nerve cells. Prog Brain Res 31:265–273

Crowne DP, Radcliffe DD (1975) Some characteristics and functional relations of the electrical activity of the primate hippocampus and hypotheses of hippocampal function. In: Isaaacson RI, Pribram KH (eds) The hippocampus, Vol II, Plenum, New York, pp. 185–206

Crowne DP, Konow A, Drake KJ, Pribram KH (1972) Hippocampal electrical activity in the monkey during delayed spatial alternation problems. Electroencephalogr Clin Neurophysiol 33:567–577

Crutcher KA, Madison R, Davis JW (1981) A study of the rat septo-hippocampal pathway using anterograde transport of horseradish peroxidase. Neuroscience 6: 1961–1973

Dalton AJ (1969) Discriminative conditioning of hippocampal electrical activity in curarised dogs. Commun Behav Biol 3:283–287

Dalton AJ, Black AH (1968) Hippocampal electrical activity during the operant conditioning of movement and refraining from movement. Commun Behav Biol (Part A) 2:267–273

Deadwyler SA, West M, Lynch G (1979) Activity of dentate granule cells during learning: differentiation of perforant path input. Brain Res 169:29–43

Deadwyler SA, West MO, Christian EP (1982) Neural activity in the dentate gyrus of the rat during the acquisition and peformance of simple and complex sensory discrimination learning. In: Woody CD (ed) Conditioning: representation of involved neural functions. Plenum, New York, pp. 63–73

Descartes R (1637/1964) Discours de la methode. In: Descartes: Philosophical writings. (Translated from French by Anscombe E, Geach PT), Nelson,Sydney

Destrade C (1982) Two types of diencephalically-driven RSA (theta) as a means of studying memory formation in mice. Brain Res 234:486–493

Destrade C, Ott T (1980) Blockade of high frequency rhythmical slow activity by intrahippocampal injection of a glutamic acid antagonists. Neurosci Lett 18:73–78

Destrade C, Ott T (1982) Is a retrosplenial (cingulate) pathway involved in high frequency hippocampal rhythic slow activity (theta)? Brain Res 252:29–37

Detari L, Vanderwolf CH (1987) Activity of identified cortically projecting and other basal forebrain neurons during large slow waves and cortical activiation in anaesthetized rats. Brain Res 437:1–8

Deupree D, Coppock W, Willer H (1982) Pretraining septal driving of hippocampal rhythmic slow activity facilitates acquisition of visual discrimination. J Comp Physiol Psychol 96:557–562

Douglas RM, McNaughton BL, Goddard GV (1983) Commissural inhibition and facilitation of granule cell discharge in fascia dentata. J Comp Neurol 219:285–294

Efremova TM, Trush VD (1973) Power spectra of cortical electric activity in the rabbit in relation to conditioned reflexes. Acta Neurobiol Exp 33:743–755

Eichenbaum H, Kuperstein M, Fagan A, Nagode J (1987) Cue sampling and goal approach correlates of hippocampal unit activity in rats performing an odor discrimination task. J Neurosci 7:716–732

Eidelberg E, White JC, Brazier MAB (1959) The hippocampal arousal pattern in rabbits. Exp Neurol 1:483–490

Elazar Z, Adey WR (1967a) Spectral analysis of low frequency components in the electrical activity of the hippocampus during learning. Electroencephalogr Clin Neurophysiol 23:225–240

Elazar Z, Adey WR (1967b) Electroencephalographic correlates of learning in subcortical and cortical structures. Electroencephalogr Clin Neurophysiol 23:306–319

Ellison GD, Humphrey GL, Feeney DM (1968) Some electrophysiological correlates of classical and instrumental behaviour. J Comp Physiol Psychol 66:340–348

Ewer RF(1967) Ethology of mammals. Logos, London

Eysenck MW (1982) Attention and arousal. Springer, Berlin Heidelberg New York

Fantie BD, Goddard GV (1982) Septal modulation of the population spike in the fascia dentata produced by perforant path stimulation in the rat. Brain Res 252:227–237

Feder R, Ranck JB (1973) Studies on single neurons in dorsal hippocampal formation and septum in unrestrained rats. Part II Hippocampal slow waves and theta cell firing during bar pressing and other behaviours. Exp Neurol 41:532–555

Feenstra BWA, Holsheimer J (1979) Dipole-like neuronal sources of theta rhythm in dorsal hippocampus, dentate gyrus and cingulate cortex of urethane-anaesthetized rats. Electroencephalogr Clin Neurophysiol 47:532–538

Ferino F, Thierry AM, Glowinski J (1987) Anatomical and electrophysiological evidence for a direct projection from Ammon's horn to the medial prefrontal cortex in the rat. Exp Brain Res 65:421–426

Ferraya Moyano H, Molina JC (1980) Axonal projections and conduction properties of olfactory peduncle neurons in the rat. Exp Brain Res 39:241–248

Fleischhauer K, Wartenberg H (1967) Elektronenmikroskopische Untersuchungen über das Wachstum der Nervenfasern und über das Auftreten von Markscheiden im Corpus Callosum der Katze. Z Zellforsch Mikrosk Anat 83:568–581

Ford RD, Colom LV, Bland BH (1989) The classification of medial septum-diagonal band cells as theta-on or theta-off in relation to hippocampal EEG states. Brain Res 493:269–282

Fox SE, Ranck JB (1975) Localization and anatomical identification of theta and complex spike cells in dorsal hippocampal formation of rats. Exp Neurol 49:299–313

Fox S, Ranck JB (1981) Electrophysiological characteristics of hippocampal complex-spike cells and theta cells. Exp Brain Res 41:399–410

Fox SE, Wolfson S, Ranck JB (1983) Investigating the mechanisms of hippocampal theta rhythms: Approaches and progress. In: Seifert W (ed) Neurobiology of the Hippocampus. Academic Press, London, pp. 303–319

Fox SE, Wolfson S, Ranck JB (1986) Hippocampal theta rhythm and the firing of neurons in walking and urethane anaesthetised rats. Exp Brain Res 62:495–508

Frederickson CJ, Whishaw IQ (1977) Hippocampal EEG during learned and unlearned behaviour in the rat. Physiol Behav 18: 597–603

Frederickson CJ, Smylie CS, Howell G, Lenig R (1978) Movement dependent and movement independent hippocampal RSA in cat. Brain Res Bull 3:559–562

Frederickson CJ, Lenig RB, Frederickson MH (1980) Hippocampal RSA in cats during cued and non-cued delayed response performance. Behav Neural Biol 28:383–391

Frederickson CJ, Frederickson MH, Lewis C, Howell GA, Smylie C, Wright CG (1982) Hippocampal EEG in normal mice and in mice with congenital vestibular defects. Behav Neural Biol 34:121–131

Fregnac Y, Shulz D, Thorpe S, Bienenstock E (1988) A cellular analogue of visual cortical plasticity. Nature 333:367–370

Frotscher M (1985) Mossy fibres form synapses with identified pyramidal basket cells in the CA3 region of the guines pig hippocampus: a combined Golgi-electronmicroscopic study. J Neurocytol 14:245–259

Frotscher M, Schandler M, Leranth C (1986) The cholinergic innervation of the rat fascia dentata: identification of target structures on granule cells by combining choline acetyltransferase immunocytochemistry and Golgi impregnation. J Comp Neurol 243:58–70

Fujita Y, Sato T (1964) Intracellular records from hippocampal pyramidal cells in rabbit during theta rhythm activity. J Neurophysiol 27:1011–1025

Fuster JM (1981) The prefrontal cortex. Anatomy, physiology, and neuropsychology of the frontal lobe. Raven, New York

Galey D, Jeantet Y, Destrade C, Jaffard R (1983) Facilitation of memory consolidation by post-training electrical stimulation of the medial septal nucleus: is it mediated by changes in rhythmic slow activity. Behav Neural Biol 38:240–250

Gaztelu JM, Buno W (1982) Septo-hippocampal relationships during EEG theta rhythm. Electroencephalogr Clin Neurophysiol 54:375–387

Gaztelu JM, Dajas F, Sanchez-Arroyos R, Garcia-Austt E (1981) [14C] 2-deoxyglucose mapping of active hippocampal areas during theta rhythm induced by curaremetics in the rat. Neurosci Lett, Suppl. 7:S46

Gerbrandt LK, Lawrence JC, Eckhardt MJ, Lloyd RL (1978) Origin of the neocortically monitored theta rhythm in the curarised rat. Electroencephalogr Clin Neurophysiol 45:454–467

Geschwind N (1965a) Disconnexion syndromes in animals and man Part 1. Brain 88:237–294

Geschwind N (1965b) Disconnexion syndromes in animals and man Part 2. Brain 88:585–644

Gibson S, McGeer EG, McGeer PL (1970) Effects of selective inhibitors of tyrosine and tryptophan hydroxylases on self-stimulation in the rat. Exp Neurol 27:283–290

Glazer H (1972) Physostigmine and resistance to extinction. Psychopharmacology (Berlin) 26:387–394

Glazer HI (1974a) Instrumental conditioning of hippocampal theta and subsequent response persistence. J Comp Physiol Psychol 86:267–273

Glazer HI (1974b) Instrumental response persistence following induction of hippocamapl theta frequency during fixed-ratio responding in rats. J Comp Physiol Psychol 86:1156–1162

Glickman SE, Sroges RW (1966) Curiosity in zoo animals. Behaviour 26:151–188

Glotzbach SF (1975) Correlation of hippocampal theta activity and movement during slow-wave sleep in cats. Behav Biol 15:485–490

Goddard GV (1980) Component properties of the memory machine. In: The nature of thought: sssays in honour of D.O.Hebb. Lawrence Erlbaum, New Jersey

Gogolak G, Stumpf Ch, Petsche H, Sterc J (1968) The firing pattern of septal neurons and the form of the hippocampal theta wave. Brain Res 7:201–207

Goldman-Rakic PS, Selemon LD, Schwartz ML (1984) Dual pathways connecting the dorso-lateral prefrontal cortex with the hippocampal formation and parahippocampal cortex in the rhesus monkey. Neuroscience 12:719–743

Gottlieb DI, Cowan WM (1973) Autoradiographic studies of the commissural and ipsilateral association connections of the hippocampus and dentate gyrus of the rat. J Comp Neurol 149:393–421

Gralewicz C (1981) Hippocampal electrophysiological correlates of an avoidance behaviour in the cat. Electroencephalogr Clin Neurophysiol 52:306–315

Grastyan E, Vereckei L (1974) Effects of spatial separation of the conditioned signal from the reinforcement: a demonstration of the conditioned character of the orienting response or the orientational character of conditioning. Behav Biol 10:121–146

Grastyan E, Lissak K, Madarasz I, Donhoffer H (1959) Hippocampal electrical activity during the development of conditioned reflexes. Electroencephalogr Clin Neurophysiol 11: 409–430

Grastyan E, Karmos G, Vereczkey L, Kellenyi L (1966) The hippocampal electrical correlates of the homeostatic regulation of motivation. Electroencephalogr Clin Neurophysiol 21:34–53

Gray JA (1969) Sodium amobarbital and the effects of frustrative nonreward. J Comp Physiol Psychol 69:55–64

Gray JA (1970) Sodium amobarbital, the hippocampal theta rhythm and the partial reinforcement extinction effect. Psychol Rev 77:465–480

Gray JA (1971) Medial septal lesions, hippocampal theta rhythm and the control of vibrissal movement in the freely moving rat. Electroencephalogr Clin Neurophysiol 30:189–197

Gray JA (1972) Effects of septal driving of the hippocampal theta rhythm on resistance to extinction. Physiol Behav 8:481–490

Gray JA Ball GG (1970) Frequency-specific relation between hippocampal theta rhythm, behaviour and amobarbital action. Science 168:1246–1248

Gray JA, McNaughton N (1983) Comparison between the behavioural effects of septal and hippocampal lesions: a review. Neurosci Biobehav Rev 7:119–188

Gray JA, Araujo-Silva MT, Quintano L (1972) Resistance to extinction after partial reinforcement training with blocking of the hippocampal theta rhythm by septal stimulation. Physiol Behav 8:497–502

Green JD, Arduini AA (1954) Hippocampal electrical activity in arousal. J Neurophysiol 17:533–547

Green JD, Maxwell DS, Schindler WJ, Stumpf C (1960) Rabbit EEG "theta" rhythm: its anatomical source and relation to activity in single neurons. J Neurophysiol 23:403–420

Green KF, Rawlins JND (1979) Hippocampal theta in rats under urethane: generators and phase relations. Electroencephalogr Clin Neurophysiol 47:420–429

Green RA, Weinberger NM (1983) Long-term potentiation in the magnocellular medial geniculate nucleus of the anaesthetized cat. Brain Res 256:138–142

Gustafsson B, Wigstrom H (1986) Hippocampal long-lasting potentiation produced by pairing single volleys and brief conditioning tetani in separate afferents. J Neurosci 6:1575–1582

Habets AMMC, Lopes da Silva FH, de Quartel FW (1981) Autoradiography of the olfactory-hippocampal pathway in the cat with special reference to the perforant path. Exp Brain Res 38:257–265

Halgren E, Babb TL, Crandall PH (1978) Human hippocampal formation EEG desynchronization during attentiveness and movement. Electroencephalogr Clin Neurophysiol 44:778–781

Harper RM (1971) Frequency changes in hippocampal electrical activity during movement and tonic immobility. Physiol Behav 7:55–58

Haug FMS (1967) Electron microscopical localization of the zinc in hippocampal mossy fibre synapses by modified sulphide silver procedure. Histochemie 8:335–368

Hayashi H, Iijima S, Sugita Y, Teshiwa Y, Tashiro T, Matsuo R, Yasochima A, Hishikawa Y, Ishihara T (1986) Appearance of frontal mid-line theta rhythm during sleep and its relation to mental activity. Electroencephalogr Clin Neurophysiol 66:66–70

Heath CJ, Jones EG (1971) The anatomical organization of the suprasylvian gyrus of the cat. Ergeb Anat Entwicklungsgesch 45:1–64

Hebb DO (1949) The organization of behaviour. John Wiley, New York

Herreras O, Solis JM, Munoz MD, Martin del Rio R, Lerma J (1988a) Sensory modulation of hippocampal transmission. I Opposite effects on CA1 and dentate gyrus synapses. Brain Res 461:290–302

Herreras O, Solis JM, Herranz AS, Martin del Rio R, Lerma J (1988b) Sensory modulation of hippocampal transmission. II Evidence for a cholinergic locus of inhibition in the Schaffer-CA1 synapse. Brain Res 461:303–313

Hilgard ER, Bower GH(1966) Theories of learning. Appleton-Century-Crofts, New York

Hill AJ (1978) First occurrence of hippocampal spatial firing in a new environment. Exp Neurol 62:282–297

Hirano T (1984) Unit activity of the septo-hippocampal system in classical conditioning with rewarding brain stimulation. Brain Res 295:41–49

Hirsh R (1974) The hippocampus and contextual retrieval of information from memory: a theory. Behav Biol 12:421–444

Hjorth-Simonsen A (1972) Projection of the lateral part of the entorhinal area to the hippocampus and fascia dentata. J Comp Neurol 146:219–232

Holmes JE, Adey WR (1960) Electrical activity of the entorhinal cortex during conditioned behaviour. Am J Physiol 199:741–744

Holmes JE, Beckman J (1969) Hippocampal theta rhythm used in predicting feline behaviour. Physiol Behav 4:563–565

Holsheimer J (1982) Generation of theta activity (RSA) in the cingulate cortex of the rat. Exp Brain Res 47:309–312

Holsheimer J, Boer J, Lopes da Silva FH, Van Rotterdam A (1982) The double dipole model of theta rhythm generation: simulation of laminar field potential profiles in dorsal hippocampus of the rat. Brain Res 235:31–50

Holsheimer J, Stok CJ, Lopes da Silva FH (1983) Theta rhythm related hippocampal cell discharges in the urethane anaesthetised rat: evidence for a predominant entorhinal input. Electroencephalogr Clin Neurophysiol 55:464–467

Holt L, Gray JA (1983) Septal driving of the hippocampal theta rhythm produces a long-term, proactive and non-associative increase in resistance to extinction. Q J Exp Psychol [B] 35:97–118

Honzik CH (1936) The sensory basis of maze learning in rats. Comp Psychol Monogr 13 (4 serial No. 64)

Hotson JR, Prince DA (1980) Effects of EGTA on the calcium-activated after hyperpolarization in hippocampal CA3 pyramidal cells. Science 210:1125–1124

Houser CR, Crawford RP, Barber RP, Salvaterra PM, Vaughn JE (1983) Organization and morphological characteristics of cholinergic neurons: an immunocytochemical study with monoclonal antibody to choline acetyltransferase. Brain Res 266:97–119

Hume D (1748/1977) An enquiry concerning human understanding. Steinberg E (ed) Hackett, Indianapolis

Hunter WS (1929) The sensory control of the maze habit in the white rat. J Genet Psychol 36:505–537

Hunter WS (1930) A further consideration of the sensory control of the maze habit in the white rat. J Genet Psychol 38:3–19

Huston JP, Brozek G (1974) Spectral analysis of hippocampal slow wave activity during feeding and drinking in the rabbit. Physiol Behav 12:819–824

Ibata Y, Desiraju T, Pappas GD (1971) Light and electronmicroscopic study of the projection of the medial septal nucleus to the hippocampus of the cat. Exp Neurol 33:103–122

Ino T, Itoh K, Kamiya H, Shigemoto R, Akiguchi I, Mizuno N (1988) Direct projection of non-pyramidal neurons of Ammon's horn to the supramammilary region in the cat. Brain Res 460:173–177

Insausti R, Amaral DG, Cowan WM (1987a) The entorhinal cortex of the monkey: II Cortical afferents. J Comp Neurol 264:356–395

Insausti R, Amaral DG, Cowan WM (1987b) The entorhinal cortex of the monkey. III Subcortical afferents. J Comp Neurol 264:396–408

Irle E, Markowitsch HJ (1982a) Connections of the hippocampal formation, mammilary bodies, anterior thalamus and cingulate cortex. Exp Brain Res 47:79–94

Irle E, Markowitsch HJ (1982b) Widespread cortical projections of the hippocampal formation in the cat. Neuroscience 7:2637–2647

Irmis F, Radil-Weiss T, Lat J, Krekule I (1970) Interindividual differences in hippocampal theta activity during habituation. Electroencephalogr Clin Neurophysiol 28:24–31

Irmis F, Lat J, Radil-Weiss T (1971) Individual differences in hippocampal EEG during rhombencephalic sleep and arousal. Physiol Behav 7:117–119

Ito M (1966) Hippocampal electrical correlates of self-stimulation in the rat. Electroencephalogr Clin Neurophysiol 21:261–268

Iwai E, Yukie M (1988) A direct projection from hippocampal field CA1 to ventral area TE of the infero-temporal cortex in the monkey. Brain Res 444:397–401

Jack JJ, Noble D, Tsien RW (1975) Electric current flow in excitable cells. Clarendon, Oxford, pp. 218–223

James W (1890) The principles of psychology. Holt, New York

John ER, Shimococki M, Bartlett F (1969) Neural readout from memory during generalization. Science 164:1534–1536

Jones EG, Powell TPS (1970) An anatomical study of converging sensory pathways within the cerebral cortex of the monkey. Brain 93:793–820

Joseph JA, Engel BT (1981) Somatomotor related hippocampal rhythmic slow activity in the monkey (Macaca mulatta). Physiol Behav 26:865–872

Jung R, Kornmuller AE (1938) Eine Methodik der Ableitung lokalisierter Potentialschwankungen aus subcorticalen Hirngebieten. Arch Psychiat Nervenkr 109:1–30

Kamp A, Lopes da Silva FH, Storm van Leeuwen W (1971) Hippocampal frequency shifts in different behavioural stuations. Brain Res 31:287–294

Kant I (1781/1934) Critique of pure reason (J.M.D.Meiklejohn, trans.), Dent, London

Karmos G, Grastyan E, Losonczy H, Vereczkey L, Grosz J (1965) The possible role of the hippocampus in the organization of the orientation reaction. Acta Physiol Hung 26:131–141

Kawamura K (1973) Corticocortical fibre connections of the cat cerebrum. III The occipital cortex. Brain Res 51:41–60

Kelso SR, Ganong AH, Brown TH (1986) Hebbian synapses in hippocampus. Proc Natl Acad Sci USA 83:5326–5330

Kemp IR, Kaada BR (1975) The relation of hippocampal theta activity to arousal, attentive behaviour and somato-motor movements in unrestrained cats. Brain Res 95:323–343

Kimble DP (1964) Hilgard and Marquis's "Conditioning and learning". Methuen, London

Kimble DP (1970) Hippocampus and internal inhibition. Psychol Bull 79:285–295

Kimsey RA, Dyer RS, Petri HL (1974) Relationship between hippocampal EEG, novelty and frustration in the rat. Behav Biol 11:561–568

Kirk IJ, McNaughton N (1989) Separate reticular influences on hippocampal EEG in CA1 and dentate gyrus. Paper presented at 7th Australasian Winter Conference on Brain Research, Queenstown, New Zealand, August, 1989.

Klemm WR (1971) EEG and multiunit activity in limbic and motor systems during movement and immobility. Physiol Behav 7:337–343

Klemm WR (1972) Effects of electrical stimulation of brainstem reticular formation on hippocampal theta rhythm and muscle activity in unanaesthetized, cervical- and midbrain-transected rats. Brain Res 41:331–344

Knoll J (1956) Experimental studies on the higher nervous activity of animals. VI. Further studies on active reflexes. Acta Physiol Hung 12:65–92

Kohler C (1984) A projection from the deep layers of the entorhinal area to the hippocampal formation in the rat brain. Neurosci Lett 56:13–19

Kohler C, Shipley MT, Srebro B, Harkmark W (1978) Some retrohippocampal afferents to the entorhinal cortex, cells of origin as studied by the HRP method in the rat and mouse. Neurosci Lett 10:115–120

Kohler C, Chan-Palay V, Wu J-Y (1984) Septal neurons containing glutamic acid decarboxylase immunoreactivity project to the hippocampal region in the rat brain. Anat Embryol (Berl) 169:41–44

Kohler W (1938) In: Ellis WD (ed) A source book of Gestalt psychology Routledge and Kegan Paul, London

Kohonen T (1977) Associative memory. Springer, Berlin Heidelberg New York

Kolb B, Whishaw IQ (1977) Effects of brain lesions and atropine on hippocampal and neocortical electroencephalograms in the rat. Exp Neurol 56:1–22

Komisaruk BR (1970) Synchrony between limbic system theta activity and rhythmical behaviour in rats. J Comp Physiol Psychol 70:482–492

Konig FR, Klippel RA (1967) The rat brain: A stereotaxic atlas of the forebrain and lower parts of the brainstem. Williams and Wilkins, New York

Konopacki J, MacIver MB, Bland BH, Roth SH (1987a) Carbachol-induced EEG "theta" activity in hippocampal brain slices. Brain Res 405:196–198

Konopacki J, MacIver MB, Bland BH, Roth SH (1987b) Theta rhythm in hippocampal slices; relation to synaptic responses of dentate neurons. Brain Res Bull 18:25–27

Konopacki J, Bland BH, Roth SH (1987c) Phase shifting of CA1 and dentate EEG rhythms in hippocampal formation slices. Brain Res 417:399–402

Konopacki J, Bland BH, MacIver MB, Roth SH (1987d) Cholinergic theta rhythm in transected hippocampal slices; independent CA1 and dentate generators. Brain Res 436:217–222

Konopacki J, Bland BH, Roth SH (1988) Carbachol-induced EEG "theta" in hippocampal formation slices: evidence for a third generator of theta in CA3c area. Brain Res 451:33–42

Konorski J, Santibanez HHG, Beck J (1968) Electrical hippocampal activity and heart rate in classical and instrumental conditioning. Acta Biol Exp 28:169–185

Korchin SJ (1964) Anxiety and cognition. In: Cognition: theory, research, promise. Papers read at the martin Scheerer Memorial Meetings on Cognitive Psychology, University of Kansas, May 1962. Scheerer C (ed), Harper and Row, New York pp. 58–78

Kramis R, Vanderwolf CH (1980) Frequency-specific RSA-like hippocampal patterns elicited by septal, hypothalamic and brainstem electrical stimulation. Brain Res 192:383–398

Kramis R, Vanderwolf CH, Bland BH (1975) Two types of hippocampal slow activity in both the rabbit and the rat: relations to behaviour and effects of atropine, diethyl ether, urethane and pentobarbital. Exp Neurol 49:58–85

Kramis RC, Routtenberg A (1969) Rewarding brain stimulation, hippocampal activity and foot-stomping in the gerbil. Physiol Behav 11:7–11

Kramis RC, Routtenberg A (1977) Dissociation of hippocampal EEG from its behavioural correlates by septal and hippocampal electrical stimulation. Brain Res 125:37–49

Krayniak PF, Siegel A, Meibach RC, Fruchtman D, Scrimenti M (1979) Origin of the fornix in the squirrel monkey. Brain Res 160:401–411

Krnjevic K, Ropert N (1982) Electrophysiological and pharmacological characteristics of facilitation of hippocampal population spikes by stimulation of the medial septum. Neuroscience 7:2165–2184

Krnjevic K, Reiffenstein RJ, Ropert N (1981) Disinhibitory action of acetylcholine in the rat's hippocampus: extracellular observations. Neuroscience 6:2465–2474

Krnjevic K, Dalkara T, Yim C (1986) Synchronization of pyramidal cell firing by ephaptic currents in hippocampus in situ. Adv Exp Med Biol 203:413–423

Krnjevic K, Ropert N, Casullo J (1988) Septohippocampal disinhibition. Brain Res 438:182–192

Kunkel DD, Lacaille J-C, Schwartzkroin PA (1988) Ultrastructure of strtaum lacunosum-moleculare interneurons of hippocampal CA1 region. Synapse 2:382–394

Kuperstein M, Eichenbaum H, VanDeMark T (1986) Neural group properties in the rat hippocampus during theta rhythm. Exp Brain Res 61:438–442

Kurtz RG (1975) Hippocampal and cortical activity during sexual behaviour in the female rat. J Comp Physiol Psychol 89:158–169

Kurtz RG, Adler NT (1973) Electrophysiological correlates of copulatory behaviour in the male rat. J Comp Physiol Psychol 84:225–239

Kuypers HGJM, Szwarcbart MK, Mishkin M, Rosvold HE (1965) Occipitotemporal corticocortical connections in the rhesus monkey. Exp Neurol 11:245–262

Lacaille JC, Schwartzkroin PA (1988) Stratum lacunosum-moleculare interneurons of hippocampal CA1 region. I. Intracellular response characteristics, synaptic responses and morphology. J Neurosci 8:1400–1410

Lacaille JC, Schwartzkroin PA (1988) Stratum lacunosum-moleculare interneurons of hippocampal CA1 region. II. Intrasomatic and intradendritic recordings of local circuit synaptic interactions. J Neurosci 8:1411–1424

Lacaille J-C, Mueller AL, Kunkel DD, Schwartzkroin PA (1987) Local circuit interactions between oriens/alveus interneurons and CA1 pyramidal cells in hippocampal slices: electrophysiology and morphology. J Neurosci 7:1979–1993

Landfield PW (1976)Computer-determined EEG patterns associated with memory-facilitating drugs and with ECS. Brain Res Bull 1:9–17

Landfield PW (1977) Different effects of posttrial driving or blocking of the theta rhythm on avoidance learning in rats. Physiol Behav 18:439–445

Landfield PW, McGaugh JL, Tusa RJ (1972) Theta rhythm: a temporal correlate of memory storage processes in the rat. Science 175:87–89

Lang M, Lang W, Diekmann V, Kornhuber HH (1987) The frontal theta rhythm indicating motor and cognitive learning. In: Johnson R, Rohrbaugh JW, Parasuraman R (eds) Current trends in event-related potential research (EEG Suppl 40), Elsevier, Amsterdam, pp. 322–327

Laurberg S (1979) Commissural and intrisic connections of the rat hippocampus. J Comp Neurol 184:685–708

Lee KH, Chung K, Chung JM, Coggeshall RE (1986) Correlation of cell body size, axon size and signal conduction velocity for individually labelled dorsal-root gangion cells in the cat. J Comp Neurol 243:335–346

Lee KS (1982) Sustained enhancement of evoked potentials following brief high-frequency stimulation of the cerebral cortex in vitro. Brain Res 239:617–623

Leichnetz GR, Astruc J (1975a) Preliminary evidence for a direct projection of the prefrontal cortex to the hippocampus in the squirrel monkey. Brain Behav Evol 11:355–364

Leichnetz GR, Astruc J (1975b) Efferent connections of the orbitofrontal cortex in the marmoset (*Saguinus oedipus*). Brain Res 84:169–180

Leichnetz GR, Astruc J (1976) The efferent projections of the medial prefrontal cortex in the squirrel monkey (*Saimiri sciurens*). Brain Res 109:455–472

Leitenberg H (1965) Is time-out from positive reinforcement an aversive event? A review of the experimental evidence. Psychol Bull 64:428–444

Leranth C, Frotscher M (1987) Cholinergic innervation of hippocampal GAD- and somatostaitin-immunoreactive commissural neurons. J Comp Neurol 261:33–47

Leung L-WS (1984a) Pharmacology of theta phase shift in the hippocampal CA1 region of freely moving rats. Electroencephalogr Clin Neurophysiol 58:457–466

Leung L-WS (1984b) Theta rhythm during REM sleep and waking: correlations between power,phase and frequency. Electroencephalogr Clin Neurophysiol 58:553–564

Leung L-WS (1984c) Model of gradual phase shift of theta rhythm in the rat. J Neurophysiol 52:1051–1065

Leung L-WS, Borst JGG (1987) Electrical activity of the cingulate cortex.1 Generating mechanisms and relations to behaviour. Brain Res 407:68–80

Leung LS, Yim C (1986) Intracellular records of theta rhythm in hippocampal CA1 cells of the rat. Brain Res 367:323–327

Leung LS, Lopes da Silva FH, Wadman WJ (1982) Spectral characteristics of the hippocampal EEG in the freely moving rat. Electroencephalogr Clin Neurophysiol 54:203–219

Levy WB, Steward O (1983) Temporal contiguity requirements for long-term potentiation/depression in the hippocampus. Neuroscience 8:791–797

Leyhausen P (1965) The communal organization of solitary mammals. Symp Zool Soc Lond 14:249–263

Leyhausen P (1979) Cat behaviour: the predatory and social behaviour of domestic and wild cats. Tonkin BA (trans), Garland STPM Press, New York

Lieb J, Sclabassi R, Crandall P, Buchness R (1974) Comparison of the action of diazepam and phenobarbital using EEG-derived power spectra obtained from temporal lobe epileptics. Neuropharmacology 13:769–784

Linsker R (1986a) From basic network principles to neural architecture: emergence of spatial oponent cells. Proc Natl Acad Sci USA 83:7508–7512

Linsker R (1986b) From basic network principles to neural architecture: emergence of orientation-selective cells. Proc Natl Acad Sci USA 83:8390–8394

Linsker R (1986c) From basic network principles to neural architecture: emergence of orientation columns. Proc Natl Acad Sci USA 83:8779–8783

Lissak K, Grastyan E (1960) The changes of hippocampal electrical activity during conditioning. In: Jasper EE, Smirnov GD (eds) The Moscow colloquium on electroencephalography of higher nervous activity. Electroencephalogr Clin Neurophysiol (Suppl) 13:271–279

Lomo T (1971) Patterns of activation in a monosynaptic cortical pathway: the perforant path input to the dentate area of the hippocampus. Exp Brain Res 12:18–45

Lopes da Silva FH, Kamp A (1969) Hippocampal theta frequency shifts and operant behaviour. Electroencephhalogr Clin Neurophysiol 26:133–143

Lynch G, Smith RL, Mensah P, Cotman C (1973) Tracing the dentate gyrus mossy fiber system with horseradish peroxidase histochemistry. Exp Neurol 40:516–524

Lynch G Rose G, Gall, C (1978) Anatomical and functional aspects of septo-hippocampal projections. In: Functions of thr septo-hippocampal system. CIBA Symp. No. 58, Elsevier, Amsterdam, pp. 5–20

Lysakowski A, Wainer BH, Rye DB, Bruce G, Hersh LB (1986) Cholinergic innervation displays strikingly different laminar preferences in several cortical areas. Neurosci Lett 64:102–118

Macadar AW, Chalupa LM, Lindsley DB (1974) Differentiation of brainstem loci which affect hippocampal and neocortical electrical activity. Exp Neurol 43:499–514

Macadar O, Roig JA, Monti JM, Budelli R (1970) The functional relationship between septal and hippocampal unit activity and hippocampal theta rhythm. Physiol Behav 5:1443–1449

MacIver MB, Harris DP, Konopacki J, Roth SH, Bland BH (1986) Carbachol-induced rhythmical slow wave activity recorded from dentate granule neurons in vitro. Proc West Pharmacol Soc 29:159–161

MacKintosh NJ (1974) The psychology of animal learning. Academic Press, London

Macrides F (1975) Temporal relationship between hippocampal slow waves and exploratory sniffing in hamsters. Behav Biol 14:295–308

Macrides F, Eichenbaum H, Forbes WB (1982) Temporal relationship between sniffing and limbic theta rhythm during odor discrimination reversal learning. J Neurosci 2:1705–1717

Madison DV, Lancaster B, Nicoll RA (1987) Voltage-clamp analysis of cholinergic action in the hippocampus. J Neurosci 7:133–141

Marr D (1969) A theory of cerebellar cortex. J Physiol (Lond) 202:437–470

Marr D (1970) A theory of cerebral neocortex. Proc R Soc Lond (Biol) 176:161–234

Marr D (1971) Simple memory. Philos Trans R Soc (Biol) 262:23–81

Martin SM, Moberg GP, Horowitz JM (1975) Glucocorticoids and the hippocampal theta rhythm in loosely restrained, unanaesthetised rabbits. Brain Res 93:535–542

Maru E, Takahashi LK, Iwahara S (1979) Effects of median raphe nucleus lesions on hippocampal EEG in freely moving rat. Brain Res 163:223–234

Matthews DA, Salvaterra PM, Crawford GD, Houser CR, Vaughn JE (1987) An immunocytochemical study of choline acetyltransferase-containing neurons and axon terminals in normal and partially deafferented hippocampal formation. Brain Res 402:30–43

Mayer Ch, Stumpf Ch (1958) Die Physostigminwirkung auf die Hippocampustatigkeit nach Septumlasionen. Naunyn Schmiedeberg's Arch Pharmacol 234:490–500

McFarland WL, Teitelbaum H, Hedges EK (1975) Relationship between hippocampal theta activity and running speed in rat. J Comp Physiol Psychol 88:324–328

McLennan H, Miller JJ (1974) The hippocampal control of neuronal discharge in the septum of the rat. J Physiol (Lond) 237:607–624

McLennan H, Miller JJ (1976) Frequency related inhibitory mechanisms controlling rhythmic activity in the septal area. J Physiol (Lond) 254:827–841

McNaughton BL, Barnes CA, O'Keefe J (1984) The contributions of position, direction and velocity to single unit activity in the hippocampus of freely-moving rats. Exp Brain Res 52:41–49

McNaughton BL, Morris RGN (1987) Hippocampal synaptic enhancement and information storage within a distributed memory system. Trends Neurosci 10:408–415

Meibach RC, Siegel A (1977a) Subicular projections to posterior cingulate cortex in rats. Exp Neurol 57:264–274

Meibach RC, Siegel A (1977b) Efferent connections of the septal area in the rat: an analysis utilizing retrograde and anterograde transport methods. Brain Res 119:1–20

Mellgren SI, Srebro B (1973) Changes in acetylcholinesterase and distribution of degenerating fibres in the hippocampal region after septal lesions in the rat. Brain Res 52:19–36

Merzenich MM, Kaas JH, Wall JT, Nelson RJ, Sur M, Felleman D (1983a) Topographic reorganization of somatosensory cortical areas 3b and 1 in adult monkeys following restricted deafferentation. Neuroscience 8:33–55

Merzenich MM, Kaas JH, Wall JT, Sur M, Nelson RJ, Felleman D (1983b) Progression of change following median nerve section in the cortical representation of the hand in areas 3b and 1 in adult owl and squirrel monkeys. Neuroscience 10:639–665

Mesulam M-M, Van Hoesen GW, Pandya DN, Geschwind N (1977) Limbic and sensory connections of the inferior parietal lobule (area PG) in the rhesus monkey: a study with a new method for horseradish peroxidase histochemistry. Brain Res 136:395–414

Miller MW, Vogt BA (1984) Direct connections of rat visual cortex with sensory, motor and association areas. J Comp Neurol 226:184–202

Miller R (1981) Meaning and purpose in the intact brain. Oxford University Press, Oxford

Miller R (1987) Representation of brief temporal patterns, Hebbian synapses, and the left hemisphere dominance for phoneme recognition. Psychobiology 15:241–247

Miller R (1988) Cortico-striatal and cortico-limbic circuits: a two-tiered model of learning and memory functions. In: Markowitsch HJ (ed) Information processing by the brain. Views and hypotheses from a physiological-cognitive perspective. Hans Huber Press, Bern, pp. 179–198

Miller R (1989) Cortico-hippocampal interplay: self-organizing phase-locked loops for indexing memory. Psychobiology 17:115–128

Miller RE (1974) Social and pharmacological influences on the non-verbal communication of monkeys and men. In: Krames L, Pliner P, Alloway T (eds) Advances in the study of communication and affect Vol 1, Nonverbal communication Plenum, New York, pp. 77–101

Milliaressis E, Bouchard A, Jacobowitz DM (1975) Strong positive reward in median raphe: specific inhibition by p-chlorophenylalanine. Brain Res 98:194–201

Milner TA, Loy R, Amaral DG (1983) An anatomical study of the development of the septo-hippocampal projection in the rat. Dev Brain Res 8 (Brain Res 284):343–371

Misgeld U, Frotscher M (1986) Post-synaptc GABAergic inhibition of non-pyramidal neurons in guinea pig hippocampus. Neuroscience 19:193–206

Mishkin M, Petri HL (1984) Memories and habits: Some implications for the analysis of learning and retention. In: Squires LR, Butters N (eds) Neuropsychology of memory, Guildford, New York, pp. 287–296

Mitchell SJ, Ranck JB (1980) Generation of theta rhythm in medial entorhinal cortex of freely moving rats. Brain Res 189:49–66

Mitchell SJ, Rawlins JNP, Steward O, Olton DS (1982) Medial septal area lesions disrupt theta rhythm and cholinergic staining in medial entorhinal cortex and produce impaired radial arm maze behaviour in rats. J Neurosci 2:292–302

Mitzdorf U, Singer W (1978) Prominent excitatory patterns in cat visual cortex (area 17 and area 18): a current source density analysis of electrically evoked potentials. Exp Brain Res 33:371–394

Mizuki Y, Masotoshi T, Isozaki H, Nishijima H, Inanaga K (1980) Periodic appearance of theta rhythm in the frontal midline area during performance of a mental task. Electroencephalogr Clin Neurophysiol 49:345–351

Mizuki Y, Tanaka M, Isozaki H, Inanaga K (1981) Effects of centrally acting drugs on the frontal midline theta activity in man. Electroencephalogr Clin Neurophysiol 52:70P

Mizuki Y, Tanaka M, Inanaga K (1982) Effects of centrally acting drugs on the frontal midline theta activity. Electroencephalogr Clin Neurophysiol 54:28P

Moise SL, Costin A (1974) Hippocampal, hypothalamic and lateral geniculate activity during visual discrimination in the monkey. Physiol Behav 12:835–841

Monmaur P (1982) Hippocampal theta rhytms from CA1 and dentate generators during paradoxical sleep of the rat: differential alterations after septal lesions. Physiol Behav 28:467–471

Monmaur P, Thompson MA (1983) Topographic organization of septal cells innervating the dorsal hippocampal formation of the rat: special reference to both CA1 and dentate theta generators. Exp Neurol 82:366–379

Monmaur P, Thompson MA (1985) Hippocampal-dentate theta disturbance after selective CA1 pyramidal cell damage in the rat. Brain Res 328:301–311

Monmaur P, Thompson MA (1986) Spatial distribution of hippocampal-dentate theta rhythm following colchicine injection into the hippocampal formation of the rat. Brain Res 365:269–277

Monmaur P, Houcine O, Delacour J (1979) Experimental dissociation between wakefulness and paradoxical sleep hippocampal theta. Physiol Behav 23:366–378

Montoya CP, Sainsbury RS (1985) The effects of entorhinal cortical lesions on type 1 and type 2 theta. Physiol Behav 35:121–126

Morales FR, Roig JA, Monti JM, Macadar O, Budelli R (1971) Septal unit activity and hippocampal EEG during the sleep-wakefulness cycle of the rat. Physiol Behav 6:563–567

Morris RGM, Black AH (1978) Hippocampal electrical activity and behaviour elicited by non-reward. Behav Biol 22:524–532

Morris RGM, Hagan JJ (1983) Hippocampal electrical activity and ballistic movement. In: Seifert W (ed) Neurobiology of the hippocampus. Academic Press, London, pp. 321–331

Mosko S, Lynch G, Cotman CW(1973) The distribution of septal projections to the hippocampus of the rat. J Comp Neurol 152:163–174

Mountcastle VB, Lynch JC, Georgopoulos A, Sakata H, Acuna C (1975) Posterior parietal association cortex of the monkey: command functions for operations within extrapersonal space. J Neurophysiol 38:871–908

Muller RU, Kubie JL (1987) The effects of changes in the environment on the spatial firing of hippocampal complex-spike cells. J Neurosci 7:1951–1968

Muller RU, Kubie JL, Ranck JB (1987) Spatial firing patterns of hippocampal complex-spike cells in a fixed environment. J Neurosci 7:1935–1950

Nafstad PHJ (1967) An electron microscope study on the termination of the perforant path fibres in the hippocampus and fascia dentata. Z Zellforsch Mikrosk Anat 76:532–542

Nauta WJH (1964) Some efferent connexions of the prefrontal cortex in the monkey. In: Warren JM, Akert K (eds) The frontal granular cortex and behaviour. McGraw Hill, New York, pp. 397–409

Nicholson C, Freeman JA (1975) Theory of current source density analysis and determination of conductivity tensor for anuran cerebellum. J Neurophysiol 38:356–368

Niki H (1974a) Prefrontal unit activity during delayed alternation in the monkey. I. Relation to direction of response. Brain Res 68:185–194

Niki H (1974b) Prefrontal unit activity during delayed alternation in the monkey. II. Relation to absolute versus relative direction of response. Brain Res 68:197–204

Nunez A, Garcia-Austt E, Buno W (1987) Intracelular theta rhythm generation in identified hippocampal pyramids. Brain Res 416:289–300

Nunez PL (1981) Electrical fields of the brain: the neurophysics of EEG. Oxford University Press, New York

Nyakas C, Luiten PGM, Spencer DG, Traber J (1987) Detailed projection patterns of septal and diagonal band efferents to the hippocampus in the rat with emphasis on innervation of CA1 and dentate gyrus. Brain Res Bull 18:533–545

O'Keefe J (1976) Place units in the hippocampus of the freely-moving rat. Exp Neurol 51:78–109

O'Keefe, J (1979) A review of the hippocampal place cells. Prog Neurobiol 13:419–439

O'Keefe J, Conway D (1978) Hippocampal place units in the freely-moving rat: why they fire when they fire. Exp Brain Res 31:573–590

O'Keefe J, Dostrovsky J (1971) The hippocampus as a spatial map. Preliminary evidence from unit activity in the freely-moving rat. Brain Res 34:171–175

O'Keefe J, Nadel L (1978) The hippocampus as a cognitive map. Clarendon Press, Oxford

O'Keefe J, Speakman A (1987) Single unit activity in the rat hippocampus during a spatial memory task. Exp Brain Res 68:1–27

Olpe HR, Klebs K, Kung E, Campiche P, Glatt A, Ortmann R, D'Amato F, Pazza MF, Mondadori C (1987) Cholinomimetics induce theta rhythm and reduce hippocampal pyramidal cell excitability. Eur J Pharmacol 142:275–283

Olton DS, Branch M, Best PJ (1978) Spatial correlates of hippocampal unit activity. Exp Neurol 58:387–409

Paiva T, Lopes da Silva FH, Mollevanger W (1976) Modulating systems of hippocampal EEG. Electroencephalogr Clin Neurophysiol 40:470–480

Palm G (1980) On associative memory. Biol Cybern 36:19–31

Palm G (1982) Neural assemblies. An alternative approach to artificial intelligence. Studies of Brain Function No 7, Springer, Berlin, Heidelberg, New York

Pandya DP, Hallett M, Mukherjee SK (1969) Intra- and interhemispheric connections of the neocortical auditory system in the rhesus monkey. Brain Res 14:49–65

Pandya DN, Kuypers HGJM (1969) Cortico-cortical association connections in the rhesus monkey. Brain Res 13:13–36

Pandya DN, Van Hoesen GW, Mesulam M-M (1979) The cortical projections of the cingulate gyrus in the rhesus monkey. Anat Rec 193:643–644

Pandya DN, Van Hoesen GW, Mesulam M-M (1981) Efferent connections of the cingulate gyrus in the rhesus monkey. Exp Brain Res 42:319–330

Pandya DN, Yeterian EH (1985) Architecture and connections of cortical association areas. In: Jones EG, Peters A (eds) Cerebral cortex Vol 4, Association and auditory cortices. Plenum, New York pp. 3–55

Panula P, Revuelta AV, Cheney DL, Wu J-Y, Costa E (1984) An immunohistochemical study on the location of GABAergic neurons in rat septum. J Comp Neurol 222:69–80

Papez JW (1937) A proposed mechanism of emotion. Arch Neurol Psychiat (Chic) 38:725–743

Parmeggiani PL (1967) On the functional significance of the hippocampal theta rhythm. In: Structure and function of the limbic system, Prog Brain Res 27:413–441

Paxinos G, Bindra D (1970) Rewarding intracranial stimulation, movement and the hippocampal theta rhythm. Physiol Behav 5:227–231

Perkel DH, Mulloney B, Budelli W (1981) Quantitative methods of predicting neuronal behaviour. Neuroscience 6:823–837

Petrides M, Pandya DN (1984) Projections to the frontal cortex from the posterior parietal region in the rhesus monkey. J Comp Neurol 228:105–116

Petsche H, Stumpf Ch (1960) Topographic and toposcopic study of the origin and spread of the regular synchronised arousal pattern in the rabbit. Electroencephalogr Clin Neurophysiol 12:589–600

Petsche H, Stumpf C, Gogolak G (1962) The significance of the rabbit's septum as a relay station between the midbrain and the hippocampus. 1.The control of hippocampal arousal activity by the septum cells. Electroencephalogr Clin Neurophysiol 14:202–211

Petsche H, Gogolak G, Van Zwieten PA (1965) Rhythmicity of septal cell discharges at various levels of reticular excitation. Electroencephalogr Clin Neurophysiol 19:25–33

Piaget J (1955) The child's construction of reality. (Translated from French by Margaret Cook), Routledge & Kegan Paul, London

Pickenhain L, Klingberg F (1967) Hippocampal slow wave activity as a correlate of basic behavioural mechanisms in the rat. In: Adey WR, Tokizane T (eds) Structure and function of limbic mechanisms. Prog Brain Res 27:218–227

Pond FJ, Schwartzbaum JS (1972) Interrelationships of hippocampal EEG and visual evoked responses during appetitive behaviour in rats. Brain Res 43:119–137

Porter R, Adey WR, Brown TS (1964) Effects of small hippocampal lesions on locally recorded potentials and on behaviour performance in the cat. Exp Neurol 10:216–235

Powell EW (1963) Septal efferents revealed by axonal degeneration in the rat. Exp Neurol 8:406–422

Powell DA, Joseph JA (1974) Autonomic-somatic interaction and hippocampal theta activity. J Comp Physiol Psychol 87:978–986

Preobrazhenskaya LA (1974) Hippocampal electrical activity and different types of conditioned reflexes in dogs. Acta Neurobiol Exp 34:409–422

Preobrazhenskaya LA (1985) Dynamics of the spectral composition of the hippocampal theta rhythm in the dog during switching over of heterogeneous instrumental reflexes. Zh Vyssh Nerv Deyat 35:658–667

Radil-Weiss T, Zernicki B, Michalski A (1976) Hippocampal theta activity in the acute pretrigeminal cat. Acta Neurobiol Exp 36:517–534

Radulovacki M, Adey WR (1965) The hippocampus and the orienting reflex. Exp Neurol 12:68–83

Raisman G (1966) The connections of the septum. Brain 89:317–348

Rall W (1962) Electrophysiology of a dendritic neuron model. Biophys J 2(2):145–167

Ranck JB (1973) Studies on single neurons in dorsal hippocampal formation and septum in unrestrained rats. Part 1: Behavioural correlates and firing repertoires. Exp Neurol 41:461–531

Ranck JB (1976) Behavioural correlates and firing repertoires of neurons in the septal nuclei in unrestrained rats. In: De France JF (ed) The septal nuclei, Plenum, New York pp. 423–462

Rappelsburger P, Pockberger H, Petsche H (1981) Current source density analysis: methods and application to simultaneously recorded field potentials of rabbit visual cortex. Pflügers Arch 389:159–170

Rawlins JNP, Green KF (1977) Lamellar organization in the rat hippocampus. Exp Brain Res 28:335–344

Remahl S, Hildebrand C (1982) Changing relation between onset of myelination and axon diameter range in developing feline white matter. J Neurol Sci 54:33–45

Restle F (1957) Discrimination of cues in mazes: a resolution of the "place-vs-response" question. Psychol Rev 64:217–228

Ribak CE, Vaughn JE, Saito K (1978) Immunocyochemical localization of glutamic acid decarboxylase in neuronal somata following colchicine inhibition of axonal transport. Brain Res 140:315–332

Robinson TE (1980) Hippocampal rhythmic slow activity (RSA; theta): a critical analysis of selected studies and discussion of possible species differences. Brain Res Rev 2 (Brain Res 203):69–101

Robinson TE, Vanderwolf CH, Pappas BA (1977) Are the dorsal noradrenergic bundle projections from the locus coeruleus important for neocortical or hippocampal activation? Brain Res 138:75–98

Room P, Lohman AHM (1984) Cortical affernet to the parahippocampal gyrus in the cat. Anat Rec 208:151A-152A

Rose AM, Hattori T, Fibiger HC (1976) Analysis of the septo-hippocampal pathway by light and electronmicroscopic autoradiography. Brain Res 108:170–174

Rosene DL, Van Hoesen GW (1977) Hippocampal efferents reach widespread areas of cerebral cortex and amygdala in the rhesus monkey. Science 198:315–317

Routtenberg A (1968) Hippocampal correlates of consummatory and observed behaviour. Physiol Behav 3:533–535

Routtenberg A, Kramis RC (1968) Hippocampal correlates of aversive midbrain stimulation. Science 160:1363–1365

Rowntree CI, Bland BH (1986) An analysis of cholinoceptive neurons in the hippocampal formation by direct microinfusion. Brain Res 362:98–113

Rudell AP, Fox SE (1984) Hippocampal excitability related to the phase of theta rhythm in urethanized rats. Brain Res 294:350–353

Rudell AP, Fox S, Ranck JB (1980) Hippocampal excitability phase-locked to the theta rhythm in walking rats. Exp Neurol 68:87–96

Ruth RE, Collier TJ, Routtenberg A (1982) Topography between the entorhinal cortex and the dentate septotemporal axis in rats: I. Medial and intermediate entorhinal projecting cells. J Comp Neurol 209:69–78

Ruth RE, Collier TJ, Routtenberg A (1988) Topographical relationship between the entorhinal cortex and the septo-temporal axis of the dentate gyrus in rats. II Cells projecting from lateral entorhinal subdivisions. J Comp Neurol 270:506–516

Rye DB, Wainer BH, Mesulam M-M, Mufson EJ, Saper CB (1984) Cortical projections arising from the basal forebrain: a study of cholinergic and non-cholinergic components employing combined retrograde tracing and immunohistochemical localization of choline acetyltransferase. Neuroscience 13:627–643

Sadowski B, Longo VG (1962) Electroencephalographic and behavioural correlates of sn instrumental reward conditioned response in rabbits. A physiological and pharmacological study. Electroencephalogr Clin Neurophysiol 14:465–476

Sainsbury RS (1970) Hippocampal activity during natural behaviour in the guinea pig. Physiol Behav 5:317–324

Sainsbury RS, Montoya CP (1984) The relation between type 2 theta and behaviour. Physiol Behav 33:621–626

Sakai K, Sano K, Iwahara S (1973) Eye movements and hippocampal theta activity. Electroencephalogr Clin Neurophysiol 34:547–549

Sakanaka M, Shiosaka S, Takagi H, Senba E, Takutsuki K, Inagaki S, Yabuuchi H, Matsuzaki T, Tohyama M (1980) Topographic organization of the projection from the forebrain subcortical areas to the hippocampal formation of the rat. Neurosci Lett 20:253–257

Sano K, Iwahara S, Semba K, Sano A, Yamazaki S (1973) Eye movements and hippocampal theta activity in rats. Electroencephalogr Clin Neurophysiol 35:621–625

Saunders RC, Rosene DL (1988) A comparison of the efferents of the amygdala and the hippocampal formation in the rhesus monkey: I. Convergence in the entorhinal, prorhinal and perirhinal cortices. J Comp Neurol 271:153–184

Schaffer K (1892) Beutrag zum Histologie der Ammonshornformation. Archiv fur Mikroskopische Anatomie 39:611–632

Schiebel ME, Schiebel AB (1965) The response of reticular units to repetitive stimuli. Arch Ital Biol 103:279–297

Schwartzbaum JS (1975) Interrelationship among multiunit activity of the midbrain reticular formation and lateral geniculate nucleus, thalamocortical arousal and behaviour in rats. J Comp Physiol Psychol 89:131–157

Schwartzbaum JS, Kreinick CJ (1973) Interrelationships of hippocampal electroencephalogram, visually evoked response and behavioural reactivity to photic stimuli in rats. J Comp Physiol Psychol 85:479–490

Schwartzkroin PA, Kunkel DD (1985) Morphology of identified interneurons in the CA1 region of guinea pig hippocampus. J Comp Neurol 232:205–218

Schwartzkroin PA, Mathers LH (1978) Physiological and morphological identification of a non-pyramidal hippocampal cell type. Brain Res 157:1–10

Schwartzkroin PA, Stafstrom CE (1980) Effects of EGTA on the calcium-activated afterhyperpolarization in hippocampal CA3 pyramidal cells. Science 210:1125–1126

Schwertfeger WK (1979) Direct efferent and afferent connections of the hippocampus with the neocortex in the marmoset monkey. Am J Anat 156:77–82

Schwertfeger WK, Buhl E (1986) Various types of non-pyramidal hippocampal neuron project to the septum and contralateral hippocampus. Brain Res 385:146–154

Segal M (1973) Flow of conditioned response in limbic telencephalic system of the rat. J Neurophysiol 36:840–854

Segal M (1977) Afferents to the entorhinal cortex of the rat studied by the method of retrograde transport of horseradish peroxidase. Exp Neurol 57:750–765

Segal M (1988) Synaptic activation of cholinergic receptors in rat hippocampus. Brain Res 452:89–86

Segal M, Olds J (1972) Behaviour of units in hippocampal circuit of the rat during learning. J Neurophysiol 35:680–690

Segal M, Olds J (1973) Activity of units in the hippocampal circuit of the rat during differential classical conditioning. J Comp Physiol Psychol 82:195–204

Seki M, Zyo K (1984) Anterior thalamic afferents from the mammillary body and the limbic cortex in the rat. J Comp Neurol 229:242–256

Seltzer B, Van Hoesen GW (1979) A direct inferior parietal lobule projection to the presubiculum in the rhesus monkey. Brain Res 179:157–161

Semba K, Komisaruk BR (1978) Phase of theta wave in relation to different limb movements in awake rats. Electroencephalogr Clin Neurophysiol 44:61–71

Seress L, Ribak CE (1983) GABAergic cells in the dentate gyrus appear to be local circuit and projection neurons. Exp Brain Res 50:173–182

Shaban VM (1969) Temporal relationships between hippocampal evoked potential and unit activity. Neirofiziologiya 1:285–292

Shipley MT (1975) The topographical and laminar organization of the presubiculum's projection to the ipsi- and contralateral entorhinal cortex in the guinea pig. J Comp Neurol 160:127–146

Shute CC, Lewis PR (1967) The ascending cholinergic reticular system: neocortical, olfactory and subcortical projections. Brain 90:497–520

Sikes RW, Chronister RB, White LE (1977) Origin of the direct hippocampus-anterior thalamic bundle in the rat: a combined horseradish peroxidase-Golgi analysis. Exp Neurol 57:379–395

Sinclair BR, Seto N-MG, Bland BH (1982) Theta-cells in CA1 and dentate layers of hippocampal formation: relations to slow wave activity and motor behaviour in freely-moving rabbit. J Neurophsyiol 48:1214–1225

Sorensen KE (1980) Ipsilateral projection from the subiculum to the retrosplenial cortex in the guinea pig. J Comp Neurol 193:893–911

Sorensen KE, Shipley MT (1979) Projections from the subiculum to the deep layers of the ipsilateral presubicular and entorhinal cortices in the guinea pig. J Comp Neurol 188:313–334

Spinelli DN, Jensen FE (1979) Plasticity: the mirror of experience. Science 203:75–77

Stanfield BB, Caviness VS, Cowan WM (1979) The organization of certain afferents to the hippocampus and dentate gyrus in normal and reeler mice. J Comp Neurol 185:461–484

Stent GS (1973) A physiological mechanism for Hebb's postulate of learning. Proc Natl Acad Sci USA 70:997–1001

Steward O (1976) Topographic organization of the projections from the entorhinal area to the hippocampal formation of the rat J Comp Neurol 167:285–314

Stewart DJ, Vanderwolf CH (1987a) Hippocampal rhythmical slow activity following ibotenic acid lesions of the septal region. I. Relations to behaviour and effects of atropine and urethane. Brain Res 432:88–100

Stewart DJ, Vanderwolf CH (1987b) Hippocampal rhythmical slow activity following ibotenic acid lesions of the septal region. II. Changes in hippocampal activity during sleep. Brain Res 423:101–108

Storm-Mathisen J (1972) Glutamate decarboxylase in the rat hippocampal region after lesions of the afferent fibre systems. Evidence that the enzyme is localized in intrinsic neurones. Brain Res 40:215–235

Storm-Mathisen J (1977) Localization of transmitter candidates in the brain: the hippocampal formation as a model. Prog Neurobiol 8:119–181

Storm van Leeuwen W, Kamp A, Kok ML, De Quartel F, Lopes da Silva FH, Tielen AM (1967) Relations entre les activités éléctriques du chien, son comportement et sa direction d'attention. Actual Neurophysiol 7:167–186

Stumpf Ch (1959) Die Wirkung von Nicotin auf die Hippocampustätigkeit des Kaninchens. Naunyn Schmiedeberg's Arch Pharmacol 235:421–436

Stumpf Ch, Petsche H, Gogolak G (1962) The significance of the rabbit's septum as a relay station between the midbrain and the hippocampus. II The differential influence of drugs upon both the septal cell firing pattern and the hippocampus theta activity. Electroencephalogr Clin Neurophysiol 14:212–219

Sutherland NS, MacKintosh NJ (1971) Mechanisms of animal discrimination learning. Academic Press, New York

Swadlow HA, Waxman SG (1976) Variations in conduction velocity and excitability folowing single and multiple impulses of visual callosal axons in the rabbit. Exp Neurol 53:128–150

Swanson LW (1976) An autoradiographic study of the efferent connections of the preoptic region in the rat. J Comp Neurol 167:227–256

Swanson LW (1978) The anatomical organization of septo-hippocampal projections. In: Functions of the septo-hippocampal system, CIBA Symp no. 58. Elliott K, Whelan J (eds), Elsevier, Amsterdam, pp. 25–48

Swanson LW, Cowan WM (1977) An autoradiographic study of the organization of the efferent connections of the hippocampal formation in the rat. J Comp Neurol 172:49–84

Swanson LW (1981) A direct projection from ammon's horn to the prefrontal cortex in the rat. Brain Res 217:150–154

Swanson LW, Wyss JM, Cowan WM (1978) An autoradiographic study of the organization of intrahippocampal association pathways in the rat. J Comp Neurol 181:681–716

Swanson LW, Cowan WM (1979) The connections of the septal region in the rat. J Comp Neurol 186:621–655

Swanson LW, Sawchenko PE, Cowan WM (1981) Evidence for collateral projections by neurons in Ammon's horn, the dentate gyrus, and the subiculum: a multpile retrogade labelling study in the rat. J Neurosci 1:548–559

Teitelbaum H, McFarland WL (1971) Power spectral shifts in hippocampal EEG associated with conditioned locomotion in the rat. Physiol Behav 7:545–549

Thatcher RW, Krause PJ, Hrybyk M (1986) Cortico-cortical associations and EEG coherence: a two compartment model. Electroencephalogr Clin Neurophysiol 64:123–143

Thorlacius-Ussing O (1987) Zinc in the anterior pituitary of the rat: a histochemical and analytic work. Neuroendocrinology 45:233–242

Tolman EC (1932) Purposive behaviour in man and animals. Appleton-Century-Crofts, New York

Tolman EC, Ritchie BF, Kalish D (1946) Studies in spatial learning. I Orientation and the short-cut. J Exp Psychol 36:13–24

Torii S (1961) Two types of pattern of electrical activity induced by stimulation of hypothalamus and surrounding parts of rabbit's brain. Jpn J Physiol 11:147–156

Torii S, Sugi S (1960) Electrical activity of hippocampus in unrestrained rabbits. Folia Psychiat Neurol Jpn 14:95–103

Totterdell S, Hayes L (1987) Non-pyramidal hippocampal projection neurones: a light and electronmicroscropic study. J Neurocytol 16:477–485

Tranel D, Brady DR, Van Hoesen GW, Dimasio AR (1988) Parahippocampal projections to posterior auditory association cortex (area Tpt) in Old-World monkeys. Exp Brain Res 70:406–416

Turner MB (1965) Philosophy and the science of behaviour. Appleton-Century-Crofts, New York

Urban I, Lopes da Silva FH, Storm Van Leewen W, De Wied D (1974) A frequency shift in the hippocampal theta activity: an electrical correlate of central action of ACTH analogues in the dog. Brain Res 69:361–365

Vanderwolf CH (1969) Hippocampal electrical activity and voluntary movement in the rat. Electroencephalogr Clin Neurophysiol 26:407–418

Vanderwolf CH (1975) Neocortical and hippocampal activation in relation to behaviour: effects of atropine, eserine, phenothiazines and amphetamine. J Comp Physiol Psychol 88:300–323

Vanderwolf CH, Baker GB (1986) Evidence that serotonin mediates non-cholinergic neocortical low voltage fast activity, non-cholinergic hippocampal rhythmical slow activity and contributes to intelligent behaviour. Brain Res 374:342–356

Vanderwolf CH, Cooley RK (1974) Hippocampal electrical activity during long-continued avoidance performance: effects of fatigue. Physiol Behav 13:819–823

Vanderwolf CH, Leung L-WS (1983) Hippocampal rhythmic slow activity: a brief history and the effects of entorhinal lesions and phencyclidine. In: Siefert W (ed) Neurobiology of the hippocampus, Academic Press, London, pp. 275–302

Vanderwolf CH, Kramis R, Robinson TE (1978) Hippocampal electrical activity during waking behaviour and sleep: analyses using centrally acting drugs. In: Functions of the Septo-hippocampal system. CIBA Symp. No. 58, Elsevier, Amsterdam, pp. 199–221

Vanderwolf CH, Leung L-WS, Cooley RK (1985) Pathways through cingulate, neo- and entorhinal cortices mediate atropine-resistant hippocampal rhythmic slow activity. Brain Res 347:58–73

Van Hoesen GW, Pandya DN (1975a) Some connections of the entorhinal (area 28) and perirhinal (area 35) cortices of the rhesus monkey. II. Frontal lobe afferents. Brain Res 95:25–38

Van Hoesen GW, Pandya DN (1975b) Some connections of the entorhinal (area 28) and perirhinal (area 35) cortices of the rhesus monkey. III Efferent connections. Brain Res 95:39–59

Van Hoesen GW, Rosene DL(1979) Subicular input from temporal cortex in the rhesus monkey. Science 205:608–610

Van Hoesen GW, Pandya DN, Butters N (1972) Cortical afferents to the entorhinal cortex of the rhesus monkey. Science 175:1471–1473

Vertes RP (1979) Brainstem gigantocellular neurons: patterns of activity during behaviour and sleep in the freely moving rat. J Neurophysiol 42:214–228

Vertes RP (1981) An analysis of ascending brainstem systems involved in hippocampal synchronization and desynchronization. J Neurophysiol 46:1140–1159

Vertes RP (1988) Brainstem afferents to the basal forebrain in the rat. Neuroscience 24:907–935

Vinogradova O (1975) Functional organization of the limbic system in the process of registration of information: facts and hypothesis. In: Isaacson RI, Pribram KH (eds) The hippocamous, Vol. 2 (Neurophysiology and behaviour), Plenum, New York pp. 3–70

Vinogradova OS, Braznik ES (1978) Neuronal aspects of septo-hippocampal relations. In: Functions of the septohippocampal system, CIBA symposium, No 58, Elliott K, Whelan J (eds) Elsevier, Amsterdam, pp. 145–177

Vinogradova OS, Zolotukhina LI (1972) Sensory characteristics of the neurons in the medial and lateral septal nuclei. Zh Vyssh Nerv Deyat 22:1260–1269

Vinogradova OS, Braznik ES, Karanov AM, Zhadina SD (1980) Neuronal activity of the septum following various types of deafferentation. Brain Res 187:353–368

Vogt BA (1985) Cingulate cortex In: Jones EG, Peters A (eds) Cerebral cortex, Vol 4 (Association and auditory cortices), Plenum, New York pp. 89–149

Vogt BA, Miller MW (1983) Cotical connection between rat cingulate cortex and visual motor and post-subicular cortices. J Comp Neurol 216:192–210

Vogt BA, Pandya DN (1987) Cingulate cortex of the rhesus monkey II Cortical afferents. J Comp Neurol 262:271–289

Von der Malsburg C (1973) Self-organization of orientation sensitive cells in the striate cortex. Kybernetik 14:85–100

Wainer BH, Levey AI, Rye DB, Mesulam M-M, Mufson EJ (1985) Cholinergic and non-cholinergic septo-hippocampal pathways. Neurosci Lett 54:45–52

Walter DO, Adey WR (1963) Spectral analysis of electroencephalograms during learning in the cat, before and after subthalamic lesions. Exp Neurol 7:481–503

Watanabe K, Kawana E (1980) A horseradish peroxidase study on the mammillothalamic tract in the rat. Acta Anat (Basel) 108:394–401

Waxman SG, Swadlow HA (1976) Ultrastructure of visual callosal axons in the rabbit. Exp Neurol 53:115–127

West JR, Hodges CA, Black AC (1981) Distal infrapyramidal granule cell axons possess typical mossy fibre morphology. Brain Res Bull 6:119–124

West JR, Van Hoesen GW, Kosel KC (1982a) A demonstration of hippocampal mossy fibre axon morphology using the anterograde transport of horseradish peroxidase. Exp Brain Res 48:209–216

West MO, Christian E, Robinson JH, Deadwyler SA (1982b) Evoked potentials in the dentate gyrus reflect the retention of past sensory events. Neurosci Lett 28:319–324

Wetzel W, Ott T, Matthies H (1977) Posttraining hippocampal rhythmic slow activity ("theta") elicited by septal stimulation improves memory consolidation in rats. Behav Biol 21:32–40

Whishaw IQ (1972) Hippocampal electroencephalographic activity in the Mongolian gerbil during natural behaviours and in rat during wheel running and conditioned immobility. Can J Psychol 26:219–239

Whishaw IQ, Sutherland RJ (1982) Sparing of rhythmic slow activity (RSA) in two hippocampal generators after kainic acid CA3 and CA4 lesions. Exp Neurol 75:711–728

Whishaw IQ, Vanderwolf CH (1971) Hippocampal EEG and behaviour: effects of variation in body temperature and relation of EEG to vibrissae movement, swimming and shivering. Physiol Behav 6:391–397

Whishaw IQ, Vanderwolf CH (1973) Hippocampal EEG and behaviour: changes in amplitude and frequency of RSA (theta rhythm) asociated with spontaneous and learned movement patterns in rats and cats. Behav Biol 8:461–484

Whishaw IQ, Bland BH, Bayer SA (1978) Postnatal hippocampal granule cell agenesis in the rta: effects on two types of rhythmical slow activity (RSA) in two hippocampal generators. Brain Res 146:249–268

Whishaw IQ, Flannigan KP, Schallert T (1982) An assessment of the state hypothesis of animal "hypnosis" through an analysis of neocortical and hippocampal EEG in spontaneously immobile and hypnotized rabbits. Electroencephalogr Clin Neurophysiol 54:365–374

Wickens JR (1988) Electrically coupled but chemically isolated synapses: dendritic spines and calcium in a rule for synaptic modification. Prog Neurobiol 31:507–528

Wiggins RC, Bissell AC, Durham L, Samorajski T (1983) The corpus callosum during post-natal undernourishment and recovery: a morphometric analysis of myelin and axon relationships. Brain Res 328:51–57

Wigstrom H, Gustaffson B (1985) On long-lasting potentiation in the hippocampus: a proposed mechanism for its dependance on coincident pre- and post-synaptic activity. Acta Physiol Scand 123:519–522

Willey JT, Maeda G, Rafuse D (1975) Antidromic units in the prepyriform cortex driven by olfactory peduncular volleys. Brain Res 92:132–136

Wilson CL, Motter BC, Lindsley DB (1976) Influences of hypothalamic stimulation upon septal and hippocampal electrical activity in the cat. Brain Res 107:55–68

Winson J (1972) Interspecies differences in the occurrence of theta. Behav Biol 7:470–487

Winson J (1974) Patterns of hippocampal theta rhythm in the freely moving rat. Electroencephalogr Clin Neurophysiol 36:291–301

Winson J (1976a) Hippocampal theta rhythm. I. Depth profiles in the curarised rat. Brain Res 103:57–70

Winson J (1976b) Hippocampal theta rhythm II. Depth profiles in the freely moving rabbit. Brain Res 103:71–79

Winson J (1978) Loss of hippocampal theta rhythm results in spatial memory deficit in the rat. Science 201:160–162

Winson J, Abzug C (1978) Neuronal transmission through hippocampal pathways dependant on behaviour. J Neurophsyiol 41:716–732

Witter MP, Groenwegen HJ (1984) Laminar origin and septotemporal distribution of entorhinal and perirhinal projections to the hippocampus in the cat. J Comp Neurol 224:371–385

Witter MP, Griffioen AW, Jorritsma-Byham B, Krijnen JLM (1988) Entorhinal projections to the hippocampal CA1 region in the rat: an underestimated pathway. Neurosci Lett 85:193–198

Wyss JM (1981) An autoradiographic study of the efferent connections of the entorhinal cortex in the rat. J Comp Neurol 199:495–512

Yamaguchi Y, Yoshii N, Miyamoto K,Itoigawa N (1967) A study of the invasive hippocampal theta-waves to the cortex. Prog Brain Res 27:281–292

Yerkes RM, Dodson JD(1908) The relation between strength of stimulus and rapidity of habit formation. J Comp Neurol Psychol 18:459–482

Yokota T, Fujimori B (1964) Effects of brain-stem stimulation upon hippocampal electrical activity, somatomotor reflexes and autonomic functions. Electroencephalogr Clin Neurophysiol 16:375–382

Yoshii N, Shimokochi M, Miyamoto K, Ito M (1967) Studies on the neural basis of behaviour by continuous frequency analysis of EEG. Prog Brain Res 21A:217–250

Young GA (1976) Electrical activity of the dorsal hippocampus in rats operantly trained to lever press and to lick. J Comp Physiol Psychol 90:78–90

Zin R, Conforti N, Feldman S (1977) Sensory responsiveness of single cells in the medial septal nucleus in intact and hypothalamic-deafferented rats. Exp Neurol 54:7–23

Index

T = Table; F = Figure; N = Footnote

Studies of Brain Function

Volumes already published in the series:

1 *W. Heiligenberg*
Principles of Electrolocation and Jamming Avoidance in Electric Fish

2 *W. Precht*
Neuronal Operations in the Vestibular System

3 *J. T. Enright*
The Timing of Sleep and Wakefulness

4 *H. Braak*
Architectonics of the Human Telencephalic Cortex

5 *H. Collewijn*
The Oculomotor System of the Rabbit and Its Plasticity

6 *M. Abeles*
Local Cortical Circuits

7 *G. Palm*
Neural Assemblies

8 *J. Hyvärinen*
The Parietal Cortex of Monkey and Man

9 *E. Zrenner*
Neurophysiological Aspects of Color Vision in Primates

10 *U. Bässler*
Neural Basis of Elementary Behavior in Stick Insects

11 *G. A. Orban*
Neuronal Operations in the Visual Cortex

12 *M. Heisenberg, R. Wolf*
Vision in Drosophila

13 *Yu. I. Arshavsky, I. M. Gelfand, G. N. Orlovsky*
Cerebellum and Rhythmical Movements

14 *G. Roth*
Visual Behavior in Salamanders

15 *U. Windhorst*
How Brain-like is the Spinal Cord?

16 *Jos J. Eggermont*
The Correlative Brain

17 *R. Miller*
Cortico-Hippocampal Interplay and the Representation of Contexts in the Brain

18 *V. Braitenberg, A. Schüz*
Anatomy of the Cortex: Statistics and Geometry

GPSR Compliance

The European Union's (EU) General Product Safety Regulation (GPSR) is a set of rules that requires consumer products to be safe and our obligations to ensure this.

If you have any concerns about our products, you can contact us on ProductSafety@springernature.com

In case Publisher is established outside the EU, the EU authorized representative is:

Springer Nature Customer Service Center GmbH
Europaplatz 3
69115 Heidelberg, Germany

Batch number: 10620141

Printed by Printforce, the Netherlands